MATHEMATICS RESEARCH DEVELOPMENTS

SELF-ORGANIZATION

THEORIES AND METHODS

Mathematics Research Developments

Additional books in this series can be found on Nova's website
under the Series tab.

Additional e-books in this series can be found on Nova's website
under the e-book tab.

Computational Mathematics and Analysis Series

Additional books in this series can be found on Nova's website
under the Series tab.

Additional e-books in this series can be found on Nova's website
under the e-book tab.

MATHEMATICS RESEARCH DEVELOPMENTS

SELF-ORGANIZATION

THEORIES AND METHODS

WENJUN ZHANG
EDITOR

nova publishers
New York

For permission to use material from this book please contact us:
Telephone 631-231-7269; Fax 631-231-8175
Web Site: http://www.novapublishers.com

Library of Congress Cataloging-in-Publication Data

Self-organization : theories and methods / editors, WenJun Zhang.
pages cm
Includes bibliographical references and index.
ISBN: 978-1-62618-865-5 (hardcover)
1. Self-organizing systems. I. Zhang, WenJun, 1963-
Q325.S437 2013
003--dc23

2013016527

Published by Nova Science Publishers, Inc. †New York

CONTENTS

PREFACE

Self-organization is a universal mechanism in nature. In the past thirty years, numerous phenomena, theories and methods on self-organization have been founded around the world. The book, Self-organization: Theories and Methods, is published to present recent achievements in theories and methods of self-organization. This book includes such theories and methods of self-organization as ant algotithms, particle swarm algorithm, artificial neural network, motion and migration algorithms, self-adaptive Kalman Filter, finite state approximation, etc. Chapters are contributed by more than 20 scientists from China, Italy, Spain, Japan, Russia, Serbia, India, Turkey, in the areas of mathematics, computational science, artificial intelligence, aeronautics and astronautics, automation and control, and life sciences. It will provide researchers with various aspects of the latest advances in self-organization. It is a valuable reference for the scientists, university teachers and graduate students in mathematics, natural science, engineering science, and social science.

The main contents are included as follows:

Chapter 1-In this chapter, concepts, theories, and methods of self-organization are briefly overviewed.

Chapter 2-Classification is a data mining functionality that assigns instances in a collection to target classes. The goal of classification is to accurately predict the target class for an unlabelled sample by learning from instances described by a set of attributes and a class label. Conventional classification methods specified in literature are less efficient in classifying very small datasets with repeated attribute values. To overcome this drawback, the study explores a classification method by mining similar patterns among the instances in very small datasets. Classification by Mining Patterns (CMP) algorithm is proposed to predict the class label of the unlabelled sample by mining similar patterns among the instances in the dataset. To mine similar patterns, the instances in the dataset belonging to the same class label are grouped. The instances in each group that differ by one attribute value are merged. Such merged instances form mined patterns. To predict the class label of the unlabelled samples, the attribute values in the mined patterns are compared with the corresponding attribute values of the unlabelled sample. The count of number of attribute values in the mined patterns matching with that of the unlabelled sample gives the attribute match count. The mined patterns having the maximum attribute match count are grouped and the majority class label of the mined patterns is predicted as the class label of the unlabelled sample. To choose the attributes in the mined patterns for comparison, an Ant Colony Optimization based Feature Selection Mined Patterns (ACOFSMP) is proposed. Simulation results are shown to prove that the CMP algorithm is efficient for classifying very small datasets with repeated attribute values.

Chapter 3-This chapter aims to present a study on auto-organisation in the context of Ant Colony Optimisation (ACO) metaheuristic. Heuristic optimisation algorithms should decide how to share efforts between exploration of newer solutions, and the exploitation of those previously discovered. In classical ACO algorithms this decision is taken beforehand, by setting up the algorithm parameters values. Usually it is necessary to change them every time a new problem is tackled. The purpose is to introduce in ACO a self-organisation dynamics that allows to balance the exploitation and the exploration. Inspired by the biological studies on the organisation of the real ant colonies, two kinds of ants are defined: one dedicated to the exploration, and the other assigned to the exploitation. Besides, it is defined a population dynamics to establish a relationship between both populations. This allows to self-regulate the populations between them according to the quality of the solutions found. The ultimate outcome is the Ant Colony Extended (ACE), a new algorithm with the capability of self-organise, which eliminates the necessity of using parameters. The chapter is focused on the population dynamics and its influence on the algorithms behaviour. The results are promising and encourage further investigation of how to apply this approach to other heuristic optimisation algorithm.

Chapter 4-In this chapter, a generalization of the popular and widely used Particle Swarm Optimization (PSO) algorithm is presented. This novel optimization technique, named Generalized PSO (GPSO), is inspired by linear control theory. It overcomes some typical flaws of the classical PSO, enabling direct control over the key aspects of particle dynamics during the optimization process. The basic idea of this algorithm with its detailed theoretical and empirical analysis is presented, and parameter-tuning schemes are proposed. GPSO is also compared to the classical PSO and Genetic Algorithm (GA) on a set of benchmark problems. The presented results demonstrate the effectiveness of the proposed algorithm. Finally, two practical engineering applications of the GPSO algorithm are described, in the area of optimal gear design and architecture and urban design.

Chapter 5-Artificial neuronets (AN), especially with error back-propagation (BP) training algorithms, have been widely investigated and applied in various science and engineering fields. However, BP-type neuronets, which are self-adaptive systems, have shown some inherent weaknesses, such as, the possibility of being trapped into local minima, the difficulty in choosing appropriate learning rate, and most importantly, the inability to determine the optimal neuronet structure. To solve the inherent weaknesses of AN, lots of improvements for BP-type algorithms have been investigated. However, as researchers (including the authors) realize and experience quite frequently, the inherent weaknesses of BP-type neuronets still exist. In this chapter, differing from others algorithmic improvements on the training procedure, the authors way about the problem solving exploits some elegant structure design, parameter-setting, pseudoinverse and numerical optimization techniques. In other words, a new type of AN using linearly-independent or orthogonal polynomials as activation functions, is presented and analyzed by us (the authors). These finally lead us to propose a weights and structure determination (WASD) method, which is based on a weights-direct-determination (WDD) method, for the authors presented feed forward AN. Based on the authors' previous work, single-input neuronets equipped with the WASD method have successfully overcome the above weaknesses. To investigate and verify two- or multiple-input neuronets equipped with this method, the authors firstly put forward various novel neuronets based on different activation functions. Then, corresponding WASD algorithms are proposed for the presented neuronets. For better performance (e.g., more efficiency and conciseness in self-organizing systems), the authors further propose pruning techniques in the neuronet structure determination. Finally, based on various target functions, numerical results further substantiate the efficacy of the proposed neuronets equipped with the corresponding WASD algorithms, which shows the better performance in terms of training (or say, approximation or learning), generalization (or say, testing or validation) and prediction.

Chapter 6-Humans produce adaptive locomotion through dynamic interactions among the nervous system, the musculoskeletal system, and the environment in a self-organized manner. A human musculoskeletal system has more degrees of freedom (DOFs) than necessary for locomotion and humans solve the redundancy problem in some way to establish locomotion.

It has been suggested that individual DOFs are not manipulated independently, but some DOFs are functionally connected by object tasks to reduce the number of DOFs. These relationships among DOFs appear as low-dimensional structures in the DOFs. This chapter shows such low-dimensional structures in joint movements and muscle activities (kinematic and muscle synergies) during human locomotion based on the analysis of measured data. In addition, it shows a constructive approach using a computer simulation with a neuromusculoskeletal model based on anatomical and physiological findings to examine the functional roles of the low-dimensional structures that generate adaptive locomotor behaviors through dynamic interactions among the nervous system, the musculoskeletal system, and the environment.

Chapter 7-This chapter discusses the self-organized movement of a set of some individuals. It is adopted that individuals perceive reality as its projection on coordinate system that is firmly attached to the individual. The movement of individuals is interpreted as the movement of points in this coordinate system according to the algorithms by which each individual is equipped. The projection of reality and other individuals simultaneously changing the parameters of algorithms and simultaneously impose constraints on the movement of these points. The set of individuals moves so that each individual performs arbitrary motion within the allowed constraints on one of two ways: a) individuals follow one or more leaders, b) individuals preserve system configuration during movement, i.e. not change some typical magnitude of set. Motion of individuals or of set of individuals from one position to another always takes place according to the functions of natural impulses. Short was given comparison between function of the natural impulse and other custom-fit functions.

Chapter 8-Habitat diversity is an important mechanism for species migration and distribution. A stochastic model for species migration in heterogeneous environment is described in this chapter. This model is used to describe various dynamics types in biological process such as periodic oscillation, monotonic increase and decline, and fluctuation. The species extinction likelihood in migrated area largely determines the dynamics and pattern of species migration. There is a significant difference in species migration between heterogeneous and homogeneous habitats. Higher fitness to species will lead to a faster migration and higher percentage of migrated areas. Existence of boundary areas may retard species migration.

Chapter 9-In this chapter a Robust Self-Adaptive Kalman Filter (RSAKF) algorithm with the filter gain correction is developed for the case of sensor/actuator malfunctions. The proposed RSAKF utilizes time variable factors in order to reduce the effect of the faults on the estimation procedure. In this sense, the procedures with single and multiple factors for the adaptation of the filter are presented. In the first case, the filter is adapted by using single adaptive factor as a corrective term on the filter gain while in the second one, an adaptive matrix built of multiple adaptive factors is used to fix the relevant term of the Kalman gain matrix, individually. After chosing the efficent method of adaptation, an overall concept for the RSAKF is proposed. In this concept, the filter detects the type of the fault, either in the sensors or actuators, and after the fault isolation it applies the required adaptation process such that the estimation characteristic is not deteriorated. Effectiveness of the proposed filters is investigated via simulations for the state estimation problem of an UAV. The results of the presented algorithms are compared for different types of sensor/actuator faults and in this context recommendations about their utilization are given.

Chapter 10 – This chapter takes into account the problem of designing regular approximations for the stochastic pushdown computing. It is proven that stochastic pushdown automata accepting with finite cut point and for which equivalent non-dirty context-free grammars exist admit arbitrarily accurate regular approximations. The author addresses the problem of designing the uniform representation of the stochastic (context-free/pushdown) computing characterizing it in terms of converging sequences of finite state approximations.

I hope this book will help readers understand recent advances in theories and methods of self-organization.

WenJun Zhang, Professor
Sun Yat-sen University, China
International Academy of Ecology
and Environmental Sciences
e-mail: wjzhang@iaees.org;
zhwj@mail.sysu.edu.cn
December 22, 2012

In: Self-organization: Theories and Methods
Editor: WenJun Zhang

ISBN: 978-1-62618-865-5

Chapter 1

AN OVERVIEW ON THEORIES AND METHODS OF SELF-ORGANIZATION

WenJun Zhang*
Sun Yat-sen University, Guangzhou, China
International Academy of Ecology and Environmental Sciences, Hong Kong

ABSTRACT

Self-organization is a universe mechanism in nature. Self-organizing systems are the systems which can evolve and improve the organizational behavior or structure by themselves. In a self-organizing system, the system evolves automatically to form an ordered structure based on some compatible rules. Without any external instruction and force, the self-organizing system emerges only from the interactions between the basic components of the system. In this chapter, theories and methods of self-organization are briefly overviewed.

Keywords: self-organization; principles; theories; methods

1. Mechanisms and Principles of Self-Organization

Organization can be classified into two basic categories, i.e., self-organization and external-organization. The major difference between the two categories of organizations is whether the organizational instructions/forces come from outside the system or from inside the system. The organization with organizational instructions/forces from inside the system is self-organization. In other word, a system is called self-organizing system if there is any specific intervention from outside in the process of evolution (Foerster, 1960; Heyligen, 2002; Zhang, 2013). The stronger a system's self-organization capacity is, the stronger the system's ability to generate and maintain new functions is.

* E-mail: zhwj@mail.sysu.edu.cn, wjzhang@iaees.org

Self-organizing systems are those systems which can evolve and improve the organizational behavior or structure by themselves. Self-organization is a process to describe the system's global state. In a self-organizing system, the system evolves automatically to form an ordered structure according to some compatible rules. Unlike other organizations, the self-organizing system arises only from the interactions between basic components of the system, without external instructions and forces. In the process of self-organization, several structural components can interact and cooperate to produce the behaviors that only a group will have. The dynamic interactions between low-level components typically include attraction and repulsion, that is, positive and negative feedbacks.

Generally, self-organization arises from the increase in complexity or information. According to the thermodynamic laws, this situation will only occur in the open systems far away from equilibrium. For most systems, this means the energy supply to the system is needed for generating and maintaining a certain mode. In an abstract sense, self-organization is a process that makes the system's entropy increase in the absence of external forces (i.e., a dynamic process that from disordered to ordered states) (Glansdorff and Prigogine, 1971; Nicolis and Prigogine, 1977).

From the perspective of systematic theory, self-organization is an irreversible dynamic process. From the view of mathematics and physics, self-organization means the dimension reduction of state space or the reduction of degrees of freedom, i.e., the system converges spontaneously to one or more steady states (i.e., attractors). In such a system, the local interactions between the basic components of the system can change the modes of system's organization, and the global behaviors of the system cannot be understood intuitively. They cannot be understood by simply observing existing laws and behaviors of between-component interactions (Zhang, 2012). The complex patterns and behaviors of self-organizing systems are thus often called "emergency". In a word, the global properties of self-organizing systems are not predictable.

Self-organization usually requires to be based on three elements (Bonabeau et al., 1999): (1) strong nonlinear and dynamic interactions, even though they do not necessarily relate to the positive or negative feedbacks, (2) a balance between development and exploration, and (3) complex and diverse interactions.

Self-organization is ubiquitous in nature and human society, which covers many fields as physics, chemistry, biology, economics and society. One of the most important examples of self-organization is the nonlinear processes in physics (Glansdorff and Prigogine, 1971; Ansari and Smolin, 2008; Brau et al., 2011). Self-organization is also chemical-related, which is often considered to be synonymous to self-assembly (Kim et al., 2006; Coleman et al., 2011; Harada, et al. 2011). It is very important for the description of biological systems at both sub-cellular level and ecosystem level (Hess and Mikhailov, 1994; Misteli, 2001; Camazine, 2003; Clyde et al., 2003; Motegi et al., 2011). It can also be found in the mathematical systems, like cellular automata (Zhang, 2012).

1.1. Thermodynamic Basis of Self-Organization

The spontaneous formation of new structures, for examples, crystallization process, Bénard phenomenon, and Belouzov-Zhabotinsky reaction (Sun and Lin, 2004), are all self-organizing processes. In other word, the formation of a structure or a mode does not need to be given any external force. The components of those systems seem to be arranged into a more ordered pattern by themselves. At first glance, self-organization violates the second law of thermodynamics. This law holds that the entropy of an independent system can only increase rather than decrease. In other word, the second law of thermodynamics means that an isolated system should evolve in a uniform, simple, difference-eliminating way, which is substantially an evolution to a low-level organization.

In the example of crystallization process, the randomly moved molecules that have been bonded into a crystal structure and thus have been fixed will transmit their kinetic energy to the liquid they are dissolved. Thus, the reduction in the entropy of crystallization process is just offset by the increase in the entropy of liquid. The entropy of the whole system has actually increased, and therefore it is coincident with the second law of thermodynamics.

For self-organizing systems, which are not in their equilibrium, it is hard to determine whether the second law of thermodynamics is true or not. Prigogine started to study the systems far away from equilibrium states since the 1950s and proposed the theory of dissipative structures (Glansdorff and Prigogine, 1971; Nicolis and Prigogine, 1977) in which the most used models to explain dynamic self-organizing process are Bénard cells and the Brusselator. The theory of dissipative structures tried to address such questions as, under what conditions a system will be able to evolve from the disordered to the ordered, and form a new, stable, and internally-dynamic structure. Such a structure must be an open system, i.e., there are energy/matter flows in the system; the system will continuously generate entropy, but at the same time the entropy will be actively dissipated from the system or output from the system. Thus, at the cost of environmental ordering, the system will be able to increase ordering of its own. The system will be able to follow the second law of thermodynamics simply by removing its excess entropy. This dissipation can be mostly found in life systems. Plants and animals obtain energy and matter by absorbing light or food with low entropy, and then output energy and matter by draining the metabolic waste of high entropy. This will reduce its internal entropy to offset the degradation process required by the second law of thermodynamics.

The output of entropy cannot explain why and how self-organization happened. Prigogine held that self-organization would mostly occur in the nonlinear systems far away from equilibrium.

1.2. Principles of Self-Organization

The first symposium on self-organization was held in 1959. In this symposium, the British cybernetic expert, Ashby, proposed "the principle of self-organization" (Ashby, 1947). He believed that a dynamic system, ignoring its classification or composition, always tends to move toward an equilibrium, or the "attractor". This theory reduced our uncertainty in the state of system and solved the problem of entropy in systematic science. This is equivalent to self-organization, which finally reaches the equilibrium and the final equilibrium can be considered to be a state with mutually adaptation of all components in the system. Another cybernetic expert, Heinz von Forster, proposed the principle of order from noise (Foerster, 1996). He believed that the larger a system is subject to random interference, paradoxically, the quicker it performs self-organization (i.e., become more ordered). This idea is very simple: The larger the state space that a system moves through is, the faster it reaches the attractor. If the system stays at its initial state, it will not reach the attractors, and self-organization will thus not occur (Foerster, 1996). Generally, there are multiple attractors in a nonlinear system. The attractor theory holds that the behavior trajectories of complex system in the state space can be represented by the dynamic equations. These dynamic equations are always determined by a set of "attractors". What attractor the system will move toward depends on the attraction domain that the initial state falls into. What attractor the system eventually reaches is uncertain. Some small fluctuations will cause the system to move toward an attractor rather than another attractor. Prigogine thus proposed a related principle, i.e., order arises from fluctuations.

1.3. Fundamental Theories of Self-Organization

So far, self-organization theory is generally believed to mainly consist of three parts, the theory of dissipative structures (Glansdorff and Prigogine, 1971; Nicolis and Prigogine, 1977), synergetics (Haken, 1978, 2004), and catastrophe theory (Saunders, 1980). However, the basic thoughts and theoretical kernel of self-organization can be derived entirely from the theory of dissipative structures and synergetics.

(1) The theory of dissipative structures. Prigogine officially proposed the theory of dissipative structures in 1969 (Nicolis and Prigogine, 1977). The theory mainly aimed to explain the exchange of matter and energy between the system and the environment and its effect on the self-organizing system. The structure established from the exchange of matter and energy between the system and the environment is a dissipative structure, such as city, life, etc. Far away from equilibrium, the openness of system, and nonlinear mechanism between different components of system, are three conditions for the formation of a dissipative structure. Far away from equilibrium refers to that the distribution of matter and energy in different areas of a system is extremely uneven. The theory of dissipative structures is mostly used to discuss the evolution of complex systems. It uses two levels of approaches, i.e., deterministic and stochastic approaches in the discussion of system evolution. Deterministic approach uses macro-physical variables to describe system dynamics and features. Stochastic approach treats macro-physical variables as the average of corresponding random variables. Analyzing random variables will not only obtain the averaged values but also help understand fluctuation characteristics of the system.

(2) Synergetics. Haken first proposed the concept of synergy in 1976. Another science on self-organization, synergetics, was thus established (Haken, 1978, 1983, 2004). About synergetics, Haken held that on the one hand, in a system many subsystems interact to produce the structure and function at the macroscopic scale; on the other hand, there are many different scientific disciplines cooperating to explore the general principles for governing self-organizing systems. The order parameters generate and govern subsystems by competition and cooperation between various subsystems. Serving of various subsystems to order parameters reinforces order parameters and further promotes the serving of subsystems to order parameters, so that the system can be spontaneously organized by itself (Haken, 1978, 2004). Competition and cooperation between order variables will result in different forms of evolutions of self-organization. Synergetics mainly discusses the coordination (i.e., synergy) mechanism between internal components of the system studied. It holds that the coordination between various components in the system is the basis of self-organizing process. Competition and cooperation between order parameters of the system are direct forces for the formation of new structures. Because of independent evolution of components in the system, various local and collaborative evolutions, as

well as random inferences by environmental factors, the actual state of system always deviates from the average. The magnitude of such a deviation is called fluctuation. When the system is in its transition from one steady state to another steady state, and the independence evolution and collaborative evolution between system components move into a balance, any small fluctuation will be amplified, and quickly spread to the whole system. The resultant giant fluctuation will promote the system moving into an ordered state. In addition, Harken proposed the concept of functional structure, i.e., the function and structure are dependent for each other. If the energy or matter flow is cut off, the physical and chemical system will lose their structure, while a biological system is mostly able to maintain a fairly long time. Such biological systems seem to combine non-dissipative and dissipative structures together (Haken, 1978, 2004).

(3) Catastrophe theory. Catastrophe theory was first proposed by the French mathematician, Thom R, in 1969. Since the 1970s, Zeeman and other scientists have further developed catastrophe theory and apply it to various aspects of physics, biology, ecology, medical science, economics and sociology, and have produced significant impacts (Zeeman, 1976). This theory was built on the basis of the stability theory. It considers a catastrophe process as a process that transit towards a new steady state through an unsteady state started from an original steady state. In the mathematical view, this means the changes of values of a set of parameters and mathematical functions that denote the states of system. Thus it is a theory to describe the phenomena that the continuous change of parameters lead to the discontinuous change of the states of the system. It treats with the systems that are almost structural stable in the state space but there is some structural instability on some point sets of measure 0. The basic characteristics of catastrophe systems include multiple steady states, reachability, jumping, lagging, and divergence (it reflects the sensitivity of evolutionary trajectory to the path of control parameters). Catastrophe theory holds that different outcomes may occur even if it is the same process that corresponds to the same controlling factors and critical values, and different new steady states may be achieved with different probabilities (Saunders, 1980). Generally, catastrophe theory does not reveal the mechanism to produce catastrophe phenomenon. It provides a reasonable mathematical model to describe the phenomenon of catastrophe in the real world, and categories various catastrophe types. In ecology, there are a lot of applications of catastrophe theory, such as the sudden outbreak or sudden collapse of biological population. In addition, there are also other theories of self-organization, for example, Eigen's super circle theory. Super circle theory is a self-organization theory on molecular evolution. However, it is a scientific hypothesis, the impact is still limited.

1.4. Properties of Self-Organization

Interacting with the environment, the self-organizing system can evolve to form new structures and functions. Unlike ordinary mechanic systems, it has its own peculiar properties. Those peculiar properties can be used as part of the definition of self-organizing systems. For example, there is no centralized control, and continuous adaptation to the changing environment, etc.

(1) Local interactions generate global ordering. The most obvious change in the self-organizing system is the formation of global ordering. Local interactions follow immediately basic physical processes, and any impact from one region to another region must first move through all intermediate regions. When it passes through intermediate regions, all the processes will be disturbed by the turmoil occurred in the intermediate regions. First we assume the system is disordered and all components of system evolve randomly. The impact of any passed will be quickly dispersed and ultimately destructed by these random turbulences. The result is thus that, starting in the chaotic state, the distant parts of system are actually independent, that is, they do not affect each other. During the process of self-organization, all the components of the system are closely linked. Understanding of the structure of a regional component will be valuable to know the structure of components of its consecutive regions.
(2) Distributed control. We tend to consider that a highly organized system is instructed and controlled by external or internal forces. This kind control is called centralized control. In a self-organizing system, the control to organization is distributed throughout the system. All components contribute to the final arrangement of the states of system. Despite some of the advantages of centralized control with respect to distributed control, centralized control must be based on distributed control at some levels. For example, the function of human brain is dispersed in the network formed by interacting neurons. Different brain regions perform specific functions, but not a neuron or a group of neurons has all the ability to control brain. This is a result of self-organization.
(3) Nonlinearity and feedback. Nonlinearity means the whole is not equal to the simple sum of its parts, i.e., superposition principle is not met. Referring to the mechanical movement, a linear phenomenon is generally manifested as smooth motion in time and space. It can be described by the functions with good performance, and the functions are continuous and differentiable. The nonlinear phenomenon is a movement from regular motion to irregular motion, with obvious jumping and intermitting features. From the view of disturbance and parameter theory, the response of a linear system is smooth and proportional changes, but nonlinear system will exhibit substantial changes in some of the key points because of the small changes in parameters, and it form and maintain spatially regular and ordered structures. Linear relationship is independent, while the nonlinear relationship is an interactive one, which makes it violate superposition principle and produce gain or loss. In the nonlinear systems, there are feedbacks between system components, and each component affects the others and other components in turn affect it. Positive feedback plays a role similar to the input so that the system's deviation increases, and the system's oscillation is thus amplified. Negative feedback causes reverse outcome compared with input's role, so that the system's output error can be reduced and the system is thus stabilized. In the complex self-organizing systems, there are often several chains of positive and negative feedbacks, so a change can be enlarged in a certain direction but suppressed in the other directions. This will result in the very complex behaviors hard to be predicted.
(4) Far away from equilibrium. Equilibrium is a special state of a system. At this state, the everywhere measurable macroscopic physical properties of the system are

uniform throughout the system (so that there is not any macroscopic irreversible process inside the system). At the equilibrium state, the system follows the first law of thermodynamics: $d_E = d_Q - pd_V$, i.e., the increment of energy inside system is equal to the absorbed heat subtracting by outward work done by the system. It is also coincident with the second law of thermodynamics: $d_S/d_t \geqslant 0$, that is, the spontaneous evolution of system is always toward the direction of entropy's increase. For the system at equilibrium state, it must abandon its extra energy; it will remain in the minimum energy state without the input of external energy. The system far away from equilibrium is more sensitive and more vulnerable to environmental changes due to its dependence on external energy input. But it is more powerful to respond changes. On the other hand, the surplus of external input energy allows the system to amplify the self-organization process, for example, offsetting small turbulences or maintaining positive feedbacks longer in the aid of strong interactions. This makes the system more vigorous and more adaptive to external changes.

(5) Systematic termination and organizational hierarchy. The interactions between individual components of self-organizing system can be, to some extent, defined as an ordered structure. However the ordering does not mean organization. Organization is an ordered structure and can achieve a particular function. In a self-organizing system, this function is to maintain a particular structure under various disturbances. The general characteristics of self-sufficiency can be understood as a closure. A process with causal relationship can be represented as a chain or a sequence: A → B → C → D → ..., where A initiates B, B initiates C, C initiates D, and so on. Overall, this will lead to a continuous change. However, there may be its own termination of a link in the chain, for example, O returns J. So the cycle of the system becomes J, K, L, M, N, O, J, K, L. Thus the corresponding arrangement of the system will always be maintained or recycled. In addition, if the loop is placed in a negative feedback region, it is relatively not affected by the impact of external interference (Foerster, 1960). In a self-organizing system, a lot of autonomous and organization-closed subsystems may generate. Those subsystems will interact in a more indirect way. They will also adapt to the structure for termination and determine subsystems at a higher level. New generated subsystems will contain the original subsystems as their components. Each self-organizing system constitutes a series of subsystems. A self-organizing system thus forms a layered structure. Each self-organizing system belongs to a high-level self-organizing system and contains low-level self-organizing systems. It interacts with other self-organizing systems at the same level. Therefore, the hierarchy is one of the characteristics of self-organizing systems.

2. Algorithms of Self-Organization

2.1. Computer and Mathematical Modeling

Because it is difficult to predict complex behaviors of self-organizing systems, computer simulation and mathematical modeling are useful methods for theoretical experiments of those systems. They also help people understand how these systems work. One method for self-organization modeling is to use nonlinear differential equations, and another method is to use cellular automata (Ballestores and Qiu, 2012; Zhang, 2012).

2.2. Intelligence Computation

Many complex problems are hard to be effectively addressed by traditional artificial intelligence technologies. Intelligence computation is a powerful technique to solve more complex problems. In the intelligence computation, computation is intelligent. It can automatically adjust parameters during the process of computation, and thus achieve optimal results (Koza, 1992). Evolutionary computation searches the optimal solution by simulating biological evolution in nature, for example, genetic algorithms, etc. Swarm intelligence algorithms are a kind of new evolutionary algorithms, which closely relate to evolutionary strategies and genetic algorithms.

2.2.1. Swarm Intelligence Algorithms

The concept, swarm intelligence, was first proposed by Hackwood and Beni (1992) in their cellular automata system. Swarm intelligence refers to that a group of unintelligent entities can cooperate to solve problems in a distributed way. They can directly or indirectly communicate by changing the local environment. These unintelligent entities behave intelligently through cooperation (Bonabeau et al., 1999; Hu and of Li, 2008; Zhang, et al., 2008). A significant feature of swarm intelligence is, although the behavior of an individual is simple, but when all individuals work together, the system will exhibit very complex behaviors. Without centralized control and global model, swarm intelligence provides a solution for distributed problems.

(1) Ant colony algorithm. Ant colony algorithm (ant algorithm) is a method for finding the optimal path in the graph. It is a probabilistic algorithm (Colomi and Maniezzo, 1991; Dorigo et al., 1996). It is proposed by Dorigo in his doctoral thesis in 1992, inspired by the behaviors of ants in the process of looking for food path (Colomi and Maniezzo, 1991). Ant colony in nature can cooperate to find the shortest path from the nest to the food, and can change strategies as circumstances change and quickly re-find the shortest path. Numerous studies found that ant colony algorithm is a self-organization algorithm. At the start of algorithm, a single artificial ant searches for solution in an unordered way. After a period of algorithm evolution, the artificial ants spontaneously tend to find some solutions close to the optimal solution, which is a process from the disordered to the ordered.

(2) Particle swarm algorithm. Particle swarm algorithm is also called Particle Swarm Optimization (PSO). PSO is an evolutionary computation based on iteration, which is proposed by Kennedy and Eberhart (1995). Particle swarm algorithm was originally a graphical simulation to preying behaviors of a flock of birds. The basic idea is inspired by their early findings on group behaviors of birds, and they thereafter used and improved the biological population model. In the particle swarm algorithm, each particle in the particle swarm is equivalent to a bird in the bird flock. They all track the currently optimal particle (which is equivalent to the bird most near the food) in the solution space, and they constantly update their position and velocity. Through continuous iteration, the algorithm reaches the optimal solution (similar to bird

finding food) (Shi and Eberhart, 1998; Eberhart and Shi, 2000; the Krink and Løvbjerg, 2002; Clerc, 2004, 2006; Zhang et al. , 2007; Niknam and Amiri, 2010).

(3) Stochastic diffusion search. In 1989, Bishop proposed stochastic diffusion search method in order to solve the problem of incentive equivalence in pattern recognition (Bishop, 1989). Stochastic diffusion search is one of the swarm intelligence algorithms. Unlike most swarm intelligence algorithms, stochastic diffusion search uses direct communication between entities (Beattie and Bishop, 1998; Nasuto et al., 1998; Myatt et al., 2004; Meyer, 2004; Meyer et al., 2006). In stochastic diffusion search, each of the entity holders holds an assumed solution about the problem to be solved, and assesses the solution partially. The successful entity directly communicates with unsuccessful entities to repeatedly test its assumption. A positive feedback mechanism is thus established so that the group can quickly converge to the optimal solution in the solution space. In the solution space the regions largely aggregated by entities are considered as candidate solutions. Through cooperation between the locally-run simple entities, the global solution can be obtained in the region with most aggregated entities. The stochastic diffusion search is a truly adaptive algorithm, because even if the optimal solution is found, there are still some entities to explore the solution space, which makes the algorithm adapt to changes of the environment (Nasuto et al., 1998).

2.2.2. Genetic Algorithms

Genetic algorithms are a kind of stochastic search algorithms that simulate the evolution of organisms (i.e., survival of the fittest, genetic mechanism). It was first proposed by Holland (1975). It aimed to explain the adaptive processes of natural and artificial systems. Main characteristic of genetic algorithm include, (1) directly operating the structural objects, (2) there is no assumptions on derivative and function continuity, (3) implicit parallelism and better search performance on global optimization, and using probabilistic optimization-searching method which can automatically obtain and guide optimized search space, and adaptively adjust the search direction without determinant rules. These properties make genetic algorithms being widely used in combinatorial optimization, machine learning, signal processing, adaptive control and artificial life. Genetic algorithms, along with adaptive systems, cellular automata, chaos theory and artificial intelligence, are considered key technologies that will have significant impacts on the future of computing technology.

2.3. Other Algorithms

In addition to the commonly used algorithms above, a variety of new algorithms have been proposed to study self-organization. Widely recognized algorithms include, fish colony algorithm (Li, 2003; Grosenick et al., 2007; Chen et al., 2009), bee colony algorithm, co-evolutionary algorithm, Memetic algorithm, hybrid optimization algorithm, bio-inspired algorithm, evolutionary programming, evolutionary strategy, parallel algorithm, etc.

ACKNOWLEDGMENTS

I am very grateful for the help of Mr KaiYu Zhao in the manuscript preparation.

REFERENCES

Ashby WR. 1947. Principles of the self-organizing dynamic system. *Journal of General Psychology*, 37(2): 125-128.

Ballestores F Jr, Qiu ZY. 2012. An integrated parcel-based land use change model using cellular automata and decision tree. *Proceedings of the International Academy of Ecology and Environmental Sciences*, 2(2): 53-69.

Beattie PD, Bishop JM. 1998. Self-Localisation in the 'Senario' autonomous wheelchair. *Journal of Intelligent and Robotic Systems*, 22: 255-267.

Bishop JM. 1989. *Anarchic Techniques for Pattem Classification*. UK University of Reading, UK.

Bonabeau E, Dorigo M, Theraulaz G. 1999. *Swarm Intelligence: From Natural to Artificial Systems*. Oxford University Press, UK.

Camazine S. 2003. *Self-Organization in Biological Systems*. Princeton University Press, USA.

Chen GZ, Wang JQ, Li CJ, et al. 2009. An improved artificial fish swarm algorithm and its applications. *Systems Engineering*, 27(12): 105-110.

Clerc M. 2004. Discrete Particle Swarm Optimization, illustrated by the Traveling Salesman Problem. In: *New Optimization Techniques in Engineering*. 219-239, Springer.

Clerc M. 2006. Particle Swarm Optimization. *ISTE* (International Scientific and Technical Encyclopedia).

Clyde DE, Corado MS, Wu X, et al. 2003. A self-organizing system of repressor gradients establishes segmental complexity in Drosophila. *Nature*, 426(6968): 849-853.

Coleman AC, Beierle JM, Stuart MC, et al. 2011. Light-induced disassembly of self-assembled vesicle-capped nanotubes observed in real time. *Nature Nanotechnology*, 6(9): 547-552.

Colomi ADM, Maniezzo V. 1991. Distributed Optimization by Ant Colonies. In: The First European conferenceon *Artificial Life*. 134-142, Elsevier, France.

Dorigo M, Maniezzo V, Colorni A. 1996. The ant system: optimization by a colony of cooperating agents. *IEEE Transactions on Systems, Man, and Cybernetics*, 26: 29-41.

Eberhart RC, Shi Y. 2000. Comparing inertia weights and constriction factors in particle swarm optimization. In: *Proceedings of the Congress on Evolutionary Computation*. 84-88, IEEE, USA.

Foerster HV. 1960. On self-organising systems and their environments. *Self Organising Systems*, 30-50.

Foerster HV. 1996. Self-organization, the progenitor of conversation and interaction theories. *Systems Research and Behavioral Science*, 13(3): 349-362.

Glansdorff P, Prigogine I. 1971. Thermodynamic Theory of Structure, Stability and Fluctuations. Wiley-Interscience, New York, USA.

Grosenick L, Clement TS, Fernald RD. 2007. Fish can infer social rank by observation alone. *Nature*, 445: 429-432.

Hackwoods S, Beni G. 1992. Self-organization of sensors for swarm intelligence. In: *Proceedings of the 1992 IEEE International Conference on Robotics and Automation*. 819-829, IEEE, Nice, France.

Haken H. 1978. *Synergetics: An Introduction*. Springer, Berlin, Germany.

Haken H. 1983. Synopsis and introduction. In: *Synergetics of the Brain* (Basar E, Flohr H, Haken H, et al., eds). Springer, Berlin, Germany983).

Haken H. 2004. Future trends in synergetics. *Solid State Phenomena*, 97-98: 3-10.

Harada A, Kobayashi R, Takashima Y, et al. 2011. Macroscopic self-assembly through molecular recognition. *Nature Chemistry*, 3(1): 34-37.

Hess B, Mikhailov A. 1994. Self-organization in living cells. *Science*, 264(5156): 223-224.

Heyligen F. 2002. The science of self-organisation and adaptivity. In: *The Encyclopedia of Life Support Systems*. UNESCO Publishing-Eolss Publisher, Oxford, UK.

Holland JH. 1975. *Adaptation in Natural and Artificial Systems*. University of Michigan Press, USA.

Hu ZG, Li J. 2008. The progress of swarm intelligence algorithms. *Control Theroy and Applications*, 27(2): 13-15.

Kennedy J, Eberhart R. 1995. Particle swarm optimization. In: *Proceedings of the IEEE Conference on NeuralNetworks*. Piscataway, NJ, USA.

Kim YK, Cook S, Tuladhar, SM, et al. 2006. A strong regioregularity effect in self-organizing conjugated polymer films and high-efficiency polythiophene:fullerene solar cells. *Nature Materials*, 5(3): 197-203.

Koza JR. 1992. *Genetic Programming: On the Programming of Computers by Means of Natural Selection*. MIT Press, Cambridge, USA.

Krink T, Løvbjerg M. 2002. The LifeCycle Model: combining particle swarm optimisation, genetic algorithms and hill climbers. In: Proceedings of Parallel Problem Solving from Nature VII (PPSN). 621-630, IEEE, USA

Li XR. 2003. *A New Intelligence Optimization Algorithm-Artificial Fish Swarm Algorithm*. Zhejiang University Press, China.

Meyer KD. 2004. *Foundation of Stochastic Diffusion Search*. University of Reading, UK.

Meyer KD, Nasuto SJ, Bishop JM. 2006. Stochastic Diffusion Search: Partial Function Evaluation in Swarm Intelligence Dynamic Optimisation. *Studies in Computational Intelligence Series*. 185-207, Springer-Verlag, Cambridge, USA.

Misteli T. 2001. The concept of self-organization in cellular architecture. *Journal of Cell Biology*, 155(2): 181-185.

Motegi F, Zonies S, Hao Y, et al. 2011. Microtubules induce self-organization of polarized PAR domains in *Caenorhabditis elegans* zygotes. *Nature Cell Biology*, 13(11): 1361-1367.

Myatt DR, Bishop JM, Nasuto SJ. 2004. Minimum stable convergence criteria for stochastic diffusion search. *Electronics Letters*, 40(2): 112-113.

Nasuto SJ, Bishop JM, Lauria S. 1998. Time complexity analysis of the stochastic diffusion search. *IFAC Symposium on Neural Computation*. 260-266, Vienna, Austria.

Nicolis G, Prigogine I. 1977. Self-Organization in Non-Equilibrium Systems. *Wiley Interscience*, New York, USA.

Niknam T, Amiri B. 2010. An efficient hybrid approach based on PSO, ACO and k-means for cluster analysis. *Applied Soft Computing*, 10(1): 183-197.

Saunders PT. 1980. *An Introduction to Catastrophe Theory*. Cambridge University Press, England.

Shi Y, Eberhart RC. 1998. A modified particle swarm optimizer. In: *Proceedings of IEEE International Conference on Evolutionary Computation*. 69-73, IEEE, Piscataway, NJ, USA.

Sun DC, Lin FY. 2004. *An Introduction to Systems Engineering*. Tsinghua University Press, Beijing, China.

Zeeman EC. 1976. Catastrophe theory. *Scientific American*, 65-70, 75-83

Zhang Q, Kang LS, Li DN. 2008. A summary for swarm intelligence algorithm and its application. *Journal of Huanggang Normal University*, 28(6): 44-48.

Zhang Q, Li CH, Liu Y, et al. 2007. Fast multi-swarm optimization with Cauchy mutation and crossover operation. In: *Lecture Notes in Computer Science*. 344-352, Springer, Berlin, Germany.

Zhang WJ. *Computational Ecology: Graphs, Networks and Agent-based Modeling*. World Scientific, Singapore, 2012

Zhang WJ. 2013. Selforganizology: A science that deals with self-organization. *Network Biology*, 3(1): 1-14.

In: Self-organization: Theories and Methods
Editor: WenJun Zhang

ISBN: 978-1-62618-865-5

Chapter 2

AN ANT COLONY OPTIMIZATION-BASED APPROACH OF FEATURE SELECTION FOR EFFICIENT CLASSIFICATION OF VERY SMALL DATASETS BY MINING PATTERNS

N. K. Sreeja*[*1] and A. Sankar[2]
Department of Computer Applications,
[1] Sri Krishna College of Technology, India
[2] PSG College of Technology, India

ABSTRACT

Classification is a data mining functionality that assigns instances in a collection to target classes. The goal of classification is to accurately predict the target class for an unlabelled sample by learning from instances described by a set of attributes and a class label. Conventional classification methods specified in literature are less efficient in classifying very small datasets with repeated attribute values. To overcome this drawback, the study explores a classification method by mining similar patterns among the instances in very small datasets. Classification by Mining Patterns (CMP) algorithm is proposed to predict the class label of the unlabelled sample by mining similar patterns among the instances in the dataset. To mine similar patterns, the instances in the dataset belonging to the same class label are grouped. The instances in each group that differ by one attribute value are merged. Such merged instances form mined patterns. To predict the class label of the unlabelled samples, the attribute values in the mined patterns are compared with the corresponding attribute values of the unlabelled sample. The count of number of attribute values in the mined patterns matching with that of the unlabelled sample gives the attribute match count. The mined patterns having the maximum attribute match count are grouped and the majority class label of the mined patterns is predicted as the class label of the unlabelled sample. To choose the attributes in the mined patterns for comparison, an Ant Colony Optimization based Feature Selection Mined Patterns (ACOFSMP) is proposed. Simulation results are shown to prove that the CMP algorithm is efficient for classifying very small datasets with repeated attribute values.

[*] E-mail: sreeja.n.krishnan@gmail.com.

Keywords: Classification, Mined Patterns, Ant Colony Optimization based feature selection

1. INTRODUCTION

Classification is the task of learning from instances which are described by a set of attributes and a class label. An unlabelled sample is an instance described by a set of attributes whose class label is to be predicted. The result of learning is a classification model that is capable of accurately predicting the class label of unlabelled samples. Some of the techniques proposed in literature for classification include Neural Networks [6], Support Vector Machine (SVM) [2], classification using trees, lazy learning methods [11]. While these existing classification methods prove their efficiency for classifying large datasets, it is found that these methods are less efficient to classify very small datasets having very few attributes with repeated attribute values. The efficiency of existing decision tree algorithms [9] has been well established for classifying relatively small data sets. However, it is computationally expensive to train. To overcome this drawback, a classification method by mining similar patterns is proposed.

In this paper, an approach is proposed to mine similar patterns among the instances in the dataset to intelligently predict the class labels of the unlabelled samples. This paper aims to classify unlabelled samples for very small datasets having very few attributes with repeated attribute values. The similar instances in each class label group in the dataset differing by one attribute value are merged to form mined patterns. A novel algorithm called Classification by Mining Patterns (CMP) [10] is proposed to classify unlabelled samples by comparing the attribute values of the unlabelled sample with the corresponding attribute values of the mined patterns derived from the dataset. The attributes for comparison is selected using an Ant Colony Optimization Feature Selection Mined Patterns algorithm. The count of the number of attributes of each mined pattern matching with the corresponding attributes of the unlabelled sample is found. This denotes the attribute match count of each mined pattern. The mined patterns having the maximum attribute match count are grouped. The class label of the unlabelled sample is predicted as the majority class label of the mined pattern having the maximum attribute match count. Experimental results are shown to illustrate that CMP is efficient for classification of very small datasets with repeated attribute values when compared to the existing classification methods. Since mined patterns are formed among instances differing in only one attribute value, CMP algorithm is not applicable for high dimensional datasets where more than one attribute values differ among the instances.

This paper is organized as follows. Section 2 describes Ant Colony Optimization. Section 3 describes Classification by Mining Patterns. Section 4 describes a case study. The experimental results are reported in section 5. A comparison of classification accuracy of CMP using ACOFSMP and classical feature selection methods is described in section 6. The comparison of classification accuracy of CMP using ACOFSMP with the existing classification methods is described in section 7. Finally, the conclusion is presented in section 8.

2. Ant Colony Optimization

A colony of ants denoting a set of computational, concurrent and asynchronous agents moves through states of the problem corresponding to partial solutions of the problem to solve [7]. They move by applying a stochastic local decision policy based on two parameters, called *trails* and *attractiveness*. By moving, each ant incrementally constructs a solution to the problem. When an ant completes a solution, or during the construction phase of the solution, the ant evaluates the solution and modifies the trail value on the components used in its solution. This pheromone information will direct the search of the future ants. Furthermore, an Ant Colony Optimization (ACO) algorithm includes two more mechanisms such as *pheromone evaporation* and, optionally, *daemon actions*. Trail evaporation decreases all trail values over time, in order to avoid unlimited accumulation of trails over some component. Daemon actions can be used to implement centralized actions which cannot be performed by single ants, such as the invocation of a local optimization procedure, or the update of global information to be used to decide whether to bias the search process from a non-local perspective [12].

3. Classification by Mining Patterns

A novel approach called Classification by Mining Patterns algorithm is proposed for classifying unlabelled samples for very small datasets having few attributes with repeated attribute values.

3.1. Mining Patterns from the Dataset

The instances in the dataset for classification is given by (X_{i1}, X_{i2},…, $X_{in,}$ C_i) where i=1,2,…, m; m is the number of instances in the dataset. X_{i1}, X_{i2},…, X_{in} denote the attribute values of the ith instance and C_i denotes the class label of the ith instance. The similar class labels of the instances in the dataset are grouped. The attribute values of the unmerged instance belonging to a class label group are compared with the corresponding attribute values of each of its successive instances in the same group. The count of the matching attributes between the compared instances is found. If the number of matches is equal to n-1, where n is the number of attributes, then the attributes among the instances are merged. This is repeated for all unmerged instances in each of the similar class label group to form mined patterns. The pseudocode for mined pattern algorithm is shown in figure 1.

3.2. CMP Algorithm for Classification

The count of each similar class labels of the mined patterns in the dataset is found. This denotes the class label count. The unlabelled sample whose class label is to be predicted is given by (x_1, x_2, x_3, …, x_n) where x_1, x_2, x_3,…, x_n are the attribute values. Initially every attribute value of the unlabelled sample is compared with the corresponding attribute values of the mined patterns. If there is a unique pattern having all the values of its attributes matching with the attributes of the unlabelled sample, the class label of the unlabelled sample is predicted as the class label of that mined pattern. If a unique match does not occur, each of the selected attribute value x_i of the mined pattern is compared with the corresponding attribute value of the unlabelled sample. Selection of attributes for comparison is found using an ACOFSMP algorithm.

```
Procedure Minedpattern(dataset)
Begin
Group the similar class labels of the instances in the dataset;
for all instances set label(instance)="NO";
m=1;
for each class label group
 for i=1 to number of instances in the group
 for j=i+1 to number of instances in the group
  if ((label(instance_i)="NO")or label(instance_j)="NO")) then
  for k=1 to count(attributes)
   if attribute_k(instance_i)= attribute_k(instance_j) then
      count=count+1;
   endif;
  end;
   if(count= n-1) then //n is the number of attribues
     pattern_m= merge(instance_i,instance_j);
      m=m+1 ;
     label(instance_j)="YES"
   endif;
  endif;
  end;
 end;
end;
for p=1 to m-1;
   return(pattern_p);
end Minedpattern;
```

Figure 1. Pseudocode for mined pattern algorithm.

The number of attributes in each mined pattern whose value matches with the corresponding attribute value of the unlabelled sample is counted and is denoted by attribute match count. This is repeated for all mined patterns. The maximum value of the attribute match count of the mined patterns is found. The mined patterns having the maximum attribute match count are grouped.

The class label of the unlabelled sample is predicted as the majority class label of the mined patterns having maximum attribute match count. If there exists more than one majority class label then the probability of the class label is found by dividing the majority class label count by the corresponding class label count. The unlabelled sample belongs to the class label having the highest probability value. The pseudocode for CMP algorithm is shown in figure 2.

```
Procedure CMP (unlabelled sample, Minedpattern, attributes)
begin
Group the similar class labels of the Minedpattern;
for i=1 to number of groups
   classlabelcount_i=count(class label_i)
end;
for each mined pattern
   for j=1 to count(attributes)
    if (attribute_j(Minedpattern)=attribute_j(unlabelled sample))then
      attribute match count= attribute match count+1;
    endif;
   end;
end;
Group Minedpattern having max(attribute match count);
  classlabel(unlabelled sample)= majority class label of Minedpattern
                        having max(attribute match count);
  if there is more than one majority class label then
   prob(class label)=count(majority class label)divided by
              corresponding classlabelcount in the Minedpattern;
   classlabel(unlabelled sample)= class label having max(prob)in
                        the Minedpattern;
  endif;
end CMP;
```

Figure 2. Pseudocode for CMP algorithm.

3.3. Ant Colony Optimization Based Feature Selection Mined Patterns algorithm

For many classification problems, irrelevant or redundant features (attributes) decrease both the classification efficiency, and the accuracy [1]. So, feature selection is necessary for most classification problems. Feature selection can be performed without reducing the accuracy of classification remarkably [3] [5].

Choosing pivotal features can not only reduce the complexity of the target problem, but also improve the performance of algorithms in many cases [4]. To improve the classification accuracy of CMP algorithm, an Ant Colony Optimization based feature selection method is proposed. An Ant Colony Optimization based Feature Selection Mined Patterns algorithm obtains an optimal subset of features for higher classification accuracy. According to this approach, an ant agent finds the solution of obtaining the optimal subset of attributes by depositing pheromone.

The pheromone deposition of the ant agent denotes the attribute values of the mined pattern whose value is compared with the corresponding attribute values of the unlabelled sample. An ant agent has a tabulist denoting the memory. Also, an ant agent has a positive and negative position to deposit pheromone. An ant agent chooses random numbers to deposit pheromone.

The random number generated for the positive position must be between 0 and n, where n denotes the number of attributes in the mined patterns. The random number generated for the negative position must be between 0 and q, where q denotes the maximum number of attributes in the pheromone deposition stored in the tabulist. The number in the positive position denotes the number of attributes in the mined patterns to be added to the pheromone deposition.

The number in the negative position denotes the number of attributes to be deleted from the pheromone deposition stored in the tabulist. The attributes in the pheromone deposition should not be repetitive. Initially, the random number in the negative position is 0. Depending on the random number in the positive position, the ant agent chooses a group of positions randomly, denoting the position of the attributes in the mined patterns. The subset of attributes represents the pheromone deposition of the ant agent. The ant agent computes the energy value by finding the classification accuracy of CMP for the attributes denoted by the pheromone deposition using leave-one-out validation method [8].

To estimate the classification accuracy, each time, one instance from the training set is used as the test instance and the remaining instances are used as the training dataset. The instances in the training dataset are merged to form mined patterns. The class of the unlabelled sample in the test dataset is found using CMP algorithm for the attributes denoted by the pheromone deposition. This is repeated for all other test datasets. The classification accuracy is the number of correctly classified instances in the test dataset divided by the total number of instances.

The energy value of the ant agent along with its pheromone deposition is stored in its tabulist. This denotes the first trail of the ant agent finding the solution. The ant agent moves to the next trail by pheromone evaporation. To update the pheromone, the ant agent again chooses two random numbers in the positive and negative position. Based on the random number present in the positive and negative position, the ant agent chooses a group of positions.

The group of positions chosen for negative position denotes the position of the attributes to be deleted from the pheromone deposition stored in the tabulist. Thus, the ant agent updates the pheromone by either adding or deleting a subset of attributes, or both based on the random number in the positive and negative position.

The energy value of the ant agent is evaluated. If the energy value is greater than or equal to the energy value of the previous trail, the tabulist is updated with the pheromone deposition and the energy value. If the energy value is less than the energy value of the previous trail, then the newly added or deleted attributes are ignored and the tabulist is not updated.

The process is repeated for 10 trails. Once the solution is found, the subset in the pheromone deposition stored in the tabulist of the ant agent denotes the attributes that are to be used for comparison by CMP algorithm to classify unlabelled samples. Thus the attribute selection for the mined patterns is made using ACOFSMP. The selected attributes are used for classification by CMP algorithm.

3.3.1. Blocks of Ant Colony Optimization Based Feature Selection Mined Patterns Algorithm

The basic blocks of ACOFSMP algorithm shown in figure 3 is explained below.

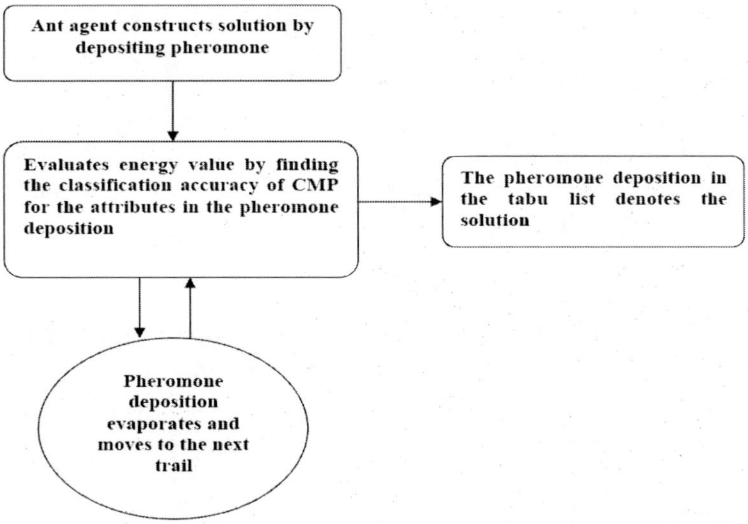

Figure 3. Blocks of ACOFSMP algorithm.

```
Procedure ACOFSMP
(No.of features n, No.of features in the tabulist q)
 begin
      j=1;
      P= Positive position;
      N= Negative position;
      (P, N)= (rnd (1, n), 0);
      Choose P features at random from n and add to the pheromonedeposition_j;
      Delete the same P features from n;
      for i=1 to number of training instances
       M_i=Minedpattern(training set_i)
       Accuracy=mean(leave-one-out(CMP(t_i, M_i, pheromonedeposition_j));
          //t_i is the test dataset
      end
      Energy_j(Ant agent)= Accuracy;
      Tabulist(Ant agent)= (Energy_j(Ant agent),pheromonedeposition_j);
      repeat
       j=j+1;
       (P, N)= (rnd (0, n), rnd(0, q));
       Choose P features at random from n and add to the pheromonedeposition_j;
       Delete the same P features from n;
       Choose N features at random from q and delete the same features from
       the pheromonedeposition_j;
       Add N features to n;
       for i=1 to number of training instances
         M_i=Minedpattern(training set_i)
         Accuracy=mean(leaveoneout(CMP(t_i,M_i,pheromonedeposition_j));
       end
       Energy_j(Ant agent)= Accuracy;
       if Energy_j(Ant agent)< Energy_j-1(Ant agent)then
         Tabulist(Ant agent)= (Energy_j-1(Ant agent),pheromonedeposition_j-1);
       else
         Tabulist(Ant agent)= (Energy_j(Ant agent),pheromonedeposition_j);
       endif;
      until (j >= 10);
      return (Tabulist(Ant agent));
    end ACOFSMP;
```

Figure 4. Pseudocode for ACOFSMP algorithm.

(i) Construct Solutions: Pheromone Deposition of the Ant Agent

To deposit pheromone, the ant agent chooses two random numbers in the positive and negative position respectively.

Depending on the random number in the positive and negative position, the ant agent chooses random positions denoting the positions of the attributes to be added or deleted.

Thus the pheromone deposition of the ant agent denotes the attributes of the mined patterns to be compared with that of the unlabelled sample.

(ii) Energy Value of the Ant Agent

The ant agent finds the attractiveness towards the solution by computing the energy value. The energy value is computed by finding the classification accuracy of CMP algorithm for the attributes denoted by the pheromone deposition of the ant agent using leave-one out method as discussed in section 3.3.

(iii) Pheromone Evaporation

The pheromone evaporation rate is given as 1-P where P=0 or 1. If the energy value of the ant agent for a subset of attributes is less than the energy value of the previous trail, then P=0 and the pheromone evaporation rate is 1. Otherwise, P=1 and the pheromone evaporation rate is 0. If the pheromone evaporation rate is 1, then the attributes added or deleted to the pheromone deposition of the ant agent in the current trail is ignored.

(iv) Selecting Attributes for Comparison

The ant agent moves through 10 trails and the pheromone deposition stored in the tabulist of the ant agent having the highest energy value is chosen as the solution.

(v) ACOFSMP Algorithm

The pseudocode for ACOFSMP algorithm is shown in figure 4. It aids in the selection of features for comparison between the patterns mined from the dataset and the unlabelled sample.

4. CASE STUDY

A case study is discussed for 9 training samples of the dataset 'contact' taken from the UCI machine learning repository. The UCI Machine Learning Repository is a collection of databases, domain theories, and data generators that are used by the machine learning community for the empirical analysis of machine learning algorithms [13]. Table 1 shows a dataset having 9 instances and 4 attributes.

4.1. Selecting Attributes Using ACOFSMP Algorithm

An ACOFSMP algorithm is used to find the optimal set of attributes for the dataset given in table 1. These attributes are used by CMP algorithm for classifying unlabelled samples. An ant agent deposits pheromone to find the optimal subset of attributes. Table 2 shows the feature selection of the mined patterns using ACOFSMP algorithm.

Table 1. Dataset

Instance Number	Age	Spectacle Prescription	Astigmatic	Tear Production Rate	Class Label
1	3	1	2	2	1
2	1	1	2	2	1
3	1	2	2	2	1
4	2	1	2	2	1
5	2	1	1	2	2
6	3	2	1	2	2
7	1	1	1	1	3
8	1	1	1	2	2
9	1	1	2	1	3

During trail-1, the ant agent chooses a random number 2 and 0 in the positive and negative position respectively. Since the positive position has a random number 2, the ant agent randomly chooses two positions {3, 4} from the training dataset. This denotes the pheromone deposition of the ant agent. Thus in trail-1, the attributes at position 3 and 4 are selected as the pheromone deposition denoting the attributes of the mined patterns to be compared. The energy value of the ant agent is computed by estimating the classification accuracy of CMP using leave-one-out method for the attribute subset {3, 4} denoted in the pheromone deposition of the ant agent. To estimate the classification accuracy of CMP using leave-one-out method, one instance of the dataset is chosen as the test dataset and the remaining instances constitute the training set. A one time validation was performed for this pair of training and test dataset and the classification accuracy is found. Thus the first test dataset containing instance 1 will be the unlabelled sample and the remaining instances from 2 to 9 constitute the training dataset. The mined patterns for these instances in the training dataset are found as described in section 3.1. The class of the unlabelled sample is found using CMP algorithm as described in section 3.2. This was repeated for remaining test datasets. The number of correctly classified unlabelled samples out of the total number of unlabelled samples gives the average classification accuracy. Thus the leave-one-out average classification accuracy of CMP algorithm for the attribute subset {3,4} is found to be 77.8 %. Therefore the energy value of the ant agent in trail-1 is 77.8. The pheromone deposition {3,4} along with the energy value 77.8 is stored in the tabulist denoting the memory of the ant agent. The ant agent moves to the next trail. To deposit pheromone in trail-2, the pheromone deposition evaporates and the ant agent chooses the numbers 1 and 1 in the positive and negative position respectively.

Therefore, the ant agent selects one attribute to be excluded from the pheromone deposition stored in the tabulist and one attribute to be included in the pheromone deposition. For the negative position, the ant agent randomly chooses position 1 and thereby the first attribute of the pheromone deposition in the tabulist is deleted. For the positive position, the ant agent chooses the attribute at position 2 to be added to the pheromone deposition. Thus the pheromone deposition {3,4} is taken from the tabulist and the attribute 3 present at position 1 in the pheromone deposition of the tabulist is deleted and attribute 2 is added to the pheromone deposition. Thus the pheromone deposition of the ant agent in trail-2 is {2, 4}. The energy value is found by estimating the leave-one-out classification accuracy of CMP which is found to be 44.4 which is less when compared to that of the earlier trail. Hence the subset of attributes denoting the pheromone deposition is ignored and the pheromone deposition {3, 4} is stored in the tabu list. The process is repeated for 10 trails. Thus the pheromone deposition {2, 3, 4} of the ant agent having the highest energy value denotes the solution. Thus the features 'Spectacle Prescription' and 'Astigmatic' and 'Tear Production Rate' in the mined patterns of the dataset are selected for comparison for the classification of unlabelled samples.

Table 2. Feature selection using ACOFSMP algorithm

Ant Agent	Positive, Negative Position	Position of features excluded from the tabulist	Position of features chosen from the instances in the training set	Pheromone Deposition	Energy Value	Tabulist	Features selected
Trail-1	2,0	-	3,4	3,4	77.8	3,4 {77.8}	{2,3,4}
Trail-2	1,1	1	2	2,4	44.4	3,4 {77.8}	
Trail-3	1,1	2	2	3,2	0	3,4 {77.8}	
Trail-4	1,0	-	1	1,3,4	22.2	3,4 {77.8}	
Trail-5	0,1	2	-	3	44.4	3,4 {77.8}	
Trail-6	1,0	-	2	2,3,4	88.9	2,3,4 {88.9}	
Trail-7	0,2	1,3	-	3	44.4	2,3,4 {88.9}	
Trail-8	0,2	1,2	-	4	44.4	2,3,4 {88.9}	
Trail-9	1,1	1	1	1,3,4	22.2	2,3,4 {88.9}	
Trail-10	0,1	2	-	2,4	44.4	2,3,4 {88.9}	

4.2. Classification of New Unlabelled Samples

The CMP algorithm predicts the class label for the unlabelled samples in the dataset given in table 3 based on the instances in the dataset in table 1. Using ACOFSMP algorithm, the attributes 'Spectacle Prescription', 'Astigmatic' and 'Tear Production Rate' of the dataset in table 1 are chosen for comparison with the unlabelled sample.

Table 3. Dataset with unlabelled samples

Unlabelled Sample Number	Age	Spectacle Prescription	Astigmatic	Tear Production Rate
1	2	2	1	1
2	3	1	2	1

The mined patterns for the dataset in table 1 are found as explained below. The instances in the dataset with similar class labels are grouped together. The instance number 1 in class label group '1' is chosen and compared with all other instances in the dataset having a class label value '1'. The count of number of attributes of instance number 1 matching with each of the other instances in the group having class label value '1' is found. If the matching count is n-1, (in this case it is 3 since the number of attributes n = 4), the attributes among the instances are merged and the patterns are derived. Hence the instances 1, 2 and 4 are merged to form the mined pattern A. This is repeated for all other unmerged instances in each class label group and the mined patterns for the dataset in table 1 are shown in table 4.

To predict the class label of the unlabelled sample number 1, each attribute of the unlabelled sample is compared with the corresponding attribute values in the mined patterns.

Since there is no unique pattern matching with all the attribute values of the unlabelled sample, the selected attributes 'Spectacle Prescription', 'Astigmatic' and 'Tear Production Rate' of the mined patterns in table 4 are compared with the corresponding attribute values of the unlabelled sample.

The count of the attribute values of each of the mined patterns matching with that of the unlabelled sample number 1 gives the attribute match count. Table 5 shows the attribute match count of each mined pattern for the unlabelled sample number 1.

Table 4. Mined Patterns derived for the dataset

Mined Patterns	Age	Spectacle Prescription	Astigmatic	Tear Production Rate	Class Label
A	3,1,2	1	2	2	1
B	1	2	2	2	1
C	2,1	1	1	2	2
D	3	2	1	2	2
E	1	1	1,2	1	3

Table 5. Matching attribute count of the unlabelled sample number '1' with the mined patterns

Mined Patterns	Age	Spectacle Prescription	Astigmatic	Tear Production Rate	Attribute match count	Class Label
A	3,1,2	1	2	2	0	1
B	1	2	2	2	1	1
C	2,1	1	1	2	1	2
D	3	2	1	2	2	2
E	1	1	1,2	1	2	3

It is observed that the mined patterns 'D' and 'E' has the highest attribute match count 2 and has the class labels '2' and '3' respectively. The probability of the unlabelled sample number '1' belonging to class label '2' is 1/2 =0.5 and that of class label '3' is 1/1=1. Thus the class label of the unlabelled sample number 1 belongs to class label '3' having the highest probability. Hence the unlabelled sample number 1 is predicted to belong to the class label '3'.

This is repeated for other unlabelled samples in the test dataset shown in table 3. Table 6 shows the predicted class label of the unlabelled samples in table 3.

Table 6. Predicted class labels of unlabelled samples

Unlabelled Sample Number	Age	Spectacle Prescription	Astigmatic	Tear Production Rate	Predicted Class
1	2	2	1	1	3
2	3	1	2	1	3

5. Experimental Results

It is shown that CMP algorithm is efficient for classification of very small datasets with small number of attributes having repeated attribute values. To show the efficiency of CMP algorithm, the experiment was carried out using very small datasets namely contact, balloon and shuttle from the UCI machine learning repository available in the website http://www.ics.uci.edu/~mlearn /databases/.

The experiment is also repeated for varying number of instances of balance dataset as shown in table 7. The efficiency of the CMP algorithm is proved using two widely used validation methods namely, 10 fold cross-validation and Full Training Full Test.

A. Full Training Full Test

According to Full Training Full Test method, the full training dataset was used for both training and testing.

The Full Training Full Test method proves the efficiency of the CMP algorithm to predict the class labels of the unlabelled samples having an exact pattern as that of the instances in the dataset.

B. 10 Fold Cross-Validation

10 fold cross-validation is to prove the efficiency of the classification model to predict the class labels of the unlabelled samples which are not already present in the dataset.

In 10 fold cross-validation, the whole training dataset was divided into ten mutually exclusive and equal-sized subsets. A subset was used as the test dataset, the other 9 subsets as a whole training dataset, one time validation was performed for this pair of training and test dataset and the classification accuracy is found.

The procedure was repeated 10 times and each subset was used as test dataset for one time. The mean classification accuracy is defined as the percentage of total number of correctly classified instances by the total number of testing instances.

Table 7. Characteristics of the dataset and classification accuracy of CMP using ACOFSMP

Data set	Number of Samples	Number of attributes	Number of classes	Attributes selected for pattern matching	Classification Accuracy (%)	
					10 fold Cross-Validation	Full Training Full Test
Contact	24	4	3	3,4	87.5	100
Balloon-adult+stretch (20)	20	4	2	3,4	100	100
Balloon-adult+stretch(16)	16	4	2	3,4	100	100
Balloon- adult-stretch(20)	20	4	2	3,4	100	100
Balloon- adult-stretch(16)	16	4	2	3,4	100	100
Balance with first 20 instances	20	4	3	1,2	96	100
Balance with first 50 instances	50	4	3	1,2,3,4	92	100
Balance with first 75 instances	75	4	3	1,2,3,4	89.3	100
Shuttle	15	6	2	1,6	93.3	100

6. Comparison of CMP Using ACOFSMP vs CMP with Classical Feature Selection Methods

The 10 fold cross-validation classification accuracy of CMP using ACOFSMP is compared with the 10 fold cross-validation classification accuracy of CMP with classical feature selection methods such as CfsSubsetEval, Principal Component Analysis (PCA), ClassifierSubsetEval and WrapperSubsetEval. It is observed from table 8 that CMP using ACOFSMP yields better results when compared to CMP using classical feature selection methods.

Also, it is observed that the classification accuracy is the same for the datasets balloon-Adult+stretch with 16 instances, balance dataset with the first 50 and 75 instances. Thus it is shown that the feature selection method using ACOFSMP is better when compared to the classical feature selection methods.

Table 8. Comparison of 10 fold cross-validation classification accuracy of CMP using ACOFSMP Vs CMP using classical feature selection methods

Dataset	10 fold cross-validation				
	CMP-ACOFSMP	CMP-CfsSubsetEval	CMP-PCA	CMP-Classifier SubsetEval	CMP-Wrapper SubsetEval
Contact	87.5	75	75	75	75
Balloon-Adult+stretch (20)	100	80	80	80	80
Balloon-Adult+stretch (16)	100	100	100	100	100
Balloon-Adult-stretch (20)	100	90	90	90	90
Balloon-Adult-stretch (16)	100	87.5	87.5	87.5	87.5
Balance with first 20 instances	96	95	96	95	95
Balance with first 50 instances	92	92	92	92	92
Balance with first 75 instances	89.3	89.3	89.3	89.3	89.3
Shuttle	93.3	86.6	86.6	86.6	86.6

7. Comparison of CMP Using ACOFSMP with Existing Classification Techniques

The classification accuracy of datasets of CMP using ACOFSMP was shown to be efficient when compared with the existing classification techniques such as classification trees, lazy learning methods and Naive Bayes methods executed using Weka 3.6.1 machine learning software package.

7.1. Comparison of CMP Using ACOFSMP vs Classification using Trees

The 10 fold cross-validation classification accuracy of CMP using ACOFSMP is compared with classification using tree methods. The experiments for classification using trees is executed in Weka 3.6.1 and compared with that of CMP.

It is observed from table 9 that CMP using ACOFSMP scales well when compared to the tree methods. It is observed from table 10 that though the Full training Full test classification accuracy of Random Tree and CMP using ACOFSMP are the same, the classification accuracy of Random Tree is less for 10 fold cross-validation when compared to CMP using ACOFSMP as observed from table 9.

Thus CMP using ACOFSMP classifies efficiently when compared to the tree methods.

Table 9. Comparison of 10 fold cross-validation classification accuracy of CMP using ACOFSMP vs Tree methods

Dataset	CMP using ACOFSMP	ID3	Random Tree	J4.8	CART
Contact	87.5	66.6	70.8	83.3	79.26
Balloon-Adult+stretch (20)	100	60	100	100	100
Balloon-Adult+stretch (16)	100	87.5	100	100	87.5
Balloon-Adult-stretch (20)	100	60	95	100	100
Balloon-Adult-stretch (16)	100	87.5	93.8	100	87.5
Balance with first 20 instances	96	95	70	75	85
Balance with first 50 instances	92	92	88	90	92
Balance with first 75 instances	89.3	88	82.6	82.7	86.6
Shuttle	93.3	73.3	93.3	93.3	93.3
Average	95.3	78.9	88.2	91.6	90.1

7.2. Comparison of CMP Using ACOFSMP vs Lazy Learning Methods

The 10 fold cross-validation classification accuracy of CMP using ACOFSMP is compared with the classification accuracy using lazy learning methods as shown in table 11 and 12. The experiments for classification using lazy learning methods are executed in Weka 3.6.1.

It is observed from table 11 that CMP using ACOFSMP outperforms the classical lazy learning methods for all datasets. It is observed from table 12 that although the Full training Full test classification accuracy of LB1 and CMP using ACOFSMP are the same, the classification accuracy of LB1 is less for 10 fold cross-validation when compared to CMP using ACOFSMP as observed from table 11.

Thus CMP using ACOFSMP has the highest classification accuracy for all datasets when compared to lazy learning approaches.

Table 10. Comparison of Full training Full test classification accuracy of CMP using ACOFSMP vs Tree methods

Dataset	CMP using ACOFSMP	ID3	Random Tree	J4.8	CART
Contact	100	62.5	100	91.7	87.5
Balloon-Adult+stretch (20)	100	60	100	100	100
Balloon-Adult+stretch (16)	100	75	100	100	100
Balloon-Adult-stretch (20)	100	60	100	100	100
Balloon-Adult-stretch (16)	100	75	100	100	100
Balance with first 20 instances	100	95	100	96	96
Balance with first 50 instances	100	92	100	92	92
Balance with first 75 instances	100	88	100	88	88
Shuttle	100	60	100	93.3	93.3
Average	100	74.2	100	95.7	95.2

Table 11. Comparison of 10 fold cross-validation classification accuracy of CMP using ACOFSMP vs Lazy learning methods

Dataset	CMP using ACOFSMP	Naive Bayes	Bayes Net	LB1	KStar	LWL
Contact	87.5	70.8	70.8	62.5	70.8	70.8
Balloon-Adult+stretch (20)	100	100	100	75	100	100
Balloon-Adult+stretch (16)	100	100	75	87.5	81.25	100
Balloon-Adult-stretch (20)	100	100	100	90	100	100
Balloon-Adult-stretch (16)	100	100	75	75	75	100
Balance with first 20 instances	96	75	85	70	80	70
Balance with first 50 instances	92	92	92	86	92	92
Balance with first 75 instances	89.3	89.3	88	85.3	89.3	88
Shuttle	93.3	93.3	93.3	93.3	86.6	93.3
Average	95.3	91.2	86.6	80.5	86.1	90.5

Table 12. Comparison of Full Training Full test classification accuracy of CMP using ACOFSMP Vs Lazy learning methods

Dataset	CMP using ACOFSMP	Naive Bayes	Bayes Net	LB1	KStar	LWL
Contact	100	91.6	87.5	100	100	100
Balloon-Adult+stretch (20)	100	100	100	100	100	100
Balloon-Adult+stretch (16)	100	100	100	100	100	100
Balloon-Adult-stretch (20)	100	100	100	100	100	100
Balloon-Adult-stretch (16)	100	100	100	100	100	100
Balance with first 20 instances	100	100	96	100	100	96
Balance with first 50 instances	100	100	92	100	100	92
Balance with first 75 instances	100	94.7	88	100	97.3	88
Shuttle	100	100	93.3	100	100	100
Average	100	98.5	95.2	100	99.7	97.3

Table 13. Comparison of 10 fold cross-validation classification accuracy of CMP using ACOFSMP Vs Ensemble methods

Dataset	CMP using ACOFSMP	Bagging with J48	Random Forest	Multiboost AB with J48	Rotation Forest
Contact	87.5	79.2	70.8	87.5	75
Balloon-Adult+stretch (20)	100	100	100	100	100
Balloon-Adult+stretch (16)	100	100	100	100	75
Balloon-Adult-stretch (20)	100	100	100	100	100
Balloon-Adult-stretch (16)	100	100	100	100	81.3
Balance with first 20 instances	96	95	80	95	85
Balance with first 50 instances	92	90	88	90	92
Balance with first 75 instances	89.3	88	85.3	86.7	88
Shuttle	93.3	53.3	93.3	66.7	93.3
Average	95.3	89.5	90.8	91.8	87.7

Table 14. Comparison of Full Training Full test classification accuracy of CMP using ACOFSMP Vs Ensemble methods

Dataset	CMP using ACOFSMP	Bagging with J48	Random Forest	Multiboost AB with J48	Rotation Forest
Contact	100	95.8	100	95.8	95.8
Balloon-Adult+stretch (20)	100	100	100	100	100
Balloon-Adult+stretch (16)	100	100	100	100	100
Balloon-Adult-stretch (20)	100	100	100	100	100
Balloon-Adult-stretch (16)	100	100	100	100	100
Balance with first 20 instances	100	95	96	100	96
Balance with first 50 instances	100	92	100	100	96
Balance with first 75 instances	100	94.6	100	97.3	96
Shuttle	100	80	93.3	80	93.3
Average	100	95.3	98.8	97.01	97.5

7.3. Comparison of CMP Using ACOFSMP vs Ensemble Methods

The 10 fold cross-validation classification accuracy of CMP using ACOFSMP is compared with the classification accuracy using Ensemble methods as shown in table 13 and 14. The experiments for classification using Ensemble methods are executed in Weka 3.6.1.

It is observed from table 13 and 14 that CMP using ACOFSMP has the highest classification accuracy for all datasets when compared to Ensemble approaches.

CONCLUSION

A CMP algorithm is proposed for efficient data classification for very small datasets having very few attributes with repeated values by matching the attributes of the unlabelled sample with the corresponding attributes of the patterns mined from the dataset. An ACOFSMP algorithm is used to select the features in the mined patterns for comparison. It is proved that CMP had better classification accuracies for very small datasets with very few dimensions having similar attribute values when compared to the existing classification techniques.

REFERENCES

Blum, A., Langley, P., "Selection of relevant features and examples in machine Learning", *Artificial Intelligence,* Vol 97, pp. 245–271,1997.

Boser, B. E., Guyon, I. M., Vapnik, V. N, "A training algorithm for optimal margin classifiers", *Fifth Annual ACM Workshop on Computational Learning Theory,* pp. 144–152, 1992.

Chebrolu, S., Abraham, A., Thomas, J, "Feature deduction and ensemble design of intrusion detection systems", *Computers and Security*, Vol. 24, pp. 295–307, 2005.

Chen, Y. H., B. Yang, J.W. Dong, A. Abraham, "Time-series forecasting using flexible neural tree model", *Information Science*, Vol. 174 (3/4), pp. 219–235, 2005.

Dash, M., Liu, H., "Feature selection for classification*", Intelligent Data Analysis.* Vol. 1 (3), pp. 131–156,1997.

Lu, H., Rudy, S. Liu, H., "Effect data mining using neural networks", *IEEE Transaction on Knowledge and Data Engineering,* Vol. 8 (6), pp. 957–961,1996.

Marco Dorigo, Thomas Stutzle, " Ant Colony optimization", *Prentice Hall of India private Limited,* NewDelhi, pp.37–38, 2005.

Payam Refaeilzadeh, Lei Tang, Huan Liu," Cross-Validation", Arizona State University, 2008.

Qinlan, J. R., "Introduction of decision trees", *Machine Learning,* Vol.1, pp.86–106,1986.

Sreeja.N.K, Sankar A "Efficient classification of datasets by mining similar patterns", *Proceedings of Indian Engineering Congress*, 2011.

Thair Nu Phyu ," Survey of Classification Techniques in Data Mining", *Proceedings of the International Multi Conference of Engineers and Computer Scientists* Vol 1, 2009.

Vittorio Maniezzo, Luca Maria Gambardella, Fabio de Luigi, Ant Colony Optimization, http://www.idsia.ch/~luca/aco2004.pdf, 2004.

http://archive.ics.uci.edu/ml/datasets.html.

In: Self-organization: Theories and Methods
Editor: WenJun Zhang
ISBN: 978-1-62618-865-5

Chapter 3

SELF-ORGANIZATION AND TASK ALLOCATION: AN APPLICATION TO ANT ALGORITHMS

***Jose B. Escario*[*], *Juan F. Jimenez*[†] *and Jose M. Giron-Sierra*[‡]**
Department of Arquitectura de Computadores y Automatica
Universidad Complutense, Spain, Madrid

Abstract

The aim of this chapter is to present an auto-organization study in the context of Ant Colony Optimization (ACO). When using heuristic optimization algorithms, it is needed to decide how to share efforts between the exploration of the latest solutions, and the exploitation of those which were previously discovered. In classical ACO algorithms, this decision is taken beforehand by setting up various algorithm parameter values. Usually it is necessary to change them every time the latest or newest problem is tackled. Our purpose with ACO is to introduce a self-organization dynamic that allows for the balance between exploitation and the exploration. Inspired by biological studies on the organization of actual ant colonies, two kinds of ants are defined: one dedicated to the exploration, and the other assigned to the exploitation. This defined population dynamics shall establish the relationship between both populations. This allows to self-regulate the populations between each of them, according to the quality of the solutions found. The ultimate outcome developed is the Ant Colony Extended (ACE), a new algorithm with the capability to self-organize, which eliminates the necessity of using parameters. This chapter is focused on the population dynamics and its influence on the algorithm's behavior. The results are promising and encourage further investigation of how to apply this approach to other heuristic optimization algorithms.

[*]E-mail address: jbescario@filos.ucm.es
[†]E-mail address: juan.jimenez@fis.ucm.es
[‡]E-mail address: gironsi@fis.ucm.es

PACS 05.45-a, 52.35.Mw, 96.50.Fm
Keywords: Self-organization, Ant Colony Optimization, Swarm Intelligence, Multi-Agent System, Ant algorithms
AMS Subject Classification: 53D, 37C, 65P

1. Introduction

The metaheuristic method for Ant Colony Optimization (ACO) was originally developed to solve combinatorial optimization problems [1]. This algorithm inspired the pheromone trail formation, which is the process that establishes the route path from a colony to a food source [2].

The pheromone trail formation enables a form of communication known as stigmergy [3, 4]: indirect communication through the environment. Using chemical tracks laid on the ground, i.e., pheromone marks, the ants are able to communicate among themselves by indicating which directions they have taken. It works as follows: when an ant has to make a decision, it looks at the environment for signs that indicate which is the best direction to take. The quality of a direction is indicated by the amount of pheromone present in the environment. If an ant is successful in the location of food, it deposits a certain amount of pheromone along the way back to the colony, to communicate the direction which has led it to succeed. Thus, an ant can provide information to the colony about which directions were successful.

If two simple rules are combined: observation and modification of the environment, then a dynamical system is obtained that is able to self-organize. At the beginning, when the ants begin to forage for food, there is no trace on the environment of which is the best directions to take. Nevertheless, there is a chance that some ants may, randomly, locate a food source, and leave the corresponding pheromone tracks in the environment. In time, the pheromone reaches a significant level. Thus, the ants begin to take these directions more often than the rest, reinforcing the pheromone tracks. Eventually, almost all the ants end up taking the same directions.

Departing from a disordered state without any activity pattern, the system converges to an ordered state, where it can now be identified as possessing specific activity patterns. It is this reason, real ant colonies are considered self-organized systems [4].

Using multi-agent techniques, Marco Dorigo proposed the first algorithm, Ant System [5], which solves combinatorial problems inspired by ant colony behavior. The algorithm uses a graph representation: the nodes are the components of the solution's sequence, and the links fully connect the nodes. Each link has associated a probability of being chosen that is represented as a pheromone value. An artificial ant, i.e., an agent, traverses the graph, from node-to-node, using the pheromone values to decide the order for visiting the nodes. The ordered list of visited nodes is the solution to the combinatorial problem. When an ant has finished building a solution, it modifies the pheromone values associated to the links, reinforcing those used in its graph path. The reinforcement is made according to the quality of the solution built.

The algorithm works on the premise with the same pheromone trail formation method used by real colonies. The directions are now links in the graph and the environment is a data structure: the pheromone table. Artificial ants can read and modify this table, as real ants do it in the environment. Like in real colonies, the artificial ants start visiting the nodes of the graph randomly. Along the search, the pheromone values on some links of the graph start to increase and the artificial ants tend to visit certain nodes more frequently. After some time, the pheromone levels reach certain threshold values and, eventually, a stable activity pattern emerges: the ants tend to build solutions by visiting the nodes almost in the same order.

Although the approach of Ant System proved promising, the comparisons done using the Traveling Salesman Problem (TSP) [6] revealed that it was not competitive with other state-of-the-art algorithms [5, 7–9]. Further research improved the results of the Ant System algorithm. Among others, the most remarkable were Marco Dorigo Ant Colony System (ACS) [1, 7] and Thomas Stüztle Max-Min Ant System [1, 10, 11]. These last two algorithms clearly outperformed the Ant System, but the research also revealed that these ant algorithms would require a more carefully tuned user-defined parameters in order to achieve a reasonable performance. This is the reason that there is an increased interest on how to automatically tune the parameters of ACO algorithms [12–14], in order to make its application easier.

It is possible to take it another step further, in this line of investigation, and study the possibility of replacing the user-defined parameters by implementing the self-organization dynamics. Thus, the behavior of the algorithm is no longer determined beforehand. Instead, it becomes an emergent property derived from its own activity in the algorithm. If the dynamics are designed correctly, the performance of the algorithm should be reasonably "independent" to the problem where it is applied, since the emerging behavior is partially determined by the problem itself. The result of this investigation is a new kind of ants algorithm: Ant Colony Extended (ACE).

Like in the case of the classical ant algorithms, ACE finds its inspiration source in how a real ant colony is organized, according to the research studies of Deborah Gordon [15,16]. The viability of this new algorithm has been studied by applying it in context to the ship maneuver optimization [17], and to the minimization of continuous functions [18].

The purpose of this chapter is to give a more detailed insight of how the dynamics of this kind works, and study the self-organization capacities that it provides. The chapter is organized as follows: section 2. contains an overview of the biological inspiration, and how it is defined using the biological ideas, a self-organization dynamics for a multi-agent system. Section 3. contains a detailed description of the ACE basic framework that will serve as the base for including the self-organization dynamics. Section 4. contains the experimental studies performed using the ACE basic framework. Section 5. contains the details in how to incorporate the self-organization properties in ACE, through the definition of a population dynamics. Section 6. contains the experimental studies that include the population dynamics in ACE. The final section 7. provides our conclusions.

2. Self-Organization and Task Performance

This section introduces how the self-organization behaviors of an ant colony allows for the emergent task allocation. Details of how this behavior can be used to design their self-organization dynamics to perform a task allocation in a multi-agent system.

2.1. Biological inspiration

The biological studies [19, 20] revealed that an ant colony self-regulates its organization through interactions – these encounters among the ants, defines their direct communication. This organization consists primarily on how to distribute the ants within the different tasks that must be undertaken. These roles are behavior-based patterns, where each role is associated to the performance of a specific task[1].

From these studies, it is possible to identify several key items that are necessary to enable this kind of self-organization:

- Task distribution: the complex behaviors of a colony arise from the performance of the ants carrying out different simple tasks. It is common to differentiate between active ants: those which are already performing a task, and inactive ants: those that are not yet engaged to any task.

- Recruitment: an ant that is performing a certain task can stimulate inactive ants to engage them to the task being performed.

- Pool of inactive workers: the inactive ants are not disperse but gather around the nest waiting to be recruited, making it easier for active ants to recruit them.

- Spontaneous activation: a mechanism that allows inactive ants to start performing any task without being recruited by already active ants.

From an algorithmic perspective, among all the biological studies, the description of the foraging activity [15, 16, 19, 20] given by Deborah Gordon is particularly interesting. Since it can serve as a basis for the design in the self-organization dynamics within an algorithm context.

The foraging activity is composed of two tasks: the localization of food sources and the exploitation of those sources. Each task is carried out by a different kind of ant:

- Patrollers are ants in charge of finding food sources.

- Forager are ants in charge of exploiting the food sources.

The foraging activity works based on *encounters* [19]:

I. It starts with the spontaneous activation of patroller ants.

II. A patroller that succeeds locating a food source, comes back to the nest. On its way back, the patroller can meet inactive ants. The outcome of this *encounter* is the recruitment of inactive ants, who then become foragers, and therefore start carrying out the exploitation of a food source. In other words, the successful patrollers recruit foragers to start performing the exploitation task.

III. Once a small group of forager ants has been activated by the patrollers, the foraging activity is self-sustained, since foragers can also recruit new foragers:

[1] Although, all ants are identical, it is common to differentiate several kinds of ants according to their different roles.

- When a forager ant has success in exploiting a food source, it comes back to nest carrying food. If a forager comes back carrying food, it stimulates new foragers to start exploiting the food source.
- In the case where a forager comes back to the nest without carrying food, there is no recruitment and the forager turns inactive.

The foraging activity can be described using a causal loop diagram [21, 22] in order to illustrate the system dynamics. The diagram is shown in figure 1, and as it can be observed, there are two feedback loops that regulates the number of active foragers. The right one is a positive feedback loop: the more foragers that are active, the more foragers will come back to the nest carrying food. Therefore, they will recruit new ones and the number of active foragers increases. The left one is a negative feedback loop: since the foragers coming back to the nest without food become inactive, the number of active foragers decreases. The two top arrows show the influence of the patrollers over the active foragers.

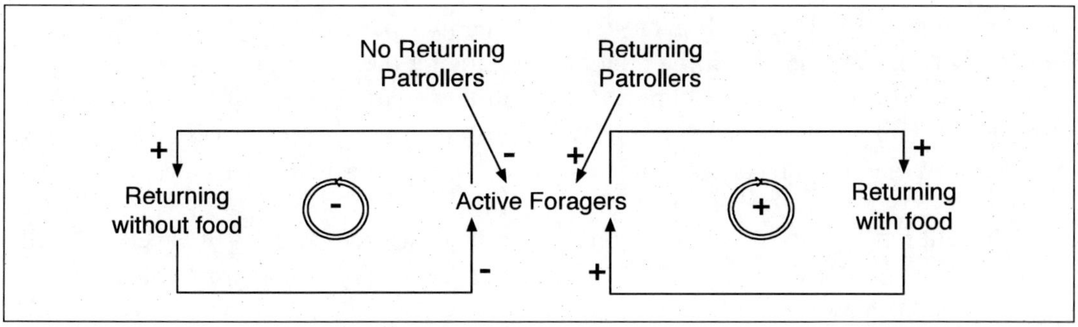

Figure 1. Causal loop diagram of the foraging activity.

Gordon describes all the foraging activity based on the concept of encounters. An encounter takes places when two ants meet, where one of the ants can stimulate the other to start performing a certain task. The encounters are what produce the recruitment, but this happens only if the generated stimulus is higher than a threshold. For instance, forager ants only start foraging if their rate of encounter with patroller ants is higher than the threshold. Sporadic encounters do not seem to stimulate forager ants, because the stimulus produced by an encounter decays over time. It can be seen as if the colony's response to the discovered food sources is "filtered" by the threshold: below this level the colony does not respond. This behavior makes sense because a high rate of encounters means that there is abundant food available, and thus the foraging activity is worthwhile. On the other hand, a low rate means that the food is scarce or too dispersed, and thus the foraging activity is not worthwhile.

The response to an stimulus using a threshold is a common way to model the division of labor and task allocation of social insects [23,24]. In these models, the stimuli are generated by the environment, remaining external to insects. For example the available food which can be perceived. Besides, these stimuli decrease as the task is being performed. Deborah Gordon proposes that the stimuli are produced by the task's performance, i.e., they are not external. It is the foraging activity by itself which produces the necessary stimuli to keep the activity going on. Therefore, these stimuli are produced according to the performance of the foraging activity itself.

For instance, let suppose there is a small food source near a colony's entrance. The food is scarce, but since it is close to the colony, it is easy to find. If the ants react to the amount of available food, i.e., external stimuli, they can ignore it, since it is not large enough. Nevertheless, if the stimuli are produced by the foraging activity, the colony probably will exploit this small food source, since it is only needed that a few ants bring some food to stimulate the foraging activity. Once the foraging activity starts, it does not depend on the available amount of food, but on the chances to find it, i.e., in the performance of the foraging activity itself.

2.2. From Biology to Multi-Agent Systems

One of the techniques for modeling complex systems and self-organization dynamics is to develop agent-based models [25, 26]. Basically, this technique involves using multi-agent systems [27], where each agent models a certain part of the system and the relationships between agents model the relationships among the different parts of the system. Thus, the computational model obtained through these simulations allows for the study in self-organization dynamics. Since ACO algorithms are multi-agent systems, the self-organization dynamics can be incorporated by using the approach of the agent-based model.

- The ant returning with or without food, can be described in terms of success or failure when accomplishing a task. For instance, a forager that returns with food, is a successful forager. Instead, a forager that returns without food, is an unsuccessful forager. The same reasoning can be applied to patrollers.

- The encounters can be defined as events, which are produced by the agents when they finish a certain task. These events increase or decrease the value of a certain signal, according to the success or failure of the agent that has carried out the task.

- The signal is what activates an agent, triggering a recruitment process.

The objective is not to imitate the ant colony behavior, but to define this behavior as a reference in the scheme of a self-organization dynamics for a multi-agent system. In virtue of this scheme, multi-agent systems are able to determine the number of agents assigned to the fulfillment of a task.

A possible scheme could be the one described by the causal loop diagram shown in figure 2. Agents doing task B can trigger the start of task A; and once the task A has started, it becomes a self-sustained process. If the performance is successful, the positive feedback loop increases the number of agents assigned to task A. While if the performance is unsuccessful, then the negative feedback loop reduces the number of agents assigned.

This kind of dynamics composes a basic module, where the task being done is regulated on the basis of a success rate. Therefore, a minimum success chance is required in order to perform a task. Notice that if the task is difficult to accomplish, it is simply ignored. In the most common situation this is not desirable. One possible solution could be to interchange the feedback loops: set the positive feedback on the fail side, and the negative feedback

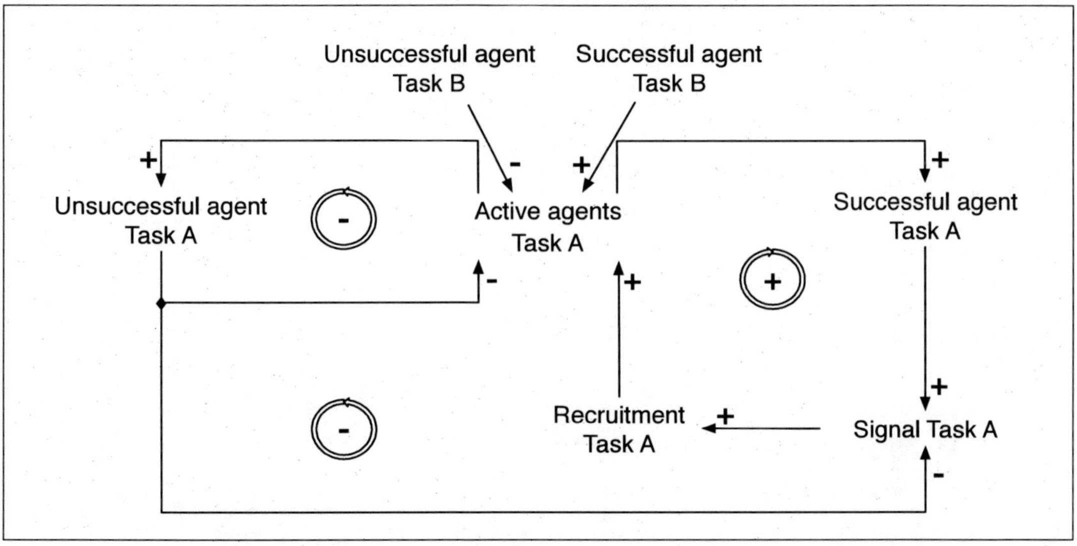

Figure 2. Causal loop diagram of a system dynamics based on success rates.

on the success side. In such a way that a basic module is obtained, which allows to adjust the number of agents according to the task's difficulty. And hence, the accomplishment of the task is ensured. Once the task is accomplished, the activity decays in order to avoid over-performing this task. This module can be easily combined to design a self-sustained system that continuously performs different tasks.

Figure 3 shows an example of a system designed that uses this module for completing two tasks (*A* and *B*). The activity begins with the spontaneous activation of the agents that start performing task *A*. If those agents are not successful, the left positive feedback loop of task *A*, tries to ensure that the task is accomplished, recruiting new agents. When those agents start to be successful, they decrease the recruitment signal for task *A*, while increase the recruitment for task *B*. This activates some agents to start performing task *B*. Notice that the activated agents on task *B* is proportional to the number of successful agents of task *A*. The positive feedback loop of task *B* tries to ensure that there are enough agents to cope with the difficulty of task *B*. When the agents assigned to task *B* are being successful, they start to decrease the number of agents assigned to task *B*, and trigger the recruitment of task *A*.

Besides ensuring the task achievement, another advantage for a system with the self-organization dynamics as just described, is able to balance the progress of the tasks performed: it rarely happens when during an extended period of time that just one task would have progressed, since a successful task cannot run indefinitely –as a module starts to fulfill its task, it tends to stop and triggers the performance of another task. Thus, the control of the system oscillates among the modules, depending of the difficulty of performing each task. Also, we notice that this kind of dynamics works as if the task completion has a fixed difficulty, or if the difficulty varies as the task is being fulfilled. This may happen either due to environmental effects or just due to the progress of the task.

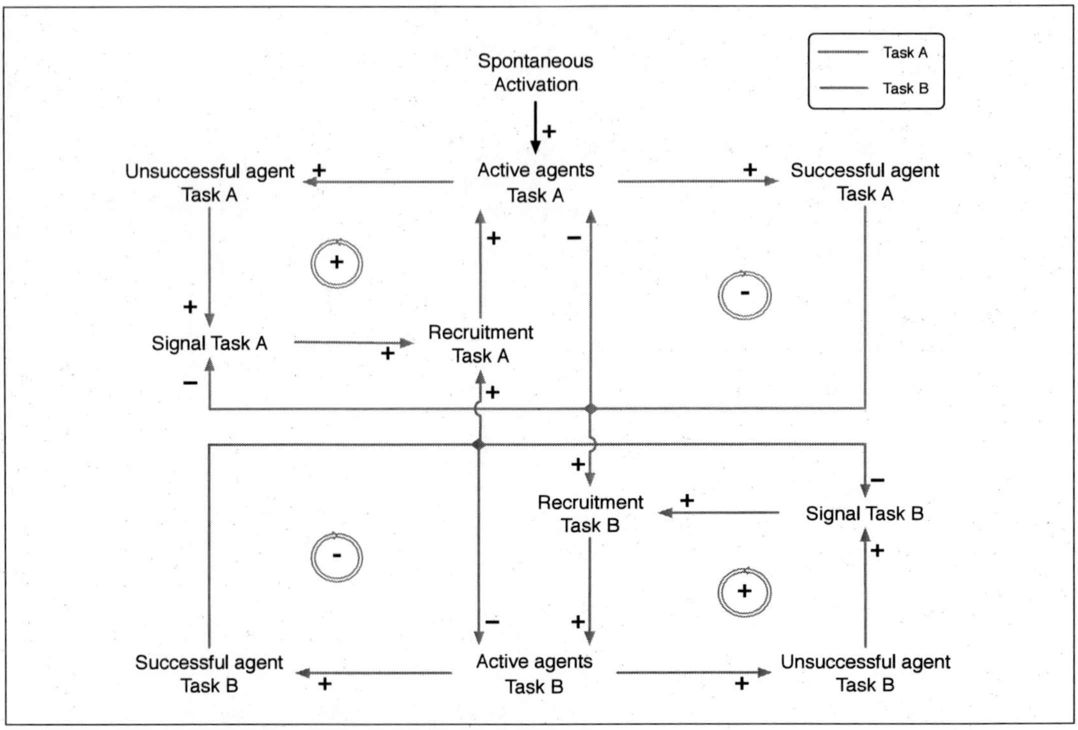

Figure 3. Example of a self-organization dynamics to accomplish two tasks.

This kind of self-organization dynamics are very attractive because they can be applied in many different contexts, in order to obtain systems that are able to auto-configure themselves to perform certain tasks. One such context is an optimization algorithm, where the dynamics can be used to perform the task of re-defining the algorithm's parameters; thus, it is possible to obtain a self-organization algorithm which configures itself while solving a problem.

3. ACE: a Basic Framework

There are two crucial aspects to define prior to the implementation of self-organization dynamics, which are the tasks to be fulfilled, and how to measure the success or failure of task's achievement. But first, it is pertinent to detail how ant algorithms work.

3.1. Information and Search

Ant algorithms can be seen as iterative search processes guided by information. The information is a codification of the steps needed to build different solutions of the problem to be solved, and this information is contained in the pheromone table.

3.1.1. Problem Representation

In the concrete case of ACE, the problem is solved by a classical approach in artificial intelligence called *solving problems by searching* [28, 29]. According to this approach, solutions are obtained by performing a search in a state space representation of the problem.

Formally, this approach requires the definition of the following elements:

- Q, a set of possible states.
- U, a set of possible actions.
- $q_0 \in Q$ the initial state.
- $\delta : Q \times U \mapsto Q$, a sucessor function.
- $goal$, a function to determine if the solution conditions are satisfied, i.e., a goal test.
- g, a cost function.

A state, q, is a symbolic expression that represents a situation in the problem's space, for example, in a robot motion planning problem the state usually represents a location. The set of actions, U, and the successor function, δ, define a relationship among the states: the way to obtain the set of all reachable states from a given one. Continuing with the previous example, if the robot is in a location q, and executes an action u (for instance, move to the right), then the function δ determines which will be the next location (state) of the robot. The state space of the problem is defined as the set of all reachable states from the initial state, q_0, by any sequence of actions. A sequence of actions from one state to another is called a path in the state space. The cost function, g, assigns a cost to a path.

The problem is solved by a searching method that moves from one state to another through the application of a certain action, until the goal test is satisfied. Notice that it is not required to build a complete state space before exploring it; the state space is explored by building it. The searching method controls how to build the state space.

Since the agents solve the problem moving within a state space, the information intended to guide the search should indicate which action must be performed at each state.

3.1.2. The Pheromone Table

According to the problem representation used, the pheromone table is an associative list, where the keys are states, and the entries are the list of pairs. Each pair consists of an action and a pheromone value, where this last represents the probability of performing the action. The scheme is represented in figure 4.

Key	Value
q_0	$[(u_0, p_0), (u_1, p_1), \ldots]$
q_1	$[(u_1, p_1), (u_2, p_2), \ldots]$

Figure 4. Pheromone table structure. The symbol q represents a state, u an action, and p a probability value

The agents can build solutions of the problem by selecting the appropriate action for each state: this selection is a random procedure based on the probability of performing each action. This probability is contained in the pheromone table for the current state. The process is called *solution construction*.

Once the agent has built a solution, for each traversed state the agent increases the probability of the actions used, and decreases the probability of the unused actions. The modification of the probabilities is done according to the quality of the obtained solution. The quality is computed using the objective function of the problem. This second process is called *pheromone update*.

3.2. Two Tasks: Exploration and Exploitation

In a search process, it is important to maintain a reasonable diversity balance, in such a way that the search does not become too disperse nor too focused.

In ant algorithms, the agents use the information of the pheromone table to perform the search, and this information determines the range of the solutions that can be found. Thus, a good search diversity can be achieved through the management of the pheromone table's information. This information refers, within this context, to the uncertainty about which action should be performed from those available for the states present in the table.

Based on this, it is possible to define two tasks:

- Exploration task: its objective is to increase the uncertainty of which action should be performed in one or more states. This can be achieved by increasing the dispersion of the probabilities distribution for the available actions of those states. If the uncertainty of which action to perform is increased, then the variety of solutions that can be found becomes larger.

- Exploitation task: it is the opposite task of the exploration, a task that reduces the uncertainty of which action to perform in one or more states, reducing the dispersion of the probabilities distribution. As the uncertainty decreases, the search tends to concentrate, because the variety of the solutions that can be found is reduced.

These two tasks can be fulfilled just by searching. The common search procedure of ACO methodology is a way to reduce the uncertainty of the pheromone table, since the agents tend to select the most probable action for a state and they increase the probability of the actions used. This means that the search process is a positive feedback loop: the search tends to increase the probability of those, already, most probable actions for each state, reducing the uncertainty.

One way to increase this uncertainty, is to allow the agents to select an action using another distribution of probabilities different than the one contained in the pheromone table, i.e., by a search using another information source. In such a way that when an agent searches using this last distribution, the actions selected for the visited states do not need to be the most probable actions in the pheromone table. Indeed, the action performed does not even need to be available in the table, if the table only contains previously used actions. In this case, the pheromone update may increase the uncertainty on which action to perform. This new distribution of probabilities can be defined through a heuristic function, which is problem dependent. If there is no available heuristic, it is always possible to use a flat distribution.

In accordance with this hypothesis, the two tasks can be seen as two different ways to search. Thus, it is possible to define two agents, each one performing a different kind of search. In order to preserve the biological inspiration, the agents are called patrollers and foragers:

- Foragers (exploitation search): agents that search for solutions using only the information provided by the pheromone table.

- Patrollers (exploration search): agents that search for solutions using the information provided by the pheromone table or a problem dependent heuristic function.

Once the tasks and the agents of the system have been identified, a criterion is needed to determine when an agent has successfully accomplished a task.

3.2.1. A Criterion to Determine the Success or Failure

As the two tasks are searches, it is possible to define the success or the failure according to the solution obtained. A possible criterion could be based on the average (μ) of the qualities of all solutions found so far. Successful agents are those that find a solution which its quality, assigned by the cost function is below μ. If this is not the case, the agent is considered unsuccessful.

This criterion makes possible to refine the results as the search progresses: if solutions are only accepted when they are better than the averaged solutions previously discovered, i.e., only successful agents are allowed to update the pheromone table. Then the value of μ decreases as more solutions are discovered. Therefore, progressively the accepted solutions must have a better quality.

This criterion can be seen as a threshold that serves to determine whether a solution is good enough or not. Since the solutions are build through a search guided by the information already contained on the pheromone table, the chances of an agent to build a solution below μ value depends of the information contained in the pheromone table.

Accuracy of the pheromone table information: according to what it has just been described, it is possible to talk about the accuracy of the pheromone table according to the chances of finding the solutions below μ, based on the information that the table contains:

- For a certain state, the information is said to be *accurate*, if the actions that are most probable of being performed serve to build solutions below μ. Thus, the whole table is said to be accurate, if for every state contained, the information tends to be accurate.

- For a certain state, the information is said to be *inaccurate*, if the actions that are most probable of being performed do not serve to build solutions below μ. In this case, the whole table is said to be inaccurate.

Notice that the value of μ is not static, instead it decreases as more solutions are discovered. Therefore, if the information of the pheromone table does not change, then the table becomes inaccurate. This relationship among μ and the accuracy of the pheromone table is key to understand the behaviors of each agent that will be discussed in later experimental sections.

3.3. An Algorithm Scheme

According to the ideas explained in the earlier sections, it is possible to define an algorithmic scheme for this basic framework. It is defined through two procedures: solution construction and pheromone update.

3.3.1. Solution Construction

This procedure defines how the agents build the solutions using the values stored in the pheromone table, and the values provided by a heuristic function. If an agent is at state q_n, the probability of selecting the action u_i is defined by equations 1 and 2.

$$P_\tau(u_i|q_n) = \frac{\tau_{u_i}^{q_n}}{\sum_{j \in N_\tau} \tau_{u_j}^{q_n}} \tag{1}$$

$$P_\eta(u_i|q_n) = \frac{\eta(u_i, q_n)}{\sum_{j \in N_\eta} \eta(u_j, q_n)} \tag{2}$$

Where $\tau_{u_i}^{q_n}$ is the pheromone value of the action u_i at state q_n, $\eta(u_i, q_n)$ is the heuristic value of the action u_i at state q_n and N_τ, N_η are the number of actions available, at state q_n; using the pheromone and the heuristic, respectively.

Notice that the heuristic function is used to provide another information source different from the pheromone table in order to select an action. This information source is defined by a probability distribution based on a priori knowledge of the problem. Therefore, the use of a heuristic function is not strictly necessary, because it is always possible to provide another information source using a flat distribution by giving the same probability to each action.

In the case where an agent, forager or patroller, reaches a state not present in the pheromone table, it uses the heuristic equation (P_η) in order to continue the search. Otherwise, depending on the kind of agent, the equations 1 and 2 are used as follows:

- Foragers search using only the pheromone equation (P_τ).
- Patrollers search employing both equations: heuristic and pheromone (P_η and P_τ), following a random process to determine which one to use:
  - If the last equation used was the pheromone one (P_τ), or the current state is an initial state (q_0)

$$\begin{cases} \chi < 1 - (1-\alpha)^{1/E} & \rightarrow P_\eta \\ \text{otherwise} & \rightarrow P_\tau \end{cases} \tag{3}$$

– If the last equation used was the heuristic one (P_η)

$$\begin{cases} \chi < 1-(1-\alpha)^{1/E}+\beta & \rightarrow P_\eta \\ \text{otherwise} & \rightarrow P_\tau \end{cases} \tag{4}$$

where χ is a random number generated from an uniform distribution, $U(0,1)$, E is the number of actions needed to build the best-so-far solution (E), $\alpha \in [0,1]$ is an user-defined parameter, $\beta < 1$ is another user-defined parameter. And the symbols "$\rightarrow P_\tau$" and "$\rightarrow P_\eta$" indicate respectively, the use of the pheromone (equation 2) or the heuristic (equation 1).

3.3.2. Pheromone Update

This procedure defines how to update the pheromone table according to the sequence built by an agent. The search starts with an empty pheromone table, its entries are filled gradually as the agents visit the states and perform new actions.

The table is updated as follows: the agents modify the pheromone values of those states visited while building a solution, increasing the value of those actions used and decreasing the values of the remaining unused actions. Pheromone values for actions belonging to non-visited states remain unchanged. The update is defined by equation 5, which is based on AntNet equations [1,30],

$$\begin{cases} \tau_{u_i}^{q_n} \leftarrow \tau_{u_i}^{q_n} + \lambda \cdot (1-\tau_{u_i}^{q_n}) & \forall u,q \in S \\ \tau_{u_j}^{q_n} \leftarrow \tau_{u_j}^{q_n} - \lambda \cdot \tau_{u_j}^{q_n} & \text{otherwise.} \end{cases} \tag{5}$$

where S is the solution built by an agent, τ_u^q is the pheromone value of selecting the action u at state q. The states q_n are those traversed by the agent while it built the solution S. The actions u_i are those used by the agent to build the solution S. And the actions u_j are those actions contained in the pheromone table that have not been used by the agent.

The term λ is the quality associated to the solution built by the agent, which is normalized between 0 and 1. The normalization is defined by the following equation:

$$\lambda = \begin{cases} \frac{g(ant)-\mu}{g(best)-\mu} & g(ant) < \mu \\ 0 & \text{otherwise} \end{cases} \tag{6}$$

where $g(ant)$ is the cost, given by the objective function, of the achieved solution, $g(best)$ is the cost of the best-so-far solution discovered, and μ is the mean of the costs for all solutions discovered so far by successful searches.

3.3.3. Pseudo-Code Scheme

The general scheme of this basic framework is shown in figure 5. For each iteration the algorithm dictates the following computations:

Solution Construction: each agent of the population performs a search step: selects an action and computes the resulting state of performing that action. Depending on the type of agent, the step can be done selecting the action using the pheromone or the heuristic.

Pheromone update: for each agent that has finished the search, the cost of the built solution is checked to see if it is below the average (μ), then the search can be considered successful and its solution is used to update the pheromone table. Next, μ and, if necessary, the best-so-far solution are updated[2].

After that, the table is updated according to the states traversed and the actions performed by the agent. It could happen that some of the actions employed for building the solution are not in the list of available actions for a state. In this case, they are added to the list.

Finally, every finished agent becomes inactive and it is retired from the active population.

Population management: since there is no population dynamics, the population is considered fixed. It is managed through two user-defined parameters: *NA* which is the number of agents of the population, and *NF* which represents how many agents must be foragers.

Once the first solution is not been indicated, a new patroller is added. This can be considered as an initialization phase. In the case when the first solution has been found, if the size of the population is lower than *NA*, it is updated by adding a new forager when their number is below the value of the parameter *NF*. Otherwise, it is added a new patroller.

4. Studies Using the Basic ACE Framework

To develop an appropriate self-organization dynamics which regulates the population, it is necessary to know how the number and kind of ants affect the algorithm's behavior. To gain more of an insight, a set of experiments were carried out that employ the basic ACE framework and the traveling salesman problem (TSP) [6]). TSP is one of the common benchmarks used to study the behavior of ant algorithms [1, 31]. All the TSP instances employed in the experiments were taken from TSPlib [32]. A suitable problem TSP representation to be used with ACE can be found in Appendix A.

The experiments are performed in batches of 16 executions, a single execution consists of building $20 \cdot 10^4$ tours. Each batch represents a different configuration of the algorithm, where the configuration is defined by the population size, the number of foragers, and how patrollers search:

- The population size is set according to the following values: 1, 2, 5, 20, and 51 agents.

[2]As μ is initialized to ∞, to accept any solution as the first one. Once the first solution has been found, μ must be reset to the value of this first solution.

```
% Initialization
samples ← 0    μ ← ∞
while termination condition not met do
   % Solution construction procedure
   for agent ∈ Population do
      if is forager(agent) then
         Pheromone Search Step (agent)    % equation 2
      else
         decision ← control policy    % equations 3,4
         if decision = true then
            Heuristic Search Step (agent)    % equation 1
         else
            Pheromone Search Step (agent)    % equation 2
         end if
      end if
   end for
   % Pheromone update procedure
   for agent ∈ Population do
      if search finish then
         if g(agent) < μ then
            samples ← samples + 1
            μ ← μ + (g(agent) − μ)/samples
            if g(agent) < g(best) then
               best ← agent
            end if
            for i ∈ {1..|U_agent|} do
               τ_{q_i} ← update(u_i, λ)    % equations 5, 6
               if u^i ∉ τ_{q_i} then
                  τ_{q_i} ← add(u_i, λ)
               end if
            end for
         end if
         Population ← retire(agent)
      end if
   end for
   % Population management
   if samples < 1 then
      Population ← add(patroller)
   else
      if size(Population) < NA then
         if number of foragers < NF then
            Population ← add(forager)
         else
            Population ← add(patroller)
         end if
      end if
   end if
end while
```

Figure 5. ACE basic framework pseudo-code scheme.

- The number of foragers is set as a fraction of the total amount of agents, that employ the following ratios: 0, 0.1, 0.2, 0.4, 0.6, 0.8, and 1.0.

- The patrollers may search either using a specific heuristic[3] or performing a blind search. In this last case, all actions have the same probability of being chosen.

The measures taken from executions that belong to the same batch are averaged to obtain a final estimation of their values. Finally, in all experiments used for the TSP instance *Eil*51 (51 cities) the algorithm parameters α and β were set to 0.99 and 0.5, respectively.

4.1. Global Behavior: Pheromone Table Information

This experiment is intended to investigate how the information of the pheromone table evolves depending on the composition of the population.

According to information theory [33, 34], the amount of information contained in the pheromone table can be measured using the concept of information entropy [33, 34]. The entropy associated to an entry in the pheromone table can be defined as,

$$H_i = \sum_{j=1}^{n} P_\tau(u_j|q_i) \cdot log(P_\tau(u_j|q_i)) \tag{7}$$

where $P_\tau(u_j|q_i)$ is the probability of selecting the action u_j at state q_i, which has n available actions. Notice that H_i measures the degree of uncertainty in making a decision. In fact, if there is only one decision available at state q_i, its probability should be $P_\tau(u|q_i) = 1$ and, consequently, the entropy of state q_i is zero.

The entropy of the whole pheromone table can be expressed as an average,

$$H_\tau = \frac{1}{N} \cdot \sum_{i=1}^{N} H_i \tag{8}$$

where N is the number of entries of the pheromone table.

The entropy (H_τ) is measured using a cumulative moving average. This allows us to obtain the average entropy along the algorithm execution ($\overline{H_\tau}$), and its evolution, i.e., how this averaged value had been reached, ($\overline{H_\tau}[i]$, where i is the number of tours built).

Evolution of the average entropy: Figure 6, shows the evolution of the average entropy ($\overline{H_\tau}[i]$) during an algorithm execution, according to the number of built tours and the ratio of foragers. First, a peak of entropy can be observed at the start of the evolution. This is due to fact that the algorithm starts with an empty pheromone table: there is no initial information available. Therefore, the entropy can only be increased with the first searches. These first searches include the initialization phase: a patroller is released at every cycle until a first solution is found. Also, several searches are performed once the initialization ends. This is due to the fact that the initialization only provides information about the visited states, and these are just a few. Hence, the performed searches, just after the initialization, are still

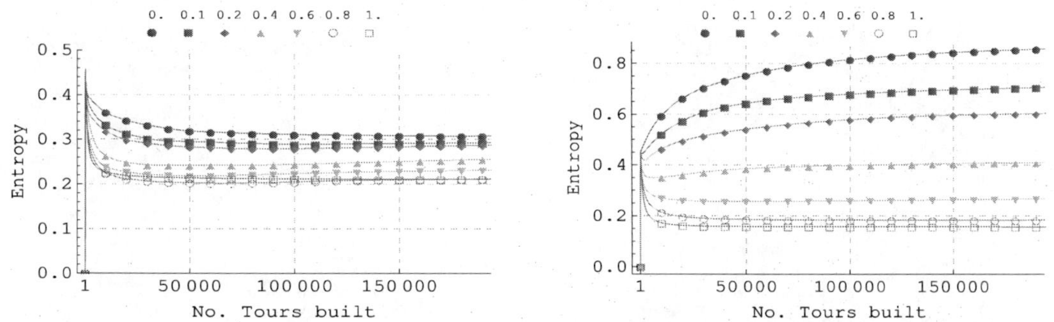

(a) Patrollers performing a heuristic search.

(b) Patrollers performing a blind search.

Figure 6. Evolution of the average entropy in the pheromone table ($\overline{H_\tau}[i]$) during an algorithm execution, according to the number of tours built and the ratio of foragers. The population size is set to 51 agents.

based on exploration (heuristic or blind search) and, accordingly, the information contained in the pheromone table keeps increasing.

As the pheromone table's information increases, the agents can use this information to guide their search. Thus, the foragers may start performing an exploitation search and the entropy can be reduced. This is what happens in the blind case, figure 6(b), when the number of foragers is appropriate to manage the uncertainty produced by the patrollers exploration ($ratio > 0.2$). It can be observed that if there are not enough foragers to cope with this uncertainty, the entropy may continue to increase, resulting to an excessive dispersion of the search.

On the other hand, when the patrollers use the heuristic function, since this is too restrictive –there is not much uncertainty on which there is action to perform– and patrollers just use it to select a few actions, therefore after the initialization phase, the entropy of the pheromone table tends to decrease to lower levels, even without foragers. This can be observed in the figure 6(a).

Comparing the results in figures 6(a) and 6(b), it can be seen that in either case, heuristic or blind search, the entropy tends to reach a steady state. This is due to the fact that the agents only modify the pheromone table when they find successful solutions –solutions below μ–. As the value of μ tends to decrease as the search progresses, increasingly it is harder to find solutions below μ. In consequence, the pheromone information tends to remain unaltered, keeping a certain level of entropy, which determines the search dispersion of the algorithm.

Average entropy: figure 7 shows the average entropy value ($\overline{H_\tau}$) of an algorithm execution, according to the foragers ratio and the population size. First, it can be observed as expected, the entropy is much higher on the case of a blind search than in the case of the heuristic search.

[3]The most common TSP heuristic function takes the nearest city as the most likely to be visited next. See Appendix A for details.

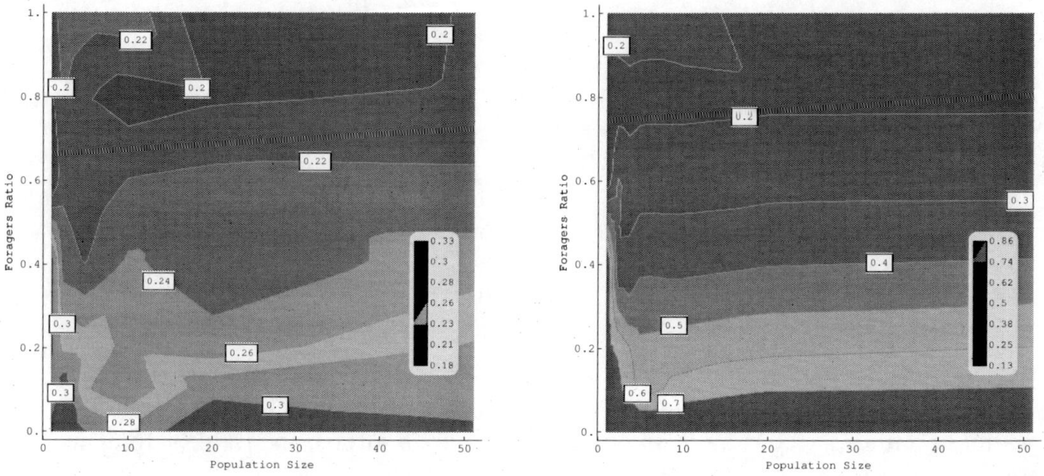

(a) Patrollers heuristic search.

(b) Patrollers blind search.

Figure 7. Average entropy of the pheromone table ($\overline{H_\tau}$) in an algorithm execution, according to the foragers ratio and the population size. (Contour lines)

The graphics show that the average entropy decreases when the number of foragers increases. This was also an expected result, because more patrollers mean more exploration and thus higher levels of entropy. Obviously, the higher the ratio of foragers, the lower the level of entropy.

The entropy level increases also when the population size increases. More agents means more patrollers which, in turn, means more exploratory searches. A low number of agents (approximately a population size below 20 agents) should be avoided because it may lead to erratic behaviors, due to the problem's degree of difficulty and the low number of available agents to solve it.

4.2. Agents Behavior: Probabilities Of Success

The previous experiment focused on the global behavior of the algorithm, i.e., the dependence of the pheromone table on the population size and composition. Nevertheless, the pheromone table also affects the behavior of single agents. The agents use the pheromone information in their search, and their probabilities of success depends on the results of that search. This mutual relationship leads to a feedback loop: where the agents depend on the pheromone table information, which in turn will depend on the agents performance. In fact, the pheromone table acts as a hub during the whole process, through linking the probability of success in a single agent with the performance of the others.

It is interesting to explore this basic relationship among the agents. The focus of this experiment is in the probability of success for foragers and patrollers, which represents the chances to find a solution below μ. This probability is measured, again, using a cumulative moving average in order to obtain the average probability along the algorithm execution, and the evolution of this average probability.

Evolution of the probability of success: figure 8 shows the evolution of the probability of success for foragers and patrollers during an algorithm execution, according to the number of built tours and the ratio of foragers.

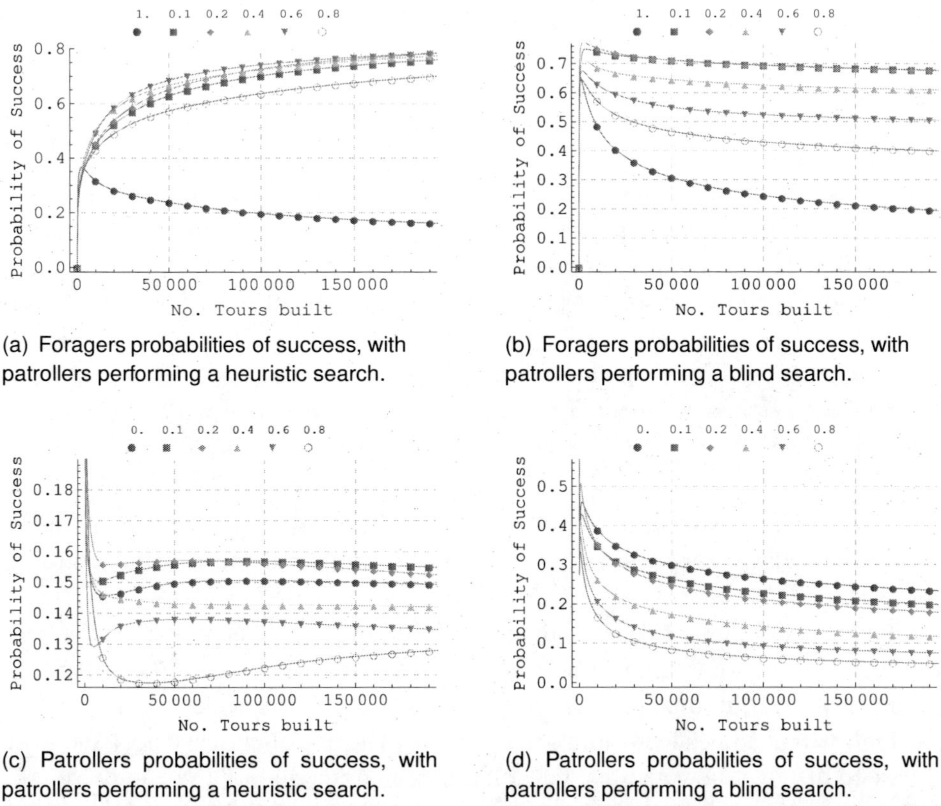

(a) Foragers probabilities of success, with patrollers performing a heuristic search.

(b) Foragers probabilities of success, with patrollers performing a blind search.

(c) Patrollers probabilities of success, with patrollers performing a heuristic search.

(d) Patrollers probabilities of success, with patrollers performing a blind search.

Figure 8. Evolution of the probabilities of success for foragers and patrollers during an algorithm execution, according to the number of tours built and the ratio of foragers. The population size is set to 51 agents.

Looking at the foragers activity, it can be observed that the probability of success decreases as the ratio of foragers increases. This is related to the accuracy of the pheromone table introduced before (section 3.2.1.). The foragers only search using the pheromone information, therefore every time that they have success, which is something relatively frequent, the value of μ decreases according to the solution built. Since the foragers do not use the complete information of the table, it happens that the μ value tends to be closer to the value of those solutions which have been built more frequently. In consequence, all the information contained in the table on how to build solutions that have not been built frequently, becomes inaccurate information: it does not serve to find solutions below μ. In time, as μ keeps decreasing, the set of solutions that can be built becomes smaller, and the amount of inaccurate information in the pheromone table increases.

When patrollers perform a heuristic search, figure 9(a), the solutions found are already close to the optimal solutions. Therefore, the value of μ can not decrease too much, and the effects of the inaccurate information are only noticeable if there are not patrollers ($ratio =$

1.0). On the other hand, when patrollers performs a blind search, figure 8(b), the solutions are far from the optimum, and thus the μ value can decrease significantly. Therefore, when the ratio of foragers is high ($ratio > 0.2$) the effects of the inaccurate information can be easily noticed, and the foragers probability of success decreases as its ratio increases.

Now for the case of patrollers, their probability of success is maximum at the beginning and decreases as the search progresses, due to the decrement of μ value. When patrollers perform a heuristic search, figure 9(c), the value of μ does not vary excessively. This is due to the fact that the heuristic allow to find solutions very close to the optimal. Thus, once the foragers decrease the uncertainty of the pheromone table, due to the first searches, the patrollers success tends to remain steady. In the case of a blind search, figure 8(d), the probability of success decreases proportionally to the difficulty of finding a solution below μ, i.e., proportional to the convergence of μ.

In both cases, as shown in figures 9(c) and 9(d), it can be observed that patrollers probability of success decreases as the ratio of foragers increases. This is due to the fact that with more foragers, the faster the value of μ decreases. Besides, the appearance of inaccurate information, which is augmented by an excessive number of foragers, also affects the patrollers probability.

Average probability of success: Figure 9 shows the average probability of success for foragers and patrollers, once the search is over, according to the foragers ratio and the population size.

It can be observed that, in general, the probability of success is higher for foragers than for patrollers. This is something to be expected, because the foragers use the pheromone information to build solutions. And the pheromone table, if it is accurate, tends to be filled with information that comes from solutions below μ. The patrollers, instead, have to employ the heuristic to discover new actions that can allow a build solution to be under μ.

The foragers, figures 9(b) and 9(a), have to reduce the uncertainty to obtain good solutions. More uncertainty in the table means a higher difficulty to fulfill their task. Therefore, when a high number of patrollers is present, foragers chances of success decreases, if there are not enough foragers to cope with the uncertainty produced by patrollers. On the other hand, an excessive number of foragers leads to the appearance of inaccurate information, which also decreases the chances of success.

In the case of patrollers, their probability of success decreases as μ decreases, and harder to find solutions below its value. As more foragers are successful, the faster the μ decreases. Therefore, the success of patrollers decreases as the ratio of foragers increases. This can be observed in both figures 9(d) and 9(c). Again, the appearance of inaccurate information also affect the patrollers. Because of this reason, in the case of a blind search, the chances of success decreases in a more proportional way to the ratio of foragers.

The study reveals that the key to maintain a reasonable performance of both agents is to set their proportion reasonably: a high number of patrollers combined with enough number of foragers. This can be observed in the case of a blind search, figure 9, where the performance of the agents remains high while their proportion tends to remain the same. The size of the population appears to have a minor impact on the performance of the algorithm, which is only affected when the population is too low.

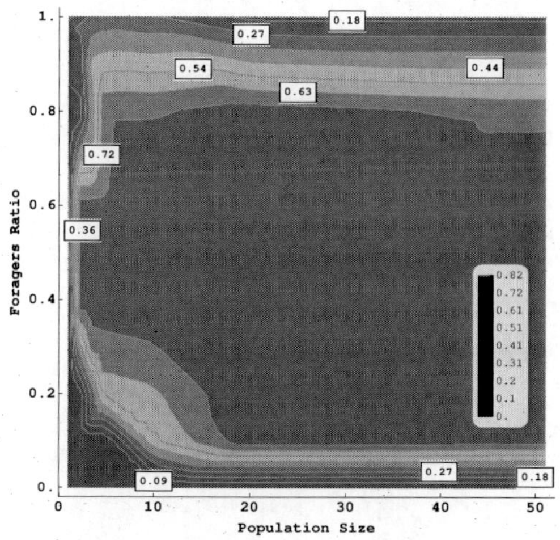

(a) Foragers probabilities of success, with patrollers performing a heuristic search.

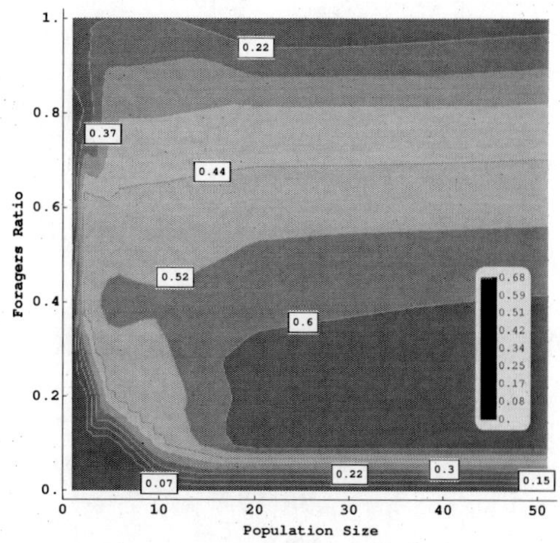

(b) Foragers probabilities of success, with patrollers performing a blind search.

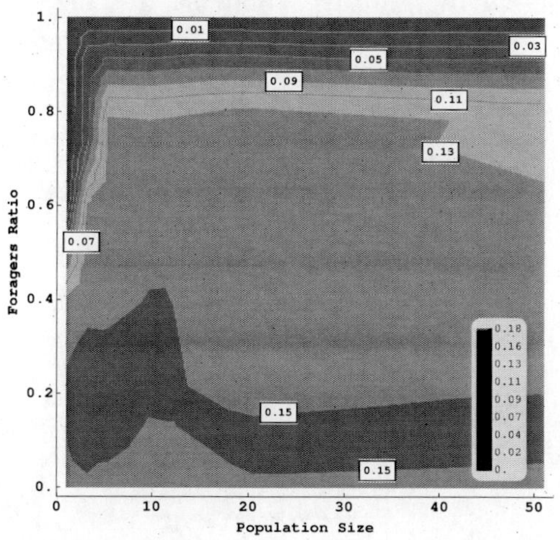

(c) Patrollers probabilities of success, with patrollers performing a heuristic search.

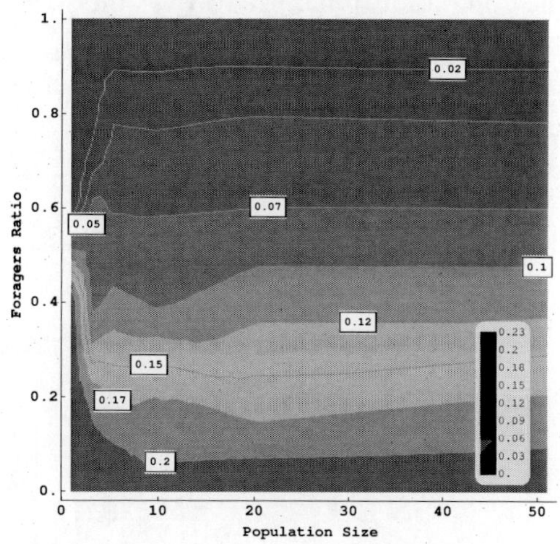

(d) Patrollers probabilities of success, with patrollers performing a blind search.

Figure 9. Average probabilities of success of foragers and patrollers in an algorithm execution, according to the foragers ratio and the population size. (Contour lines)

Based on these experiments, it is possible to identify several features with self-organization dynamics that must first be accomplished, in order to obtain a reasonable algorithm performance:

i. To keep the population large enough, in order to avoid erratic behaviors.

ii. To keep the effects on each kind of agent in equilibrium. The search should be neither excessively disperse nor be excessively concentrate.

5. ACE: Basic Framework + Population Dynamics

This section is devoted to describe population dynamics. The development of population dynamics is based on the ideas described earlier in section 2., and through the results of our previous experiments.

5.1. Schematic Description

The population dynamics follows the scheme as shown in figure 10. There is a initialization phase, which is not included in the figure, that plays the role of spontaneous activation: releasing a patroller every cycle until a first solution is found. Once the patrollers get a first solution the process is self-sustained.

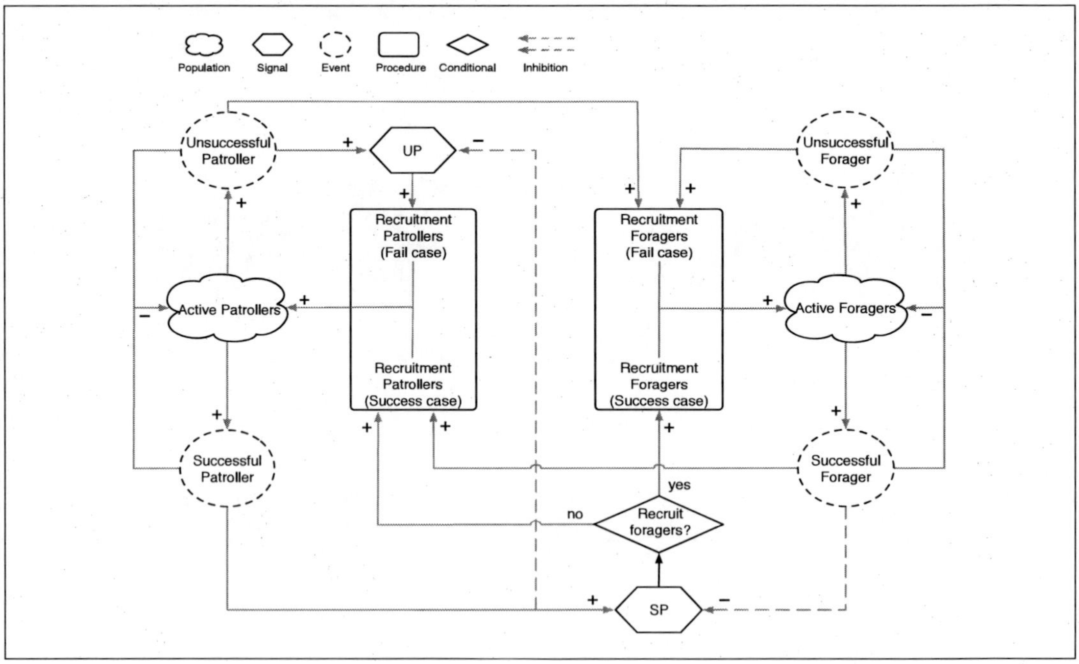

Figure 10. Ant Colony Extended population dynamics scheme represents the different parts of the system and the relationship among them.

The behavior emerges by the reaction to the events that may take place: if a patroller is unsuccessful, it triggers an event which increases the value of the signal *up*, and activated the recruitment process of the patrollers. On the other hand, if a patroller is successful, it resets the value of signal up to zero, but it does not activate the recruitment of patrollers. This serves like an inhibition mechanism that reduces the number of patrollers to the long term. Besides, the successful patroller increases the value of signal sp and activates the recruitment process of foragers.

In the case where the foragers are equivalent: if a forager is unsuccessful, it triggers an event which activates the recruitment of new foragers. If a forager is successful, it resets the value of signal sp to zero and recruits new patrollers.

The system self-organizes based on the rate of events, which allows it to adapt its behavior according to different conditions that may occur. For instance, if the patrollers are frequently unsuccessful, the system tends to recruit new patrollers. At the same time the number of forager decreases, as they finish their exploitation searches and there are not successful patrollers to stimulate the recruitment of foragers. This allows to stop the exploitation and focus the efforts on increasing the result of the exploration.

5.2. Threshold Response Functions

The experiments carried out with fixed ratios of agents in the previous section 4., showed that the foragers ratio is key to obtain a good algorithm performance: a high number of them may lead to over-exploitation. But, if there are not enough foragers, the algorithm may not be able to reduce the uncertainty caused by patrollers exploration.

The use of threshold response functions can supply a way to define a reasonable behavior for foragers. Such functions allow to model the response of an individual, to perform a certain task depending on the intensity of the stimulus. It is commonly used in biology to develop models for social insect behavior. [23,24].

There are several ways to define a function to model the threshold response. For the ACE case, the threshold response functions have the following general form:

$$T_\theta(s) = \frac{s^2}{s^2 + \theta^2}$$

where s is the stimulus value, and θ is the threshold.

The function $[0,1)$ is bounded, and its meaning is very simple: $s \ll \theta$, when the response is close to 0. For $s \gg \theta$, the response is close to 1. Therefore, the individual does not start performing the task unless the stimulus reaches some intensity, defined by the value of the threshold (θ). Using this kind of function, it is possible to define a behavior for the foragers according to the results of the previous section 4..

Over-exploitation: to define a behavior that allows the foragers reaction to the value of sp, but avoid the over-exploitation, as it is defined in a first threshold function. This function uses sp as the stimulus and the current number of foragers as the threshold: $\theta = fr[i]$, where the index i represents the current cycle of the algorithm.

$$T_{fr[i]}(sp) = \frac{sp^2}{sp^2 + fr[i]^2} \tag{9}$$

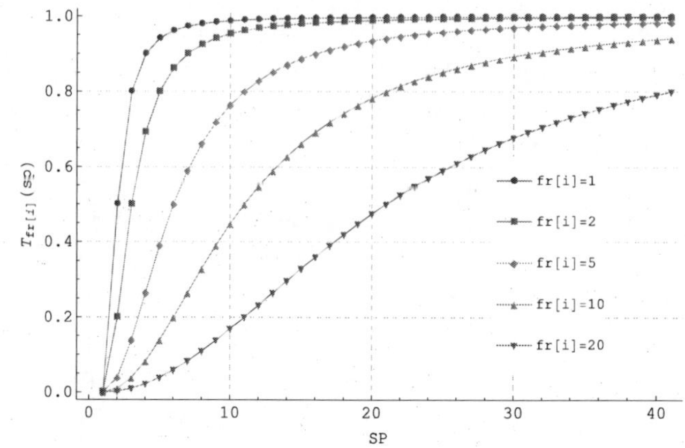

Figure 11. Foragers response ($T_{fr[i]}(sp)$) according to the sp value and the number of active foragers.

Figure 11 illustrates the obtained behavior. As more foragers become active, a greater value for the signal sp is needed to activate a new forager. This function serves to model a situation when there are already enough active foragers. In this case, only with a large number of successful patrollers, it is worth recruiting a new forager. If a successful patroller is not allowed to recruit a forager, it will keep searching. Otherwise, there would be no recruitment at all. Therefore, there would be a risk to reach populations with a low number of active agents and this is not desirable according to the experiments results.

In order to allow successful patrollers to keep searching, when the recruitment of foragers is not possible, the algorithm introduces a conditional recruitment of a new patroller by another successful patroller. The conditional recruitment is indicated in the scheme, figure 10, using a conditional module.

This conditional recruitment is modeled by the following threshold function:

$$T_{sp}(fr[i]) = \frac{fr[i]^2}{fr[i]^2 + sp^2} \tag{10}$$

Notice that the function $T_{fr[i]}(sp)$ is symmetric to $T_{sp}(fr[i])$. Thus, patrollers can recruit a forager, if it is worth doing so; otherwise, it keeps exploring. In fact, the successful patroller finishes its search and joins the pool of inactive agents. But, as it recruits a new patroller, the population of active patrollers does not change.

The relationship among these two functions, $T_{sp}(fr[i])$ and $T_{fr[i]}(sp)$, is illustrated by an example given in figure 12. The threshold function with the higher value determines the behavior: while sp does not reach the current number of active foragers $fr[i] = 10$, the patrollers just keep searching. Once sp reaches a value over the threshold, the recruitment of foragers begins.

Over-Exploration: The recruitment of new foragers by successful patrollers was designed to prevent an over-exploitation. Nevertheless, the experiments of the previous section also showed that whenever patrollers introduce a high degree of uncertainty, a large number of foragers is needed to cope with it. For these cases, the algorithm readjusts the

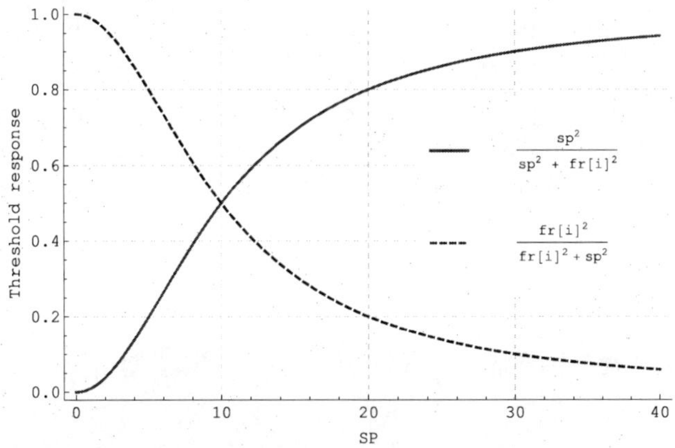

Figure 12. Threshold functions $T_{sp}(fr[i])$ and $T_{fr[i]}(sp)$, for a fixed value of $fr[i] = 10$ and different values of sp.

number of foragers, employing and additional mechanism, based on the number of active agents (patrollers and foragers) and on the degree of uncertainty that patrollers may introduce.

The degree of uncertainty introduced by patrollers can be estimated using the distance between the cost of the best solution found so far. And, the mean cost of all solution discovered so far is the value of μ. When this distance increases, there are more solutions that can be accepted as successful. Thus, patrollers often succeed. The consequence is an increment on the pheromone table uncertainty. On the other hand, when this distance decreases, it is harder to find an acceptable solution and the pheromone table uncertainty tends to remain the same.

Based on this distance criterion to estimate the uncertainty, The algorithm employs a third and last threshold function to regulate the number of foragers:

$$T_{\mu}(fr[i], pr[i]) = \frac{\left(\frac{pr[i]+fr[i]}{fr[i]}\right)^2}{\left(\frac{pr[i]+fr[i]}{fr[i]}\right)^2 + \left(\frac{\mu}{\mu - g(best)}\right)^2} \tag{11}$$

where $fr[i]$ and $pr[i]$ are the current number of forager and patrollers, respectively. At cycle i, $g(best)$ is the cost of the best solution found so far, and μ is the averaged cost of the all solutions discovered so far by successful agents.

This function, $(T_{\mu}(fr[i], pr[i]))$, regulates the number of foragers according to the distance between μ and the best solution discovered, taking into account the number of patrollers and foragers that are currently active.

When the relationship between the foragers and patrollers is below the threshold, defined by μ and $g(best)$, it means that the current number of active foragers can manage the uncertainty that patrollers may produce. Conversely, when the relationship between the foragers and patrollers is above the threshold, the foragers will not be able to manage the possible increment of the uncertainty, thus their number should be increased.

Figure 13 illustrates the function $T_{\mu}(fr[i], pr[i])$. For a fixed number of patrollers (50), it represents when to recruit foragers, based on the foragers numbers and the relationship between μ and the best solution found so far. As observed, the closer μ is to the best solution the less foragers are needed, because the patrollers will fail frequently and the consequence is there is not much probability that they will increase the uncertainty of the pheromone table. On the other hand, the patrollers chances of success increases as μ is far from the best solution, and in consequence a higher number of foragers is needed.

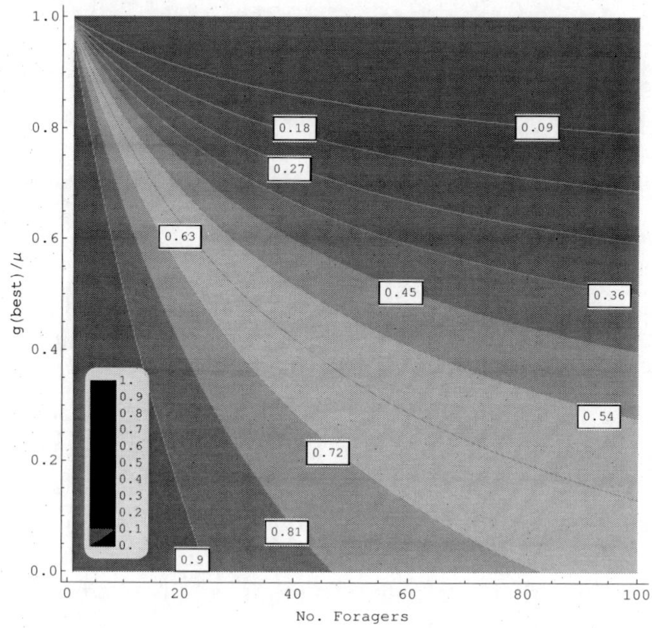

Figure 13. Value of the threshold function $T_{\mu}(fr[i], pr[i])$ according to the number of foragers $(fr[i])$, and the relationship between μ and the best solution found so far $(g(best)/\mu)$. The patrollers number $(pr[i])$ is set to 50 agents.

Finally, this function is included into the population dynamics as the mechanism that recruits new foragers in the cases where a patroller or a forager is unsuccessful can be seen in figure 10. Linking this function with the failure of a forager serves to decide when this forager must be released again, or when it is considered not worthly to do so because there is already enough active foragers. In cases where the foragers do not fail frequently, then it becomes necessary where in order to reinforce the use of this function it requires to link the recruitment of the foragers with the failure of the patrollers.

5.3. General Equations

According to this scheme presented and the threshold functions just defined, the population dynamics can be implemented as a model defined through a set of equations using the following variables:

- *RP*, number of patrollers to be activated.

- RF, number of foragers to be activated.
- up, signal to control the recruitment of patrollers.
- sp, signal to control the recruitment of foragers.
- $fr[i]$, current number of active foragers at cycle i.
- $pr[i]$, current number of active patrollers at cycle i.

The population dynamics works in two steps:

1. A first step consist in processing all agents that have ended their search at cycle i. The agents are processes sequentially, each one updates the model as it is shown here:

   (a) First, the ant updates the variables up, sp, $fr[i]$, and $pr[i]$, according to the following equations:

$$up = \begin{cases} up+1 & \text{if unsuccessful patroller.} \\ 0 & \text{if successful patroller.} \\ up & \text{otherwise.} \end{cases} \tag{12}$$

$$sp = \begin{cases} sp+1 & \text{if successful patroller.} \\ 0 & \text{if successful forager.} \\ sp & \text{otherwise.} \end{cases} \tag{13}$$

$$fr[i] = \begin{cases} fr[i]-1 & \text{if the agent is a forager.} \\ fr[i] & \text{otherwise.} \end{cases} \tag{14}$$

$$pr[i] = \begin{cases} pr[i]-1 & \text{if the agent is a patroller.} \\ pr[i] & \text{otherwise.} \end{cases} \tag{15}$$

   (b) Using the new values for ns, sc, and $fr[i]$, the ant calculates the value for variables RP and RF:

$$RP = \begin{cases} \log(up+1) & \text{if unsuccessful patroller.} \\ 1 & \text{if successful forager.} \\ T_{sp}(fr[i]) & \text{if successful patroller.} \\ & \wedge \quad T_{sp}(fr[i]) \geq T_{fr[i]}(sp) \\ RP & \text{otherwise.} \end{cases} \tag{16}$$

$$RF = \begin{cases} T_{\mu}(fr[i], pr[i]) & \text{if unsuccessful agent.} \\ T_{fr[i]}(sp) & \text{if successful patroller.} \\ & \wedge \quad T_{fr[i]}(sp) \geq T_{sp}(fr[i]) \\ RF & \text{otherwise.} \end{cases} \tag{17}$$

2. The second step takes place once that all ants have been processed. It consists of an update in the populations:

$$fr[i+1] = fr[i] + \text{round}(RF) \tag{18}$$

$$pr[i+1] = pr[i] + \text{round}(RP) \tag{19}$$

Equations 12 and 13 manage the value of the signals. It can be seen that these signals just count the number of unsuccessful patrollers since the last successful ant and the number of successful patrollers since the last successful forager, respectively. Equations 14 and 15 update the number of active foragers and patrollers, respectively. Equation 16 manages the recruitment of patrollers. And equation 17 manages the recruitment of foragers.

In order to clarify the meaning of the equations, each part is explained independently using simple simulations. These simulations use a constant rate of external events to trigger the population dynamics. For example, a "constant rate of" two unsuccessful patrollers per cycle.

Foragers recruitment. This recruitment works based on the value of signal sp. A simple simulation can be done using a external constant rate of two successful patrollers per cycle. This makes the signal sp to increase its value in two units per cycle: $sp[i+1] = sp[i] + 2$. It is also assumed that a forager needs 10 cycles to build a solution, and that all foragers are successful.

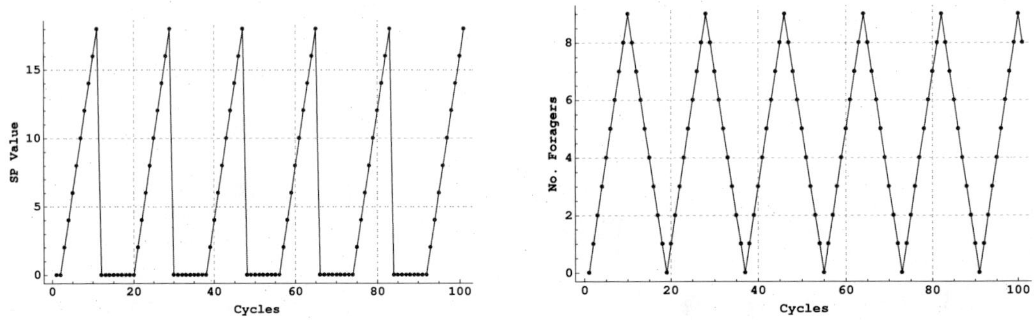

(a) Evolution of signal sp. (b) Evolution of foragers population.

Figure 14. A simple simulation of the foragers recruiting dynamics.

The result of simulating 100 cycles is shown in figure 14. As illustrated, a single forager is recruited each cycle although the signal increasing rate is two units per cycle. This is a consequence of the threshold function (equation 9) action, which limits the growing of foragers to one agent per cycle. When foragers begin to finish the construction of their solutions, they reset the value of sp (equation 13), therefore there is no recruitment of new foragers until all of them have finished their search.

Notice that the maximum amount of foragers depends on the number of cycles needed to build a solution. This is due to the fact that the system dynamics is defined through events, and the number of events depends on the number of cycles needed to build a solution.

This allows for an adjustment in the population of foragers according to the difficulty of the problem. When the algorithm is building solutions that requires a low number of cycles, the number of events is also low and the population will be small. But, if the construction of solutions takes more cycles, the number of events increases, and hence, the population size also increases. In general, the population size will be proportional to the number of cycles required to build a solution.

This basic simulation points out a fundamental requirement of the population dynamics: the algorithm should work asynchronously. Otherwise, all agents will end at the same cycle and the dynamics will not work. Notice that an agent may inhibit a signal completely. Therefore, if there is no other agents that come after him, the signal will be zero forever.

Patrollers recruitment. This recruitment works based on the value of signal *up*. This time, an external constant rate of two unsuccessful patrollers per cycle is assumed. This makes the signal *up* to increase its value by two units per cycle: $up[i+1] = up[i] + 2$. Like in the foragers simulation, it is also assumed that a patroller needs 10 cycles to build a solution, and all recruited patrollers are successful. Thus, there is a constant rate of unsuccessful patroller, given by the simulation conditions, and a variable rate of successful patroller, which are those who are recruited by the population dynamics.

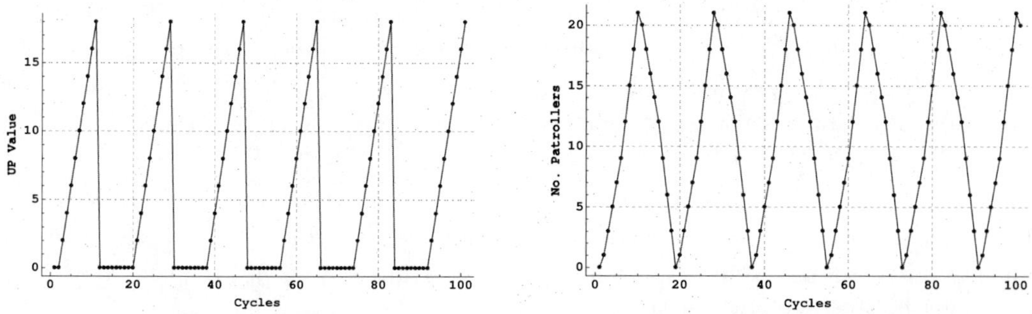

(a) Evolution of signal *up*. (b) Evolution of patrollers population.

Figure 15. A simple simulation of the patrollers dynamics.

Figure 15 shows the result of simulating 100 cycles. As observed, there is a recruitment of several patrollers per cycle. The recruitment is proportional to the logarithm of the value taken by the signal *up* (equation 16). Thus, their number will be greater than the number of foragers recruited in the previous simulation. Like in the foragers case, as soon as patrollers begin to finish the construction of their solutions, they inhibit the value of *up* (equation 12) and there is no recruitment of new patrollers until all of them have finished.

Bond between patrollers and foragers. The mutual relationship between patrollers and foragers can be illustrated running the simulation under the following conditions:

The value of signal *up* is externally increased one unit per cycle. This simulates that there is an unsuccessful patroller per cycle. These unsuccessful patrollers trigger, in turn, the recruitment of new patrollers, which are considered always successful. These patrollers increase the value of signal *sp*, that in turn triggers the recruitment of foragers. These foragers are also considered always successful. Finally, all the agents need 10 cycles to build a solution.

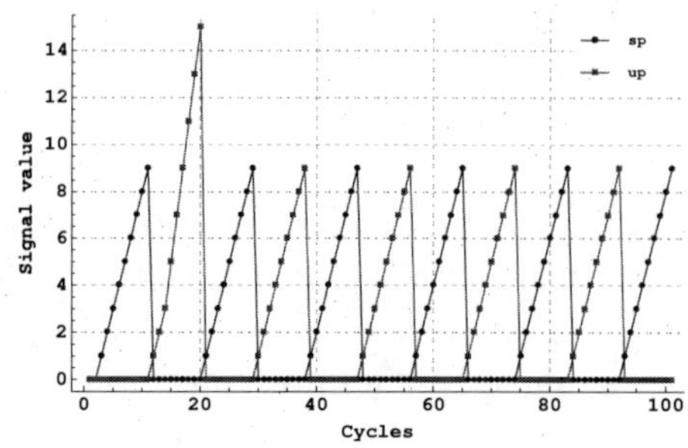

(a) Evolution of signals up and sp.

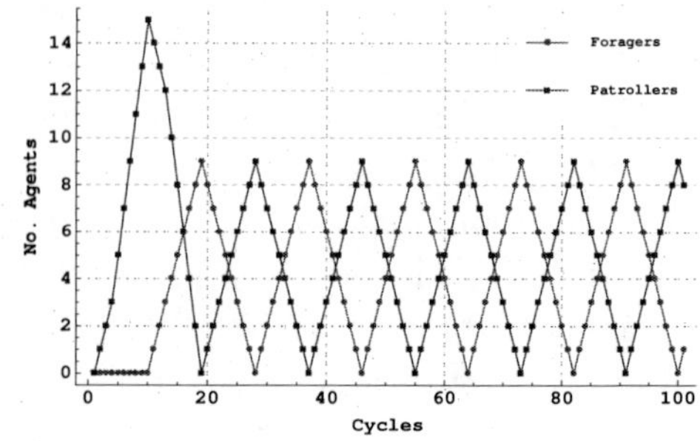

(b) Evolution of populations of patrollers and foragers.

Figure 16. A simple simulation of the both dynamics, patrollers and foragers, together.

Figure16 shows the result of simulating 100 cycles under the described conditions. First, there is a recruitment of patrollers. As these patrollers finish their searches, the signal up is set to zero (equation 12). Thus, the recruitment of patrollers stops (equation 16). Simultaneously, these successful patrollers start to increase signal sp. Therefore, there is a recruitment of foragers (equation 17) which, in turn, will recruit new patrollers once they finish their search. The result of linking both populations is an auto-regulation, alternating phases of exploration with phases of exploitation.

Short-term inhibition. The final values of signal RP, equation 16, and signal RF, equation 17, depend on the sequence of the events generated by those agents that have finished their search in the same cycle. As they are processed sequentially, the last event may inhibit the response of the previous events.

For example, when a successful foragers and one or several unsuccessful patrollers finish at the same cycle, the final value of up depends on the other agents that are processed. According to equation 16:

a) If patrollers are processed first and the forager is processed last, the forager overrides the result of patrollers and set the final value of RP to 1.

b) Conversely, if the successful forager is processed in any other position its effect is overridden by the patrollers processed later. Then, the total amount of patrollers recruited will be $log(up+1)$.

These effects are illustrated in figure 17. It compares the result of the previous simulation, where the last agent processed is a successful forager, with the results when unsuccessful patrollers are the last to be processed.

As can be observed, in one case, subfigure 17(a), the patrollers population is being altered by the foragers population, which inhibits the recruitment performed by unsuccessful patrollers. The consequence is that both populations have the same number of agents. In the other case, subfigure 17(b), patrollers operate independently of foragers. Their number is only affected by the signal up and the number of cycles needed to build a solution.

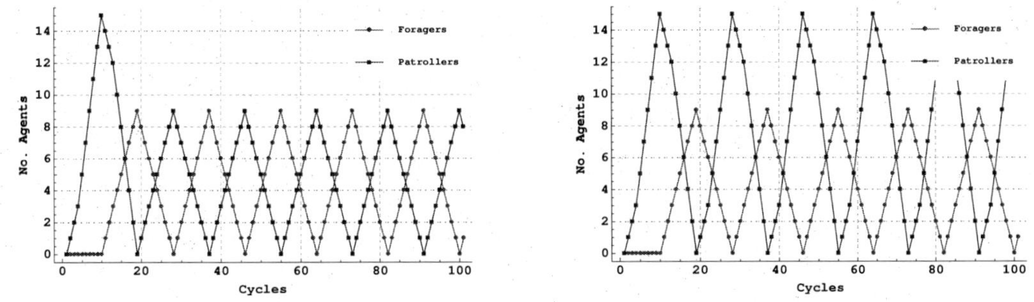

(a) Patrollers population regulated by the foragers population through the short-term inhibition.

(b) Patrollers population operating independently from the foragers population.

Figure 17. A simple example of the short-term inhibition.

This kind of inhibition is called short-term inhibition, in order to differentiate it from the inhibition produced when a signal is reset by a successful agent (see equations 12 and 13), which is called long-term inhibition. The main difference between them is that the long-term one affects the behavior of the algorithm in subsequent cycles, due to the signal resetting. In the short-term one, the behavior is modified only in the cycle where the event occurs.

Self-organization capacity. The simulations provided thus far were done under constant conditions. As a consequence, the events that regulate the population are also constant and the populations reach a stable configuration.

To explore the self-organization capacity when events are not constant, foragers can be introduced at a variable rate using equation 9. This equation introduces a foragers, only when the output value is over 0.5. This can be considered as a spontaneous activation or recruitment of foragers.

The algorithm is executed one more time, including equation 9, a constant distance between μ and $g(best)$ $((g(best)/\mu) = 0.87)$ and the same conditions already defined for previous simulations.

Figure 18 shows the result of simulating 100 cycles. It can be observed that there is first a recruitment of patrollers. At the same time, there is an introduction of foragers not recruited by successful patrollers.

When patrollers begin to finish their solutions, there are already foragers active. Thus, equation 17 does not recruit any new foragers, because the value of signal sp is below the threshold (equation 9. As patrollers can not recruit foragers, they keep searching (equation 16 and equation 10). Besides, according to the conditions imposed, patrollers are always successful. Therefore, the value of signal up is reset to 0 constantly; thereby avoiding the recruitment of new patrollers, equation 16. As there is no recruitment, neither patrollers become inactive, the patrollers population simply remains stable until a change on the foragers population takes place.

Meanwhile some foragers may finish their searches, which recruit new patrollers. This increases the number of active patrollers. As there are less foragers active, and there are more patrollers active being successful, the sp value raises quickly. This can be observed

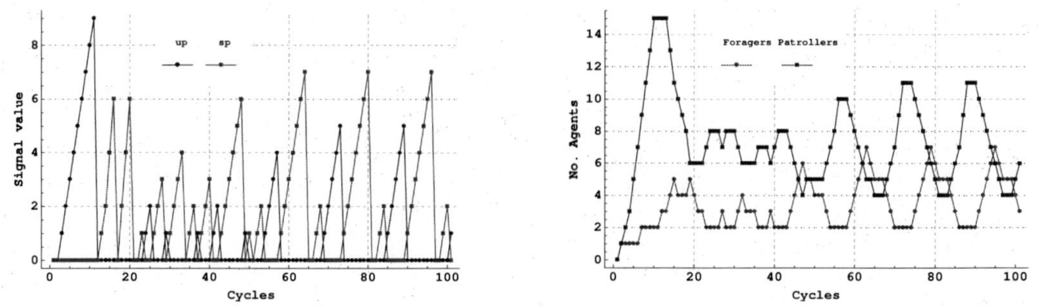

(a) Evolution of signals up and sp.

(b) Evolution of populations for patrollers and foragers.

Figure 18. A simple simulation in the self-organization for the population when there is a spontaneous recruitment of foragers using equation 9, setting $(g(best)/\mu) = 0.87$.

in the subfigure 18(a) by the behavior of signal sp. Thus, the population of patrollers varies again when some patrollers are able to recruit new foragers and become inactive.

The spontaneous recruitment of foragers is conditioned to the fact that there is no recruitment by patrollers. And, the recruitment of patrollers, in turn, is conditioned to the finalization of foragers that were spontaneously recruited. This forms a closed loop that leads to the populations by a period of oscillations, as in subfigure 18(b) approximately in cycle 20, until it is synchronized through the internal dynamics of the populations with the spontaneous recruitment of foragers.

Once the internal dynamics are synchronized with the spontaneous recruitment of foragers, subfigure 18(b) approximately in cycle 65, a stable pattern of organization emerges. At this point, it can be said that the system self-organized by the means of taking into account how the spontaneous foragers have been introduced.

A real example: the previous simulations show that the system is able to organize foragers and patrollers, synchronizing their behavior according to the events produced by each one. In a real problem, these events are produced by the real success or failure of each population. Patrollers and foragers tend to organize themselves in such a way, that both produce the same quantity of successful events. This kind of synchronization between the rates of success will balance the search, avoiding the risks of over-exploitation or over-exploration.

An example of the populations dynamics working on a real problem is shown in figures 19 and 20. The graphics represent the evolution of patrollers and foragers populations when the algorithm is applied to solve the instance ($Eil51$) of the TSP.

Figure 19 shows the evolution in the case where patrollers explore doing a blind search. It can be seen there are small oscillations in the beginning when the initialization ends and the first solutions are found. After that point, the behavior is more stable. When there is an increment of the patrollers failure rate, there are peaks that appear on their populations. These peaks of patrollers may be transferred to the foragers population, if they can increase the value of sp signal without having a forager set to zero.

Figure 20 shows the evolution when patrollers explore using an heuristic function. The behavior is similar, but now patrollers have difficulties to find solutions below μ, because the heuristic function reduces the search space. The population of patrollers is larger than in the case of a blind search, because a higher number of them is necessary to maintain the success rate of the exploration search. If the patrollers have no successes, the foragers population will be smaller.

In both cases, the population of foragers is adjusted by the threshold function 11, which increases the number of foragers according to the uncertainty in the number of patrollers that may be produced. In the case of the blind search, this threshold function helps to maintain a constant population of foragers according to this uncertainty. As observed in the figure 19, the population of foragers tends to increase as the search progresses. In the heuristic case, it is unlikely that the patrollers can produce a high increment of uncertainty. Therefore, the population of foragers is determined mainly by the success rate of patrollers. This is observed in figure 20, where the population of foragers tends to maintain the same number of foragers.

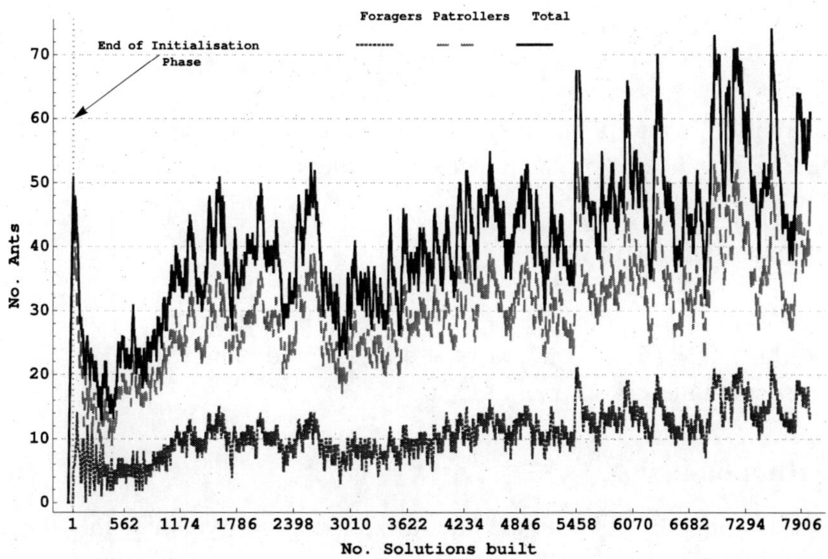

Figure 19. Evolution of the population of foragers and patrollers when the algorithm is applied to the TSP instance *Eil*51, with patrollers performing a blind search.

5.4. Pseudo-Code Scheme

The general scheme of ACE that includes the population dynamics shown in figure 21. It works exactly the same as the previous basic framework (section 2.) except for the population management. The initialization remains the same, but once the first solution is found the populations become in charge of the population dynamics. As shown in the scheme, it works as follows: for every agent that finishes its search, further updates the model (equations 12 to 17). Once all agents have been processed, the population is modified according to the recruitment equations (equations 12 to 17).

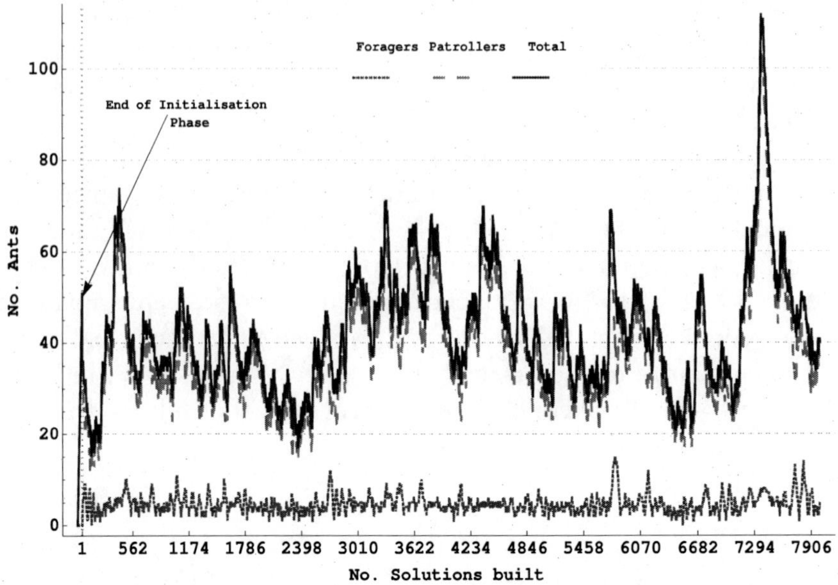

Figure 20. Evolution of the population of foragers and patrollers when the algorithm is applied to the TSP instance *Eil*51, with patrollers using a heuristic function.

6. Experimental Section

The aim of this section is to compare the effects when including the population dynamics. In order to do this, ACE (basic framework + population dynamics) is compared with the basic framework previously described (see section 3.).

6.1. Algorithm Behavior

This experiment is aimed to investigate how the population dynamics modifies the algorithm's behavior. The experiment follows the same scheme as the studies done with the basic framework (section 4.). It is measured by using a cumulative moving average, the evolution of the pheromone table information, and the probability of success for each kind of agent.

In order to facilitate the discussion, the case when patrollers use a heuristic function is analyzed separately from when performing a blind search case.

Blind search: in this case the main risk comes from the fact that patrollers explore without a heuristic. Thus, it is easy for them to increase the pheromone table uncertainty, i.e., the entropy. If this entropy reaches high levels, then the search becomes excessively dispersed. By observing the results of this experiment shown figures 22(a) and 22(b), it can be seen that the population dynamics is able to manage the number of agents, maintaining reasonable entropy levels. It avoids the higher levels of entropy, which occur when the foragers population is not large enough (forager ratio below 0.2). Hence, the search is not excessively disperse. At the same time, it avoids over-exploitation situations, which occur

```
% Initialization
samples ← 0    μ ← ∞    popdyn ← false
while termination condition not met do
  % Solution construction procedure
  for agent ∈ Population do
    if is forager(agent) then
      Pheromone Search Step (agent)    % equation 2
    else
      decision ← control policy    % equations 3,4
      if decision = true then
        Heuristic Search Step (agent)    % equation 1
      else
        Pheromone Search Step (agent)    % equation 2
      end if
    end if
  end for
  % Pheromone update procedure
  for agent ∈ Population do
    if search finish then
      if g(agent) < μ then
        samples ← samples + 1
        μ ← μ + (g(agent) − μ)/samples
        if g(agent) < g(best) then
          best ← agent
        end if
        for i ∈ {1..|U_agent|} do
          τ_qi ← update(u_i, λ)    % equations 5, 6
          if u^i ∉ τ_qi then
            τ_qi ← add(u_i, λ)
          end if
        end for
        popdyn ← true
        Population dynamics (agent)    % equations 12 to 17
      end if
      popdyn ← true
      Population dynamics (agent)    % equations 12 to 17
      Population ← retire(agent)
    end if
  end for
  % Population management
  if samples < 1 then
    Population ← add(patroller)
  else
    if popdyn = true then
      Population dynamics (rp,rf)    % equations 18, 19
      popdyn ← false
    end if
  end if
end while
```

Figure 21. ACE pseudo-code scheme.

when the numbers of foragers are excessively large (forager ratio above 0.4).

Through analyzing the behavior of the agents as shown in figures 22(d) and 22(f), it was observed that the entropy levels reached by the population dynamics are not a consequence of an increment in the foragers number, but of a decrement of the patrollers efficiency (probability of success). This is due to the fact that the short-term inhibition, carried out by successful foragers, reduces the number of active patrollers, synchronizing both populations. Hence, the patrollers probabilities of success tend to decrease. In figure 22(f), it was observed that the foragers probability of success for the case of population dynamics was similar to the success probability for the case in the basic framework with a ratio set to 0.2. Nevertheless, the patrollers probability of success is lower in the case of population dynamics than for the case in the basic framework with a ratio set to 0.2, that helps to explain the difference between the entropy levels.

The conclusion of this experiment is that the population dynamics is able to balance the population such that the entropy of the pheromone table is reasonable in order to obtain a good search performance. This balance is achieved by organizing the population in such a way that the efficiency of the patrollers decreases. This is interesting because it then makes it possible to avoid an excessive dispersion of the search, without triggering an increment of the foragers which increases the risk of over-exploitation situations.

Heuristic search: in this experiment, the patrollers explore using a TSP specific heuristic (see appendix A). Thus, it will not be easy for them to increase the pheromone table uncertainty, i.e., the entropy. This increases the risk of over-exploitation situations. As shown in figures 23(a) and 23(b), the population dynamics is able to manage the number of agents such that the entropy level is reasonable: it reaches the level of the basic framework with forager ratio set to 0.0. The heuristic is very accurate, it has almost no uncertainty as to the order for the cities visited (i.e., which action to perform at each state). Therefore, it is almost unnecessary to use forager agents.

In the behavior of the agents as indicated by subfigures 22(d) and 22(f), it can be observed that entropy levels reached by the population dynamics are a consequence of an increment in the patrollers probability of success which increases their efficiency. This increment can be explained as a consequence of the foragers behavior:

- Every time that a forager is successful it decreases the uncertainty of the pheromone table, but also decreases the value of μ. Therefore, if the foragers are numerous and they are constantly successful, then the value of μ decreases at such rate that most of the table's information becomes inaccurate: it does not serve to build solution below the μ value.

- The population dynamics is able to stop the exploitation of foragers, if the patrollers are not being successful: where there are not events to trigger the recruitment of foragers. This synchronized the performance of foragers with the success rate of patrollers, preventing to use more foragers than necessary. This allows to reduce the uncertainty of the pheromone table, but avoid the faster decrement of μ and the appearance of inaccurate information.

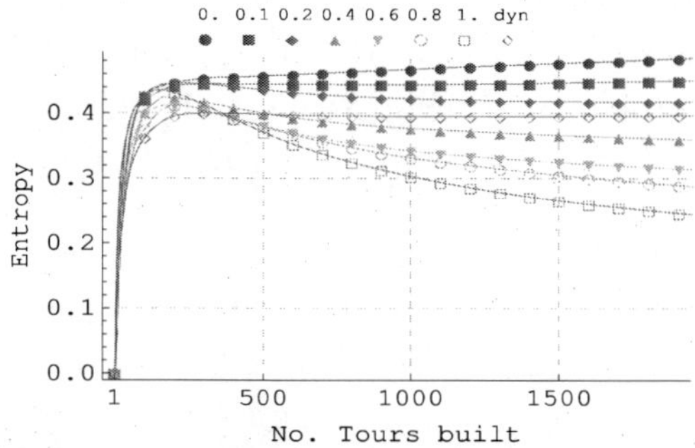

(a) Pheromone entropy evolution at early stages.

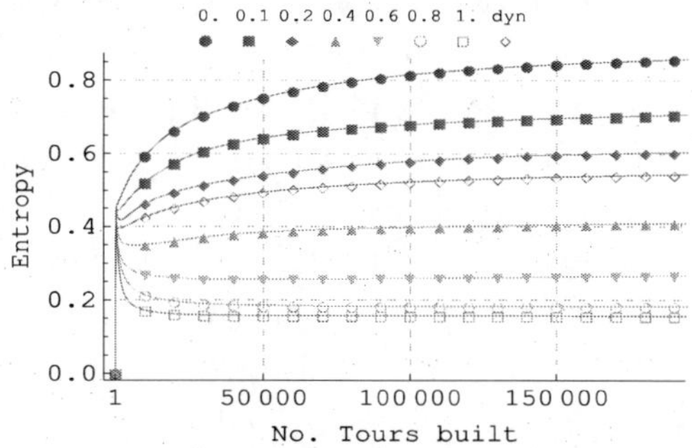

(b) Pheromone entropy complete evolution.

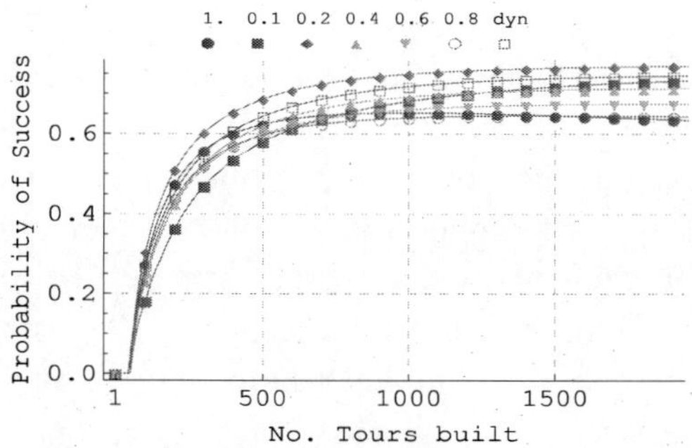

(c) Foragers success probability evolution at early stages.

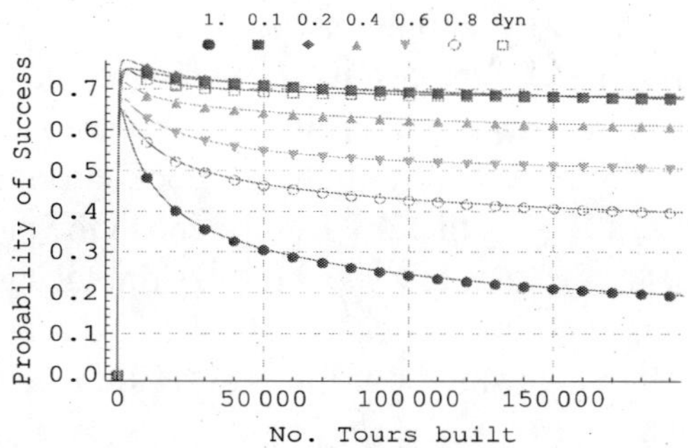

(d) Foragers success probability complete evolution.

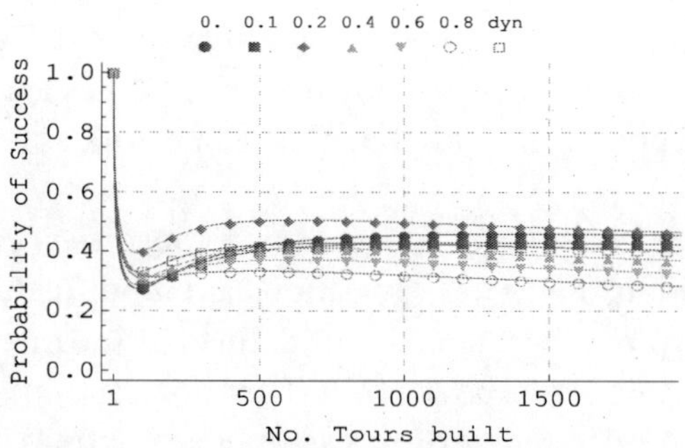

(e) Patrollers success probability evolution at early stages.

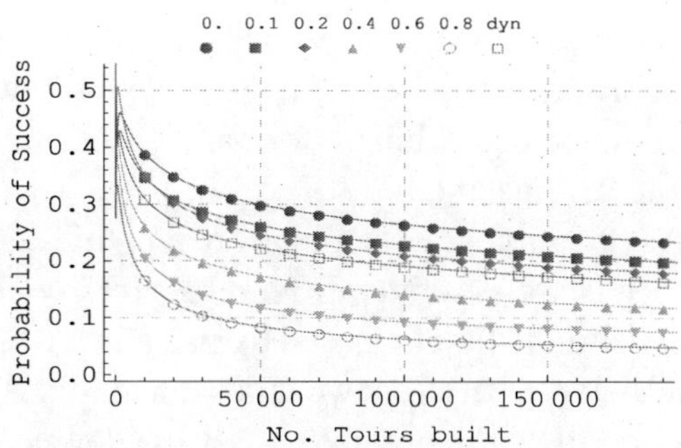

(f) Patrollers success probability complete evolution.

Figure 22. Evolution of the pheromone table's average entropy ($\overline{H_{\tau}[i]}$), and evolution of the success probabilities for foragers and patrollers during an algorithm execution. The left column shows these evolutions at early stages of the search, and the right column the complete evolutions.

To reexamine the analysis of the experiment's results, as the success rate of patrollers is low, the population dynamics releases a few foragers. These few foragers are able to reduce the uncertainty of the pheromone table, but without decreasing significantly the μ valve. This allows for an easier search for patrollers, because the pheromone table has almost no uncertainty, and practically all the information serves to find solutions below μ. That is why the patrollers probability of success is increased.

The conclusion of this experiment is that the population dynamics is able to balance the population such that the entropy of the pheromone table is reasonable in order to obtain a good search performance. This balance is achieved by organizing the population in a such a way that the efficiency of the patrollers increases. This is interesting because it encourages to keep using the foragers to reduce the uncertainty of the pheromone table, but avoid the over-exploitation situations.

To consider the results in both cases when using heuristic and blind search, it can be seen how the agents efficiency can be increased with a correct synchronization among them. In the case of a blind search, the population is organized such that the patrollers efficiency decays, allowing the foragers to reduce the entropy of the pheromone table. Conversely, in the heuristic case, the population is organized in such a way that the performance of some foragers makes it easier for patrollers to search. This is interesting because it shows how the self-organization dynamics allows the emergence of a complex relationship among the agents.

6.2. Comparative with the Basic Framework

Previous experiments show how the population dynamics modifies the internal details of the algorithm compared with the basic framework. In order to study if these changes are reflected in the algorithm performance, ACE was compared with the basic framework solving several TSP instances[4].

As the purpose of population dynamics is to provide self-organization capacities, each instance is solved under different searching conditions. These variations are modifications of the heuristic function used by patrollers to explore. The heuristic is defined as the inverse of the distance between two cities (appendix A) raised to a power of x. Therefore, modifying the value of x it is possible to set the degree of uncertainty of this heuristic function. For instance, with $x = 0$, as in the case of a blind search, where the uncertainty is maximum. As x increases, the uncertainty of the heuristic function decreases.

The experiment consists of modifying the heuristic and perform a batch of 100 independent executions for each instance. A single execution builds $n \cdot 10000$ tours, where n is the number of cities in the instance. The performance of the algorithm is measured using the error (the average deviation from the optimal value). The error is expressed in percentage according to the following formula:

$$\frac{100}{optimum} \cdot \frac{1}{n} \sum_{i}^{n} (x_i - optimum)$$

Where x_i is the distance of the best tour found in execution i

[4] ry48p, eil51, st70, ftv70, pr76, rat99, kroA100, eil101, and gr137. The number in the instance name indicates the number of cities in the instance

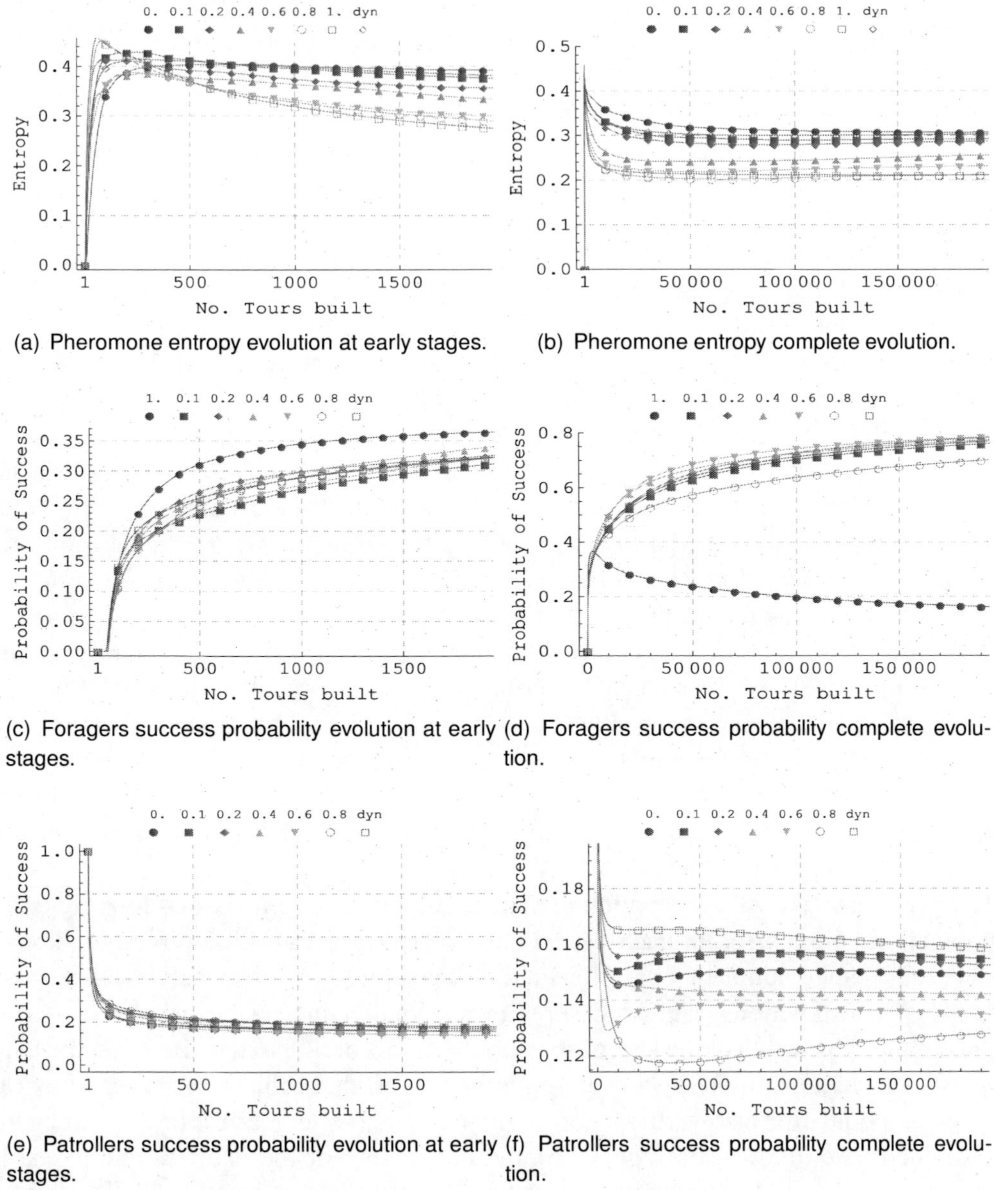

(a) Pheromone entropy evolution at early stages.

(b) Pheromone entropy complete evolution.

(c) Foragers success probability evolution at early stages.

(d) Foragers success probability complete evolution.

(e) Patrollers success probability evolution at early stages.

(f) Patrollers success probability complete evolution.

Figure 23. Evolution of the pheromone table's average entropy ($\overline{H_\tau}[i]$), and evolution of the success probabilities for foragers and patrollers during an algorithm execution. The left column shows these evolutions at early stages of the search, and the right column the complete evolutions.

Using this measure, ACE results are compared with the results of the basic framework, for different forager ratio, and a population size equal to the number of cities in the instance.

The results of the experiment are shown in figure 24. The results observed improve as the heuristic function increases its accuracy. This confirms that the heuristic employed for TSP is reliable: its recommendations can be narrowed without loosing precision in the search.

The basic framework when the heuristic is not very accurate, figures 24(a), 24(b), and 24(c), achieves the best performance with a reasonable high ratio of foragers (0.4, 0.6) in order to deal with the uncertainty produced by patrollers. In the cases where the heuristic has some accuracy, as in figures 24(d) and 24(e), the best results correspond to a lower foragers ratio: 0.2 and 0.1, respectively. Finally, when the heuristic is highly accurate the best results are obtained without any foragers, ratio=0.0.

ACE results, labeled in the figure 24 as "P. Dyn", tend to be just behind the best performance of the basic framework. When not shown as just behind the best performance, the results are very close to the optimal performance of the basic framework. This means that the self-organization capabilities provided by the population dynamics do not reach the optimal performance for each situation (for each heuristic function). This is something expected because self-organization lacks the precision of fine tuning in user-defined parameters. But for each search situation, the parameters must be tuned efficiently in order to obtain a reasonable performance. Nevertheless, the self-organization is able to obtain a high reliable performance independent of the situation. According to the results, in almost all cases, only the best parameters values reaches a better performance than the results obtained using a population dynamics. This means that the overall performance of the population dynamics is nearly as good as the performance of the best parameters values for each case.

The performance of the population dynamics can be better understood looking at how the population is organized in each case. This organization can be indirectly measured, by using the average number of patrollers and foragers used to solve each instance for the different heuristic functions. These measures are shown in figure 25. It can be observed how the algorithm is able to adapt the size of the population according to the size of the instance: large instances requires more agents as the difficulty of the problem increases. For each instance, the population also varies accordingly to the heuristic in use. For highly accurate heuristic there are almost no foragers. As the uncertainty of the heuristic increases, but still it can be considered reliable, the population dynamics raises the number of patrollers employed. Because they are not as efficient as before, thus more of them are required to perform an efficient exploration of the search space. As the heuristic becomes less reliable, the population decreases the number of patrollers and increases the number of foragers. This allows to manage the uncertainty provided by patrollers due to the use of a low accurate heuristics.

It can be observed how the population dynamics reacts to different searches situations in the direction pointed out by the results of the basic framework: when the heuristic is reliable, there is almost no need for exploitation search, thus there is no need for foragers. When the quality of heuristic decreases, more exploitation search is needed in order to avoid an excessive dispersion of the search. Thus, the number of foragers increases. This is why the results of the population dynamics are nearly as good as the results with the best values

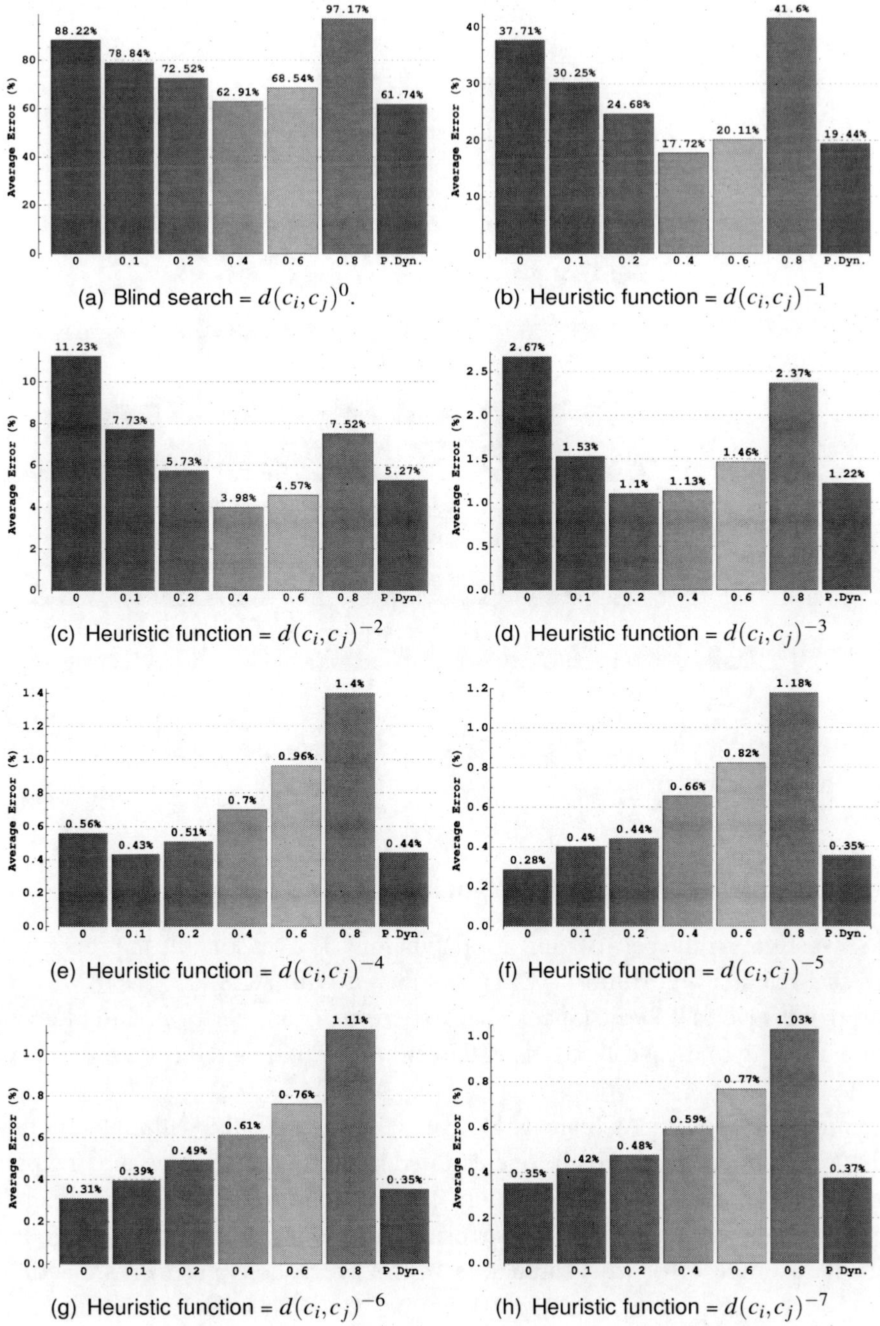

(a) Blind search = $d(c_i, c_j)^0$.

(b) Heuristic function = $d(c_i, c_j)^{-1}$

(c) Heuristic function = $d(c_i, c_j)^{-2}$

(d) Heuristic function = $d(c_i, c_j)^{-3}$

(e) Heuristic function = $d(c_i, c_j)^{-4}$

(f) Heuristic function = $d(c_i, c_j)^{-5}$

(g) Heuristic function = $d(c_i, c_j)^{-6}$

(h) Heuristic function = $d(c_i, c_j)^{-7}$

Figure 24. Average error yield by each algorithm when solving several TSP instances with different heuristic functions.

for the parameters.

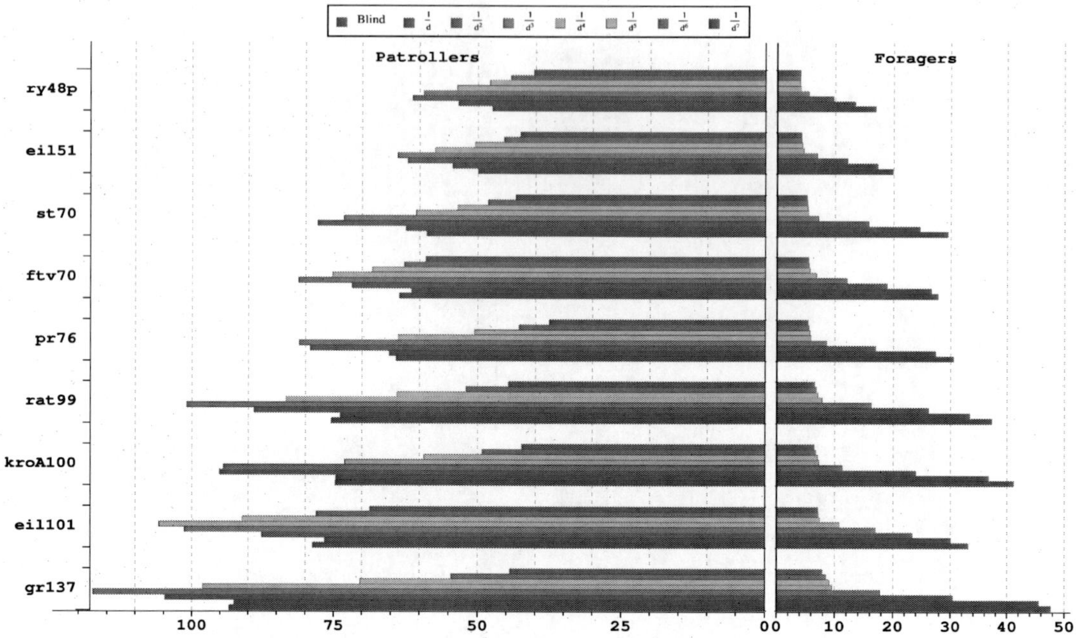

Figure 25. ACE average number of patrollers and foragers used to solve each instance for the different heuristic functions.

6.3. Comparison with other ant Algorithms

In order to investigate if the self-organization dynamics is a reasonable approach, compared with other classical ant algorithms, ACE is compared with ACS [1,7] and MMAS [1,10] algorithms, using several TSP instances. The experiments are performed implementing the algorithms according to the publications of the authors [7,10], and the source code provided by T. Stüztle [35].

The parameter values for ACS and MMAS are those suggested within the literature. For ACE the heuristic function used is the one detailed in the appendix A, where the parameters values are obtained empirically. These values can be found in table 3 of the appendix B.

The comparison is done in terms of performance: the probability of finding the optimal value, and the average error (measured like in the previous experiments, section 4.))are compared. Every algorithm is executed 1000 times for each instance. A single execution consists of running the algorithm, until it builds a fixed number of tours: $nc \cdot 10^4$, where nc is the number of cities.

Since all the algorithms are heuristic, the error analysis should be considered as the most relevant measure in the comparison. The figure 26(a) shows how the error is distributed among the different instances. As observed, ACE and MMAS have a similar performance. In some instances the ACE error is larger than MMAS (*gr*96), while in other cases MMAS error is larger than ACE (*KroA*100). The results shows that ACE has the lowest average error, but the difference is sufficiently small as to be considered negligible. On the other hand, ACS clearly has a larger error, which in several cases doubles the error values of ACE and MMAS.

Figure 26(b) shows how the probability of finding the optimal value is distributed among the different instances. There is a noticeable difference between ACE probability of finding the optimal value, and MMAS probability. For some instances, MMAS is able to find the optimal frequently, but it shows a worse performance where ACE shows a more regular performance. ACS shows the lower probability among all three algorithms.

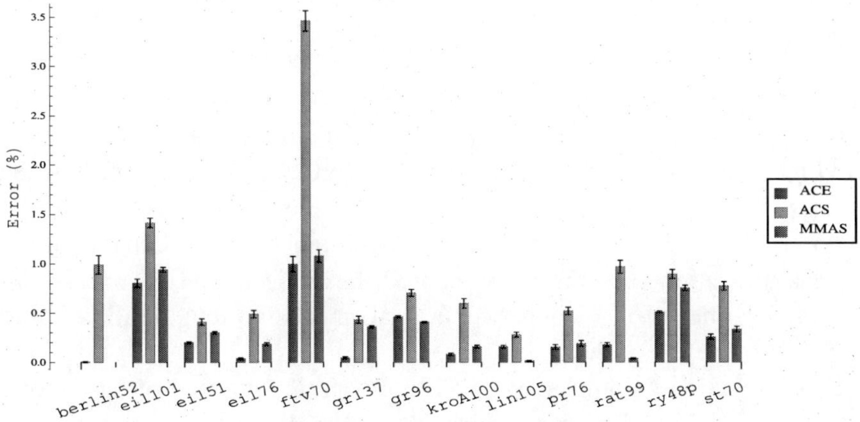

(a) Error percentage distribution

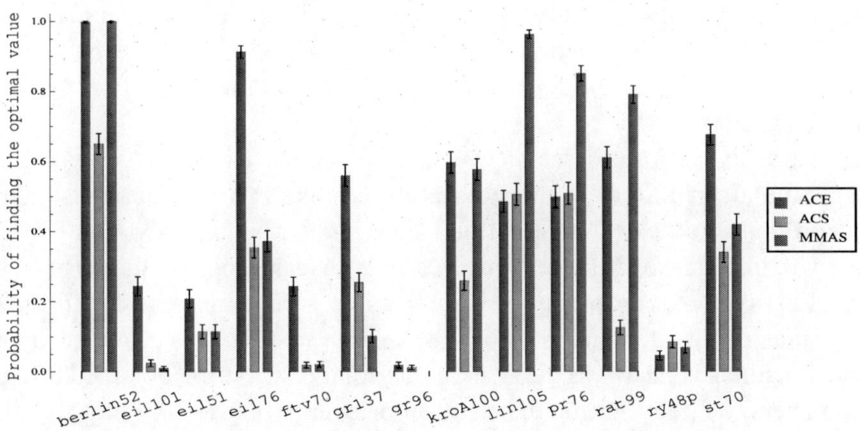

(b) Distribution of the probability of finding the optimal solution

Figure 26. Average performance yield by ACE, ACS, and MMAS when solving several TSP instances.

To account for both measures, it can be concluded that ACE performance is equivalent to those algorithms; and, depending on the scenario, it may achieve a higher performance.

Conclusion

This chapter has shown how the self-organization dynamics of real ants can be used to design multi-agent systems that resemble ants capabilities.

The self-organization dynamics of ant colonies leads to an emergent distribution of individuals, aimed to complete several tasks. This emerging task allocation arises from the interaction patterns (encounters) that take place while ants are performing their tasks. Whenever an ant interacts with another, there is a stimulus that draws one of the ants to start doing the task the other ant is already doing. This is a form of recruitment, where the chances of getting new ants depends on the performance of the recruiting ant. In this way, unsuccessful ants show a inclination to be recruited, and successful ants tend to be reluctant to shift to new tasks. This kind of dynamics makes possible that an ant colony organizes itself, according to the success rates that each ant has in performing a task.

This form of task allocation prioritizes those tasks that have more chances to be fulfilled. Reversing this dynamic, it is possible to design a self-organization mechanism that distributes a group of individuals according to the difficulty to perform each task: unsuccessful individuals recruit more partners to cope with the difficulty of performing a task. While successful individuals recruit new agents to start performing the next task. Incorporating this dynamics into a multi-agent system, can obtain a system capable of performing an efficient task allocation such that it is ensured the achievement of all tasks, balancing the progress of each one: no individual task is performed for a long period of time without making progress on the rest of the tasks.

This self-organization dynamics are used to design a new kind of ant algorithm, Ant Colony Extended. Which is an optimization algorithm designed as a multi-agent system capable to configure itself in order to solve a problem by searching. The search process is conducted using two tasks: exploration and exploitation search. The self-organization is included in the algorithm as a population dynamics. This allows to allocate the system agents accordingly to the difficulty of performing each task, and at the same time, synchronizing the progress of each task.

The population dynamics is based on events that establish a reactive system. This makes it possible for the algorithm to be able to search on unknown environments, adapting its behavior accordingly to the search situation. Thus, it can be expected that the algorithm performance would be reasonable for a large variety of searching conditions.

The traveling salesman problem has been used as a benchmark to test the algorithm. The experimental results demonstrate that the agents distribution provided by the self-organization dynamics is nearly as good as the optimal distribution obtained through user-defined parameters, i.e., setting the distribution of agents before applying the algorithm. The comparison with other classical ant algorithms (Ant Colony System and Max-Min Ant System), shows that Ant Colony Extended performance is at least equivalent to those algorithms. And, depending on the scenario, it may achieve better results.

There are adaptive techniques that can allow to automatically tune ant algorithm's parameters. This approach does not try to self-adjust them, instead, it replaces the user defined parameters by self-organization dynamics. Both approaches could be combined, thus allowing to reinforce the weak points of one with benefical techniques taken from the other.

Finally, the population dynamics paradigm is not specific to just ant colony algorithms. It can be applied to other heuristic optimization algorithms.

Appendix A: TSP State Space Representation

This preliminary section details the TSP state space representation, and the heuristic function, that will be used in the next experimental sections.

The set of possible states for a given TSP instance, will be the set for all the edges in the TSP graph. Thus, a state is represented by a pair of cities: departure city and arrival city. For a given state, an action is the selection a city to travel to; so, the set of possible actions is the set of cities.

The initial states require to define extra symbols in order to represents the initial situation: being on a city before visiting any other one. Therefore, an initial state is a pair with the external symbol as a first component, and a city as the second one.

The goal test is a function that checks if all cities have been visited. The cost function is the distance covered by a tour. The successor function gives the resulting state by the application of an action to a given state, this function handles the restrictions: if the action points to a previously visited city, the resulting state is corrected.

The heuristic function is the inverse of the distance between two cities, raised to a power of 6.

More formally, the definition is as follows:

$$\begin{aligned}
\text{TSP Cities } (C)\text{:} \quad & \{c_1, c_2, c_3, \ldots, c_n\} \\
\text{Initial states } (Q_0)\text{:} \quad & \{(\Theta, c_i) | c_i \in C\} \\
\text{Set of states } (Q)\text{:} \quad & \{(c_i, c_j) | c_i, c_j \in C\} \cup Q_0 \\
\text{Set of actions } (U)\text{:} \quad & \{c_i | c_i \in C\} \\
\text{Goal:} \quad & V = C \\
\text{Cost:} \quad & d(c_n, c_1) + \sum_{i=1}^{n-1} d(c_i, c_{i+1})
\end{aligned}$$

Sucessor function:

$$\delta((c_i, c_j), c_k) = \begin{cases} (c_j, c_k) & \text{if } c_k \notin V \\ (c_j, c_{nn}) & \text{if } c_k \in V \end{cases}$$

$$\text{where} \quad c_{nn} = nn(c_j, C \setminus V)$$

Heuristic function:

$$\eta((c_i, c_j), c_k) = \begin{cases} d(c_j, c_k)^{-6} & \text{if } c_k \notin V \\ 0 & \text{if } c_k \in V \end{cases}$$

Where Θ represents the external symbol used to define the initial states. And V is the list of cities visited so far, d is a function that gives the distance between two cities. At last, $C \setminus V$ are the cities which are in C but not in V, and *nn* is the nearest-neighbor heuristic: for a given city, and a list of cities; *nn* returns the city of the list, closest to the given one.

Before an ant starts searching, it is randomly assigned to one of the initial states. When an ant searches for a solution, it can choose the next action – next city to visit– using the heuristic or the pheromone table. Once the action has been selected, the successor function is called; if the action leads to a previously visited city, the state transition is done using the closest non-visited city.

The ACE search is divided into two tasks: exploration and exploitation. Exploration should find new possible actions for the states and exploitation should not be able to change the available actions for any state, just increase or decrease the pheromone values.

When an ant is searching using the pheromone (exploitation), and it selects a previously visited city, the successor function corrects the state transition. But the ant keeps, in its list of actions, the original selection. If the ant finishes the search successfully, it updates the pheromone table using its list of actions. In other words, the actions reinforced are always the original ones selected by the ant, and never the correction made by the successor function.

This prevents that exploitation search change for the actions available in a state that could increase the search dispersion.

For example, in an hypothetical TSP instance of five cities, the pheromone table may look like table 1. An ant could perform a search like that shown in table 2. The transition from state $(2,3)$, is done selecting the city 1. As this city has been visited, the successor function gives the proirity for the closest city non-visited: 4, which results in the state $(3,4)$. But the action responsible for that transition is 1, not 4. At last, the ant should update the table according to the following list of associations: $(\Theta,1) \rightarrow 2$, $(1,2) \rightarrow 3$, $(2,3) \rightarrow 1$, $(3,4) \rightarrow 5$, and $(4,5) \rightarrow 1$, in order to complete the cycle.

Table 1. Pheromone table example

Keys (state)	Entries (action, probability)
$(\Theta,1)$	$(2,1)$
$(1,2)$	$(3,1)$
$(2,3)$	$(1,0.5),(4,0.5)$
$(3,4)$	$(5,1)$

Table 2. Tour construction example

Search step	Visited cities	List of actions	Current state
0	[1]	[]	(Θ,1)
1	[1,2]	[2]	(1,2)
2	[1,2,3]	[2,3]	(2,3)
3	[1,2,3,4]	[2,3,1]	(3,4)
4	[1,2,3,4,5]	[2,3,1,5]	(4,5)

Table 3. Parameter values used in the ACE, ACS, and MMAS comparison

Parameter value	ACE
β	0.5
α	0.99

Parameter value	ACS	MMAS
α	1	1
β	2	2
m	10	nc
ρ	0.1	0.02
q_0	0.9	–
ξ	0.1	–
τ_0	$1/(nc \cdot L_{nn})$	$1/(\rho \cdot L_{nn})$
cl	20	20
τ_{max}	–	$1/(\rho \cdot L_{best})$
τ_{min}	–	$2 \cdot \tau_{max} \cdot \frac{(1-\sqrt[nc]{0.05})}{\sqrt[nc]{0.05} \cdot (cl+1)}$

Appendix B: Parameter Values Used in the ACE, ACS, and MMAS Comparison.

Parameters values used to perform the comparison of section 6.3.. The values are shown in table 3, where "–" means that the parameter is not present in the algorithm, cl is the number of elements in the candidate list, L_{nn} is the tour length produced by the nearest neighbor heuristic, L_{best} is the tour length of the best-so-far solution, and nc is the number of cities. The table also included the formulas used to set the limits, τ_{max} and τ_{min}, of the MMAS. At last, it must be said that MMAS uses another parameter to regulate the pheromone update that is not included in the table: the pheromone is updated using the iteration-best solution, but every 25 iterations the pheromone is updated using the best-so-far solution.

References

[1] Marco Dorigo and Thomas Stützle. *Ant Colony Optimization (Bradford Books)*. The MIT Press, June 2004.

[2] S Goss, S Aron, J. Deneubourg, and J. Pasteels. Self-organized shortcuts in the Argentine ant. *Naturwissenschaften*, 76:579–581, 1989. 10.1007/BF00462870.

[3] Eric Bonabeau. Editor's introduction: Stigmergy. *Artificial Life*, 5(2):95–96, 2012/05/08 1999.

[4] S. Camazine, J.L. Deneubourg, N.R. Franks, J. Sneyd, G. Theraula, and E. Bonabeau. *Self-Organization in Biological Systems:(Princeton Studies in Complexity)*. Princeton University Press, 2003.

[5] M. Dorigo, V. Maniezzo, and A. Colorni. Ant system: optimization by a colony of cooperating agents. *Systems, Man, and Cybernetics, Part B: Cybernetics, IEEE Transactions on*, 26(1):29–41, February 1996.

[6] D.S. Johnson and L.A. McGeoch. The traveling salesman problem: A case study in local optimization. *Local search in combinatorial optimization*, pages 215–310, 1997.

[7] M. Dorigo and L.M. Gambardella. Ant colony system: A cooperative learning approach to the traveling salesman problem. *Evolutionary Computation, IEEE Transactions on*, 1(1):53–66, 1997.

[8] T. Stützle and H.H. Hoos. Improving the ant system: A detailed report on the max–min ant system. *FG Intellektik, FB Informatik, TU Darmstadt, Germany, Tech. Rep. AIDA–96–12*, 1996.

[9] T. Stützle. Local search algorithms for combinatorial problems-analysis, algorithms and new applications. *DISKI—Dissertationen zur Künstliken Intelligenz. infix, Sankt Augustin, Germany*, 1999.

[10] Thomas Stützle and Holger H. Hoos. MAX-MIN Ant System. *Future Generation Computer Systems*, 16(8):889–914, 2000.

[11] Stutzle. MAX-MIN Ant System and local search for the traveling salesman problem. In *Evolutionary Computation, 1997., IEEE International Conference on*, pages 309–314, 1997.

[12] Manuel López-Ibáñez and Thomas Stützle. Automatic Configuration of Multi-Objective ACO Algorithms. In Marco Dorigo, Mauro Birattari, Gianni Di Caro, René Doursat, Andries Engelbrecht, Dario Floreano, Luca Gambardella, Roderich Groß, Erol Sahin, Hiroki Sayama, and Thomas Stützle, editors, *Swarm Intelligence*, pages 95–106. Springer Berlin / Heidelberg, 2010. 10.1007/978-3-642-15461-4_9.

[13] Paola Pellegrini, Thomas Stützle, and Mauro Birattari. Off-line vs on-line tuning: A study on max–min ant system for the tsp. In Marco Dorigo, Mauro Birattari, Gianni Di Caro, René Doursat, Andries Engelbrecht, Dario Floreano, Luca Gambardella, Roderich Groß, Erol Sahin, Hiroki Sayama, and Thomas Stützle, editors, *Lecture Notes in Computer Science*, pages 239–250. Springer Berlin / Heidelberg, 2010. 10.1007/978-3-642-15461-4_21.

[14] Wei-jie Yu and Jun Zhang. Pheromone-distribution-based adaptive ant colony system. In *Proceedings of the 12th annual conference on Genetic and evolutionary computation*, pages 31–38, New York, NY, USA, 2010. ACM.

[15] Deborah M Gordon. *Ants at Work: How an Insect Society is Organized*. W. W. Norton & Company, 1 edition, October 2000.

[16] D.M. Gordon. The regulation of foraging activity in red harvester ant colonies. *American Naturalist*, 159(5):509–518, 2002.

[17] Jose B. Escario, Juan F. Jimenez, and Jose M. Giron-Sierra. Optimisation of autonomous ship manoeuvres applying ant colony optimisation metaheuristic. *Expert Systems with Applications*, 39(11):10120 – 10139, 2012.

[18] Jose B. Escario, Juan F. Jiménez, and Jose M. Giron-Sierra. Ant colony extended: Search in solution spaces with a countably infinite number of solutions. In *ANTS Conference*, pages 552–553, 2010.

[19] Deborah M Gordon. *Ant Encounters: Interaction Networks and Colony Behavior (Primers in Complex Systems)*. Princeton University Press, March 2010.

[20] D.M. Gordon. Control without hierarchy. *Nature*, 446(7132):143–143, 2007.

[21] J. Aracil. *Introducción a la dinámica de sistemas*. Alianza Editorial, 1978.

[22] Jay W. Forrester. *Principles of systems*. MIT Press, 1968.

[23] E. Bonabeau, G. Theraulaz, and J.L. Deneubourg. Quantitative study of the fixed threshold model for the regulation of division of labour in insect societies. *Proceedings: Biological Sciences*, 263(1376):1565–1569, 1996.

[24] Eric Bonabeau, Marco Dorigo, and Guy Theraulaz. *Swarm Intelligence: From Natural to Artificial Systems*. Oxford University Press, USA, 1 edition, 1999.

[25] R. Axelrod. *The complexity of cooperation: Agent-based models of competition and collaboration*. Princeton University Press, 1997.

[26] J.H. Miller and S.E. Page. *Complex adaptive systems: An introduction to computational models of social life*. Princeton University Press, 2007.

[27] G. Weiss. *Multiagent systems: a modern approach to distributed artificial intelligence*. MIT press, 2000.

[28] Newell Allen and Simon Herbert. *Human problem solving*. Prentice-Hall, 1972.

[29] S.J. Russell and P. Norvig. *Artificial intelligence: a modern approach*. Prentice hall, 2010.

[30] G. Di Caro and M. Dorigo. Antnet: Distributed stigmergetic control for communications networks. *Journal of Artificial Intelligence Research*, 9:317–365, 1998.

[31] A.P. Engelbrecht. *Fundamentals of computational swarm intelligence*, volume 1. Wiley London, 2005.

[32] Gerhard Reinelt. Tsplib 95, 1995.

[33] J.R. Pierce. *An introduction to information theory: symbols, signals & noise*. Dover Publications, 1980.

[34] C.E. Shannon. A mathematical theory of communication. *Bell System Technical Journal*, 27(3):379–423, 1948.

[35] Thomas Stützle. Acotsp 1.02, 2002.

In: Self-organization: Theories and Methods
Editor: WenJun Zhang

ISBN: 978-1-62618-865-5

Chapter 4

THE GENERALIZED PARTICLE SWARM OPTIMIZATION ALGORITHM WITH APPLICATION EXAMPLES

Željko S. Kanović, Milan R. Rapaić, Zoran D. Jeličić, Milan J. Rackov, Mirna N. Kapetina and Jelena T. Atanacković-Jeličić
Faculty of Technical Sciences, Novi Sad, Serbia

ABSTRACT

In this chapter, a generalization of the popular and widely used Particle Swarm Optimization (PSO) algorithm is presented. This novel optimization technique, named Generalized PSO (GPSO), is inspired by linear control theory. It overcomes some typical flaws of the classical PSO, enabling direct control over the key aspects of particle dynamics during the optimization process. The basic idea of this algorithm with its detailed theoretical and empirical analysis is presented, and parameter-tuning schemes are proposed. GPSO is also compared to the classical PSO and Genetic Algorithm (GA) on a set of benchmark problems. The presented results demonstrate the effectiveness of the proposed algorithm. Finally, two practical engineering applications of the GPSO algorithm are described, in the area of optimal gear design and architecture and urban design.

Keywords: Genetic Algorithm (GA); Particle Swarm Optimization (PSO)

1. INTRODUCTION

Many successful optimization techniques are inspired by natural processes and phenomena. The field of global optimization has prospered much from techniques such as Genetic Algorithms (GAs) (Michalewitz, 1999), Simulated Annealing (SA) (Kirkpatrick et al., 1983), Ant Colony Optimization (ACO) (Dorigo and Blum, 2005) and others. Among these search strategies, the Particle Swarm Optimization (PSO) algorithm is relatively novel, yet well studied and proven optimizer based on the social behavior of animals moving in large groups (particularly birds) (Kennedy and Eberhart, 1995).

PSO uses a set of particles called swarm to investigate the search space, imitating the movement of birds in a flock. The position of each particle is a potential solution, characterized by the value of the optimality criterion. During the search process, particles move through the search space and eventually discover the location(s) which provide the best value of the optimality criterion.

PSO has only a few adjustable parameters and it is computationally inexpensive and very easy to implement (Ratnaweera et al., 2004; Schutte and Groenwold, 2005). Thus, it provides high calculation speed, compared to other evolutionary algorithms. However, this algorithm also has some flaws, such as premature convergence tendency and the inability to independently control various aspects of the search - oscillation frequency and the impact of personal and global knowledge.

In this chapter a modification of the original PSO algorithm will be presented, named Generalized Particle Swarm Optimization (GPSO) (Kanović et al., 2011). The idea of this modification is to consider each particle within the swarm as a second-order linear stochastic system. The authors found and explained that the stability and response properties of such a system can be directly related to its performance as an optimizer, i.e., its explorative and exploitative properties.

Based on formal analysis of response properties and stability of the control system, which represents a particle and simulates its dynamics, a new algorithm formulation with a new set of parameters is proposed, which enables more direct and independent control over key aspects of the search process. This way, one can overcome the earlier mentioned inherent flaws of the PSO, keeping at the same time all advantages of the original version of the algorithm, concerning its simplicity, small number of parameters and easy practical implementation.

Using the stability criteria well-known from the control theory, some recommendations for parameter tuning are proposed, resulting in desired properties of the algorithm, concerning its convergence, exploration and exploitation properties.

An experimental analysis of the proposed algorithm is also presented in this chapter. The algorithm is tested on a set of standard benchmark functions and the results are compared to some earlier versions of the PSO algorithm, including the modification with time-varying coefficients. The obtained results are discussed and some crucial conclusions are emphasized regarding performance of algorithms used in comparison.

This chapter will also present two engineering applications of GPSO algorithm in the field of high contact ratio (HCR) gears geometry optimization and optimal layout of buildings in a settlement in terms of maximal effect of the sunlight, i.e. shadow minimization.

In PSO, each particle in the swarm is described by its position x and velocity v. The position of each particle is a potential solution, and the best position that each particle achieved during the entire optimization process (p) is memorized. The swarm also memorizes the best position ever achieved by any of its particles (g). The position and the velocity of each particle in the k -th iteration are:

$$v[k+1] = w \cdot v[k] + cp \cdot rp[k] \cdot (p[k] - x[k]) + cg \cdot rg[k] \cdot (g[k] - x[k])$$
$$x[k+1] = x[k] + v[k+1] \qquad (1)$$

Acceleration factors cp and cg control the relative impact of the personal (local) and common (global) knowledge on the movement of each particle. In the earlier versions of the algorithm, these coefficients have constant value during the search process. Inertia factor w, which was introduced for the first time in (Shi and Eberhart, 1999), keeps the swarm together and prevents it from diversifying excessively.

Random numbers rp and rg are mutually independent and uniformly distributed on the range [0, 1].

Many improved versions of PSO emerged so far, based on velocity clamping (Shahzad et al., 2009), constriction factor approach (Clerk and Kennedy, 2002), variable inertia factor (Eberhart and Shi, 2000), time-varying acceleration coefficients (Ratnaweera et al., 2004), particle position extrapolation (Senthil Arumugam et al., 2009), sub-swarm (Jiang et al., 2007) or dynamical multi-swarm (Liang and Suganthan, 2005) strategies, knowledge-based cooperation (Jie et al., 2008) and many other, which all improve the performance of the algorithm.

In recent literature, several other modifications of the original PSO also emerged, which incorporate mutation-like operators (Senthil Arumugam and Rao, 2007; Cui et al., 2008), modify the topology of the swarm (Van der Bergh, 2001; Niu et al., 2006; DeBao and ChunXia, 2009; Jiang et al., 2010) or utilize hybridization with other techniques (Chen et al., 2010; Thangaraj et al., 2011).

However, these modifications imply more complicated calculation procedure and extended calculation time and therefore reduce the main advantage of PSO – its simplicity and high calculation speed, so they cannot be compared with previously mentioned algorithms.

Numerous studies addressed PSO both empirically and theoretically (Shi and Eberhart, 1999; Van der Berg, 2001). The effectiveness of the algorithm has also been proven for various engineering problems (Dong et al., 2005, Dimopoulos, 2007; He and Wang, 2007; Olamaei et al., 2008; Chatterjee et al., 2011; Ramesh et al., 2012; Mallipeddi et al., In press; Fang et al., 2011; Bouallegue et al., 2012; Menhas et al., 2012; Alireza, 2011; Majhi and Panda, 2011; Samanta and Nataraj, 2009). However, the theoretical justification of the PSO procedure remained long an open question.

The first formal theoretical analyses were undertaken by Ozcan and Mohan (1998), addressing the dynamics of a simplified, one-dimensional, deterministic PSO model. Clerc and Kennedy (2002) also analyzed PSO by focusing on swarm stability and the so-called "explosion" phenomenon. Jiang et al. (2007) were the first to analyze the stochastic nature of the algorithm.

Rapaić and Kanović (2009) explicitly addressed PSO with time-varying parameters, which is the most common case in practical implementations. In (Kanović et al., 2011), the idea and comprehensive analysis of a new, improved algorithm modification is presented, which overcomes in great deal the flaws of the original PSO mentioned earlier. This algorithm modification and its implementation in some engineering applications is the subject of this chapter.

2. Generalized Particle Swarm Optimization

2.1. The Basic Idea

Theoretical analyses of PSO are mostly conducted on the basis of the second-order model

$$x[k+1]-(1+w-cp\cdot rp[k]-cg\cdot rg[k])x[k]+w\cdot x[k-1]= \\ =cp\cdot rp[k]\cdot p[k]+cg\cdot rg[k]\cdot g[k] , \quad (2)$$

equivalent to the model described by (1). This equation can also be interpreted as a difference equation of the motion of a stochastic, second-order, discrete-time, linear system with two external inputs. The output of such a system is the position of the particle (x), while its inputs are personal and global best positions (p and g, respectively). In general, a second-order discrete-time system can be modeled by the recurrent relation

$$x[k+1]+a_1\cdot x[k]+a_0\cdot x[k-1]=b_p\cdot p[k]+b_g\cdot g[k], \quad (3)$$

where some or all of the parameters a_1, a_0, b_p and b_g are stochastic variables with appropriate probability distributions. Several restrictions should be imposed on these parameters in order to make (3) a successful optimizer. First, the system should be stable in the sense of control theory (Åström and Wittenmark, 1997), with stability margins growing during the optimization process. This means that particles should initially be allowed to move more freely in order to explore the search space. As the search process approaches its final stages, previously found good solutions should be exploited and the effort of the swarm should be concentrated in their vicinity. Second, the response of the system to the perturbed initial conditions should be oscillatory to allow the particles to overshoot or fly over their respective attractor points, constantly improving current best solution. Also, if both external inputs approach the same limit value as k grows, the particle position x should also converge to this limit. Finally, in the early stages of the search, the motion of the system should primarily be determined by the cognitive input p, allowing particles to behave more independently, while in later stages the social input g should prevail because the swarm should become more centralized, and global knowledge of the swarm should dominate the local knowledge of each individual particle. All of these requirements are formulated in a sequel using notions from control theory.

The characteristic polynomial of (3) is $f(z)=z^2+a_1z+a_0$. The particles' dynamics is determined by its roots, also known as the eigenvalues of the system. The system is stable if and only if the modulus ρ of the eigenvalues is less than 1. In order for the system to be able to oscillate, the roots of the characteristic polynomial must not be positive real numbers. The argument φ of the eigenvalues determines the discrete frequency of the characteristic oscillations of the system. Argument values close to π result in more frequent oscillations.

All of the previously mentioned requirements can easily be satisfied if system (3) is rewritten in the following canonical form, often used in control theory (Åström and Wittenmark, 1997):

$$x[k+1]-2\zeta\rho x[k]+\rho^2 x[k-1]=(1-2\zeta\rho+\rho^2)(c\cdot p[k]+(1-c)\cdot g[k])\,. \tag{4}$$

In (4), ρ is the eigenvalues module and ζ is the cosine of their arguments. Parameter c is introduced to replace b_p and b_g. The basic idea of GPSO algorithm is to use (4), instead of (1) or(2), in optimizer implementation, because the parameters in this equation allow a more direct and independent control of the various aspects of the search procedure. The effect of these parameters to particle trajectory in a two-dimensional space is depicted in Figure 1.

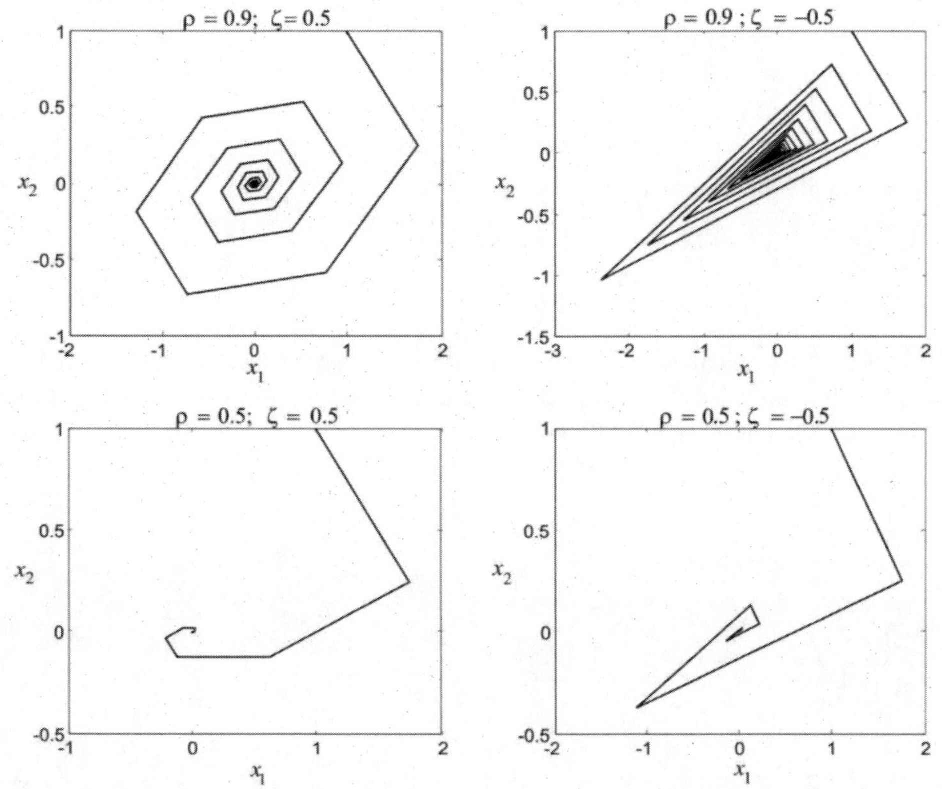

Figure 1. The effect of algorithm parameters to GPSO particle trajectories.

Both attractor points (i.e., global and personal best) are assumed to be equal to zero, for simplicity. Note that lower values of parameter ρ lead to faster convergence, while higher values result in less stable motion and slower convergence. Consequently, higher values of ρ enable the swarm to cover a wider portion of the search space, which improves exploration abilities of the algorithm, and this approach is usually used in early stages of the search. On the other hand, lower values of ρ should be applied in the later stages of the search, when search space is extensively investigated and faster convergence is preferable. Parameter ζ determines the way particles oscillate over attractor points. For ζ equal to 1, the particle motion would be non-oscillatory. For ζ equal to -1, a particle would erratically oscillate from one side of the attractor to another. Discrete frequency of oscillations, determined by ζ, has not been previously considered as a parameter in PSO dynamics. Conducting numerous numerical experiments, the authors noticed that by adjusting ζ to a value between -1 and 1, one can control oscillatory nature of particles' motion and obtain a desired behavior, in the sense of exploration properties of the algorithm. Exploitation properties, however, are mainly controlled adjusting the value of parameter *c*. Higher value of *c* favors personal (local) knowledge, while lower value of this parameter enable better exploitation of global swarm knowledge.

2.2. Convergence Analysis

Among several ways to define convergence for stochastic sequences, the notion of mean square convergence is adopted in this chapter. A stochastic sequence $x[k]$ is said to converge in mean square to the limit point a if and only if

$$\lim_{k\to\infty} \mathbf{E}(x[k]-a)^2 = 0, \tag{5}$$

where $\mathbf{E}$ denotes the mathematical expectation operator (Papoulis, 1965). Consequently, the investigation of convergence of a stochastic sequence can be replaced by the investigation of two deterministic sequences, since (5) is equivalent to

$$\lim_{k\to\infty} \mathbf{E}x[k] = a \tag{6}$$

and

$$\lim_{k\to\infty} \mathbf{E}x^2[k] = a^2. \tag{7}$$

In the sequel, $\overline{(\cdot)}$ is used instead of $\mathbf{E}(\cdot)$ to simplify notation and make it more compact.

If we assume that $p[k]$ and $g[k]$ converge to p and g respectively, GPSO's convergence is equivalent to the stability of the dynamical system (4). Let us introduce $\mu = c\cdot p + (1-c)\cdot g$ and $y[k] = x[k] - \mu$. The equation (4) leads to:

$$y[k+1] - 2\zeta\rho y[k] + \rho^2 y[k-1] = 0. \tag{8}$$

Since the inertia factor w of the classical PSO is closely related to the eigenvalue modulus, ρ is considered a deterministic parameter. The ζ factor, on the other hand, is assumed to be stochastic and independent of particle position.

Applying the expectation operator to (8) one obtains

$$\overline{y[k+1]} - 2\overline{\zeta}\rho\overline{y[k]} + \rho^2\overline{y[k-1]} = 0. \tag{9}$$

The eigenvalues of this system are

$$p_{1,2} = \rho\zeta \pm j\rho\sqrt{1-\zeta^2}, \tag{10}$$

where j is the imaginary unit, and the stability conditions are denoted as

$$0 < \rho < 1, \tag{11}$$

$$|\zeta| \leq 1. \tag{12}$$

If system (9) converges, its limit point is zero, and the mathematical expectation of any particle's position tends to μ. If personal best and global best positions are asymptotically equal, which is true for the particle achieving the global best position, then $\mu = p = g$. Now we will address the variance of $x[k]$. From (8), it is readily obtained that

$$\overline{y^2[k+1]} = 4\overline{\zeta^2}\,\rho^2\,\overline{y^2[k]} + \rho^4\,\overline{y^2[k-1]} - 4\overline{\zeta}\,\rho^3\,\overline{y[k]y[k-1]}, \tag{13}$$

$$\overline{y[k+1]y[k]} = 2\overline{\zeta}\rho\overline{y^2[k]} - \rho^2\,\overline{y[k]y[k-1]}. \tag{14}$$

Equations (13) and (14) can be considered a model of a deterministic linear system. If we define the state variables as $\alpha_1[k] = \overline{y^2[k]}$, $\alpha_2[k] = \overline{y^2[k-1]}$ and $\alpha_3[k] = \overline{y[k]y[k-1]}$, the system can be rewritten in state-space form as

$$\boldsymbol{\alpha}[k+1] = \mathbf{A}\boldsymbol{\alpha}[k], \tag{15}$$

with state vector $\boldsymbol{\alpha} = [\alpha_1\ \alpha_2\ \alpha_3]^T$ and the matrix $\mathbf{A}$ defined as

$$\mathbf{A} = \begin{bmatrix} 0 & 1 & 0 \\ \rho^4 & 4\overline{\zeta^2}\rho^2 & -4\overline{\zeta}\rho^3 \\ 0 & 2\overline{\zeta}\rho & -\rho^2 \end{bmatrix}. \tag{16}$$

The system (15) is stable if and only if all eigenvalues of $\mathbf{A}$ are less than 1 in modulus. The eigenvalues of a matrix are roots of its characteristic polynomial. The characteristic polynomial of the matrix $\mathbf{A}$ is

$$f(z)=z^3+z^2\rho^2(1-4\overline{\zeta^2})-z\rho^4(4\overline{\zeta^2}-8(\overline{\zeta})^2+1)-\rho^6. \tag{17}$$

Stability can be investigated by various methods. Jury's criterion (Åström and Wittenmark, 1997) is applied in the sequel. For a general third-order polynomial $f(z)=z^3+a_2z^2+a_1z+a_0$, the conditions under which all zeros are less than one in modulus are

$$f(1)>0,\ f(-1)<0,\ |a_0|<1,\ |a_0^2-1|>|a_0a_2-a_1|. \tag{18}$$

Applying these conditions to (17) we obtain

$$(1+\rho^2)(1-\rho^4)>4\rho^2(\overline{\zeta^2}+\rho^2(\overline{\zeta^2}-2(\overline{\zeta})^2)), \tag{19}$$

$$(1-\rho^2)(1-\rho^4)>4\rho^2(-\overline{\zeta^2}+\rho^2(\overline{\zeta^2}-2(\overline{\zeta})^2)), \tag{20}$$

$$|\rho^6|<1, \tag{21}$$

$$1-\rho^{12}>\rho^4\left|1-\rho^4+4\overline{\zeta^2}(1+\rho^4)-8(\overline{\zeta})^2\right|. \tag{22}$$

Due to (11), both (20) and (21) are identically satisfied; that is, right-hand side of (20) is always negative and its left-hand side is always positive. The stability conditions are therefore (19) and (22). Unfortunately, these conditions are not as nearly as compact as (11) and (12), and they provide no direct insight into the influence of parameters ρ and ζ on stability. However, some clarification can be obtained by introducing the helper polynomial

$$f_1(q)=\frac{f(\rho^2q)}{\rho^6}=q^3+q^2(1-4\overline{\zeta^2})-q(4\overline{\zeta^2}-8(\overline{\zeta})^2+1)-1 \tag{23}$$

and investigating its roots by means of (18):

$$f_1(1)=8((\overline{\zeta})^2-\overline{\zeta^2})>0, \tag{24}$$

$$f_1(-1)=-8(\overline{\zeta})^2<0, \tag{25}$$

$$|1|<1, \tag{26}$$

$$0 > 8\left|\overline{\zeta^2} - (\overline{\zeta})^2\right| \tag{27}$$

Condition (24) is not satisfied because the variance of ζ ($\operatorname{var}\zeta = \overline{\zeta^2} - (\overline{\zeta})^2$) is non-negative. Conditions (26) and (27) are also violated. It is therefore clear that f_1 has at least one root outside the unit circle or on its boundary at best. Since the roots of f are ρ^2 times the roots of f_1, it can be concluded that at least one root of f is greater than or equal to ρ^2. Thus, it follows that an increase in ρ has a negative influence on the convergence of both the mean value and the variance of particle positions. It is also clear that by increasing the variance of ζ, $f_1(1)$ decreases, and the right-hand side of (27) increases; therefore, the increase in the variance of ζ impedes the convergence of the algorithm.

The limit point of $\boldsymbol{\alpha}$ is $[0\ 0\ 0]^T$, and therefore, $\overline{y^2[k]}$ is zero in limit. Because $\overline{y^2[k]}$ is asymptotically equal to the variance of x, it follows that the variance of the position of any particle is asymptotically equal to zero. This concludes the proof that under conditions (11), (12), (19) and (22), GPSO system (4) converges in mean square sense.

2.3. Empirical Analysis

Proper parameter selection is crucial for the performance of PSO. The relationships between adjustable factors within classical PSO and GPSO are

$$\rho = \sqrt{w}, \tag{28}$$

$$\zeta = \frac{1 - w - cp \cdot rp - cg \cdot rg}{2\sqrt{w}}, \tag{29}$$

$$c = \frac{cp \cdot rp}{cp \cdot rp + cg \cdot rg}. \tag{30}$$

The inertia factor w keeps the particles together and prevents them from wandering too far away. Since its introduction by Shi and Eberhart (1999), it was noted that it is beneficial to gradually decrease its value. It is common to select a value close to 0.9 at the beginning of the search and a value close to 0.4 at the end. Based on (28), ρ should be decreasing from about 0.95 to about 0.6. Regarding the acceleration factors, $cp = cg = 2$ is the choice used in the original variant of the PSO algorithm (Kennedy and Eberhart, 1995). It is also common to use $cp = cg = 0.5$ as well as $cp = 2.8$ and $cg = 1.3$. It is generally noted that the choice of acceleration factors is not critical, but they should be chosen so that $cp + cg < 4$ (Schutte and Groenwold, 2005). In fact, if the last condition is violated, PSO does not converge, regardless of the inertia value (Jiang et al., 2007; Rapaić and Kanović, 2009). Ratnaweera et al. (2004) introduced the time-varying acceleration coefficients PSO (TVAC-PSO) and demonstrated that it outperforms other commonly used PSO schemes. They suggested that cp should linearly decrease from 2.5 to 0.5, while cg should simultaneously linearly increase from 0.5 to 2.5. This would correspond to changing c from about 0.8 to approximately 0.2.

It is important to note that the novelty of the GPSO algorithm does not reduce to a change in coefficient expressions. Indeed, if the coefficients of GPSO are fixed, then according to formulas (28)-(30), one can find an equivalent set of classical PSO parameters. However, if parameters vary in time, which is the case in most practical applications, linear variations in GPSO parameters are equivalent to nonlinear variations in inertia w and acceleration coefficients cp and cg. It is this nonlinearity that accounts for the performance of GPSO, as the effects of the new parameterization cannot be achieved by the linear alteration of the classical PSO parameters. One could use the classical PSO with a highly nonlinear parameter adaptation law to achieve this, but this would be impractical. In GPSO, nonlinearity has been hidden within the algorithm definition, thus allowing the same effect to be achieved by linear alteration of the newly proposed parameters. These parameters have a well-defined interpretation in terms of particle dynamics, and they allow independent control of the particles' dynamic properties, such as stability, characteristic frequency and relative impact of local to global knowledge.

Based on recommendations stated earlier, we can propose that ρ linearly decrease from 0.95 to 0.6 and c linearly decrease from 0.8 to 0.2. The proper selection of ζ proved to be the most difficult task, as this factor has no direct analogy to any of the parameters of the classical PSO. The ζ parameter equals the cosine of particle eigenvalues in the dynamic model. It is directly related to the ability of particles to oscillate around attractor points (i.e., global and personal best) and, therefore, has a crucial impact on the exploitative abilities of the algorithm. If ζ equals 1, this would prevent particles from flying over the attractor points; however, if ζ is close to -1, this would result in the desultory movement of the particles. In both cases, particles cannot explore the vicinity of the attractor points. Based on thorough empirical analysis and numerous numerical experiments conducted by the authors, it is shown that the most appropriate and robust choice is to select ζ uniformly distributed across a wider range of values, and most appropriate values for this range appeared to be from -0.9 to 0.2.

The proposed GPSO scheme is compared to TVAC-PSO developed by Ratnaweera et al. (2004), which will be denoted simply as PSO in the sequel. Among many existing PSO modifications, this algorithm is chosen for comparison because it alters the value of all three coefficients during the search, keeping at the same time the original form and implementation simplicity of PSO algorithm, which makes it similar to proposed GPSO schemes. A comparison is also made with respect to GA with linear ranking, stochastic universal sampling, uniform mutation and uniform crossover (Michalewitz, 1999), since GA is extensively used in engineering applications. The comparison has been performed using several well-known benchmark functions presented in Table 1, all of which have a minimum value of zero.

The search was conducted within a 5-dimensional search space using 100 iterations with 30 particles in the swarm. The population was initialized within a hypercube of edge 10 units. The center of the initial hypercube is displaced from the global optimal point of a particular benchmark by 100 units in each direction. The results of the experiment are presented in Table 2; values are shown for the mean, median and standard deviation of the obtained minimum after 100 consecutive runs.

Table 1. Benchmark functions used for comparison

Dixon-Price	$f(\mathbf{x})=(x_1-1)^2+\sum_{i=2}^{n} i(2x_i^2-x_{i-1})^2$
Rosenbrock	$f(\mathbf{x})=\sum_{i=1}^{n-1}\left[100(x_i^2-x_{i+1})^2+(x_i-1)^2\right]$
Zakharov	$f(\mathbf{x})=\sum_{i=1}^{n} x_i^2+\left(\sum_{i=1}^{n}0.5ix_i\right)^2+\left(\sum_{i=1}^{n}0.5ix_i\right)^4$
Griewank	$f(\mathbf{x})=\sum_{i=1}^{n}\frac{x_i^2}{4000}-\prod_{i=1}^{n}\cos\left(x_i/\sqrt{i}\right)+1$
Rastrigin	$f(\mathbf{x})=10n+\sum_{i=1}^{n}(x_i^2-10\cos(2\pi x_i))$
Ackley	$f(\mathbf{x})=20+e-20\exp(-\frac{1}{5}\sqrt{\frac{1}{n}\sum_{i=1}^{n}x_i^2})-\exp(-\frac{1}{n}\sum_{i=1}^{n}\cos(2\pi x_i))$
Michalewitz	$f(\mathbf{x})=5.2778-\sum_{i=2}^{n}\sin(x_i)\left(\sin\left(ix_i^2/\pi\right)\right)^{2m}$, $m=10$
Perm	$f(\mathbf{x})=\sum_{k=1}^{n}\left[\sum_{i=1}^{n}(i^k+0.5)\left(\left(x_i/i\right)^k-1\right)\right]^2$
Spherical	$f(\mathbf{x})=\sum_{i=1}^{n}x_i^2$

It is clear that newly proposed scheme performs better than either PSO or GA in the majority of cases; the exception is Michalewitz's benchmark, where it is consistently outperformed by GA and slightly outperformed by PSO. Note also that in many of the considered cases, GPSO algorithm shows results that are several orders of magnitudes better than those obtained by other two optimizers.

Figure 2 depicts changes in the objective value of the best, mean and worst particles within the PSO and GPSO swarms.

One can notice that the best particle converges very fast in both settings. In the GPSO swarm, however, other particles do not follow so rapidly, effectively keeping the diversity of the population sufficiently large. Figure 3 shows the maximum distance between two particles in the swarm during the optimization process. It is clear that the GPSO swarm becomes extremely diverse at certain points, spreading across the vast area of the search space, which is the main and most important effect of the newly proposed parameterization. This phenomenon is the explanation of the superior performance that GPSO exhibits in most of the analyzed cases. We must note that figure 2 and 3 depict data obtained optimizing Griewank's function, but similar behavior can be noticed for all considered test-functions.

Table 2. Results of the experiment: mean, median and standard deviation of the minimal obtained values

		PSO	GPSO	GA
Dixon-Price	mean	$9.29\cdot10^{-1}$	$2.60\cdot10^{-1}$	$1.06\cdot10^{3}$
	median	$2.50\cdot10^{-2}$	$8.25\cdot10^{-15}$	$2.94\cdot10^{2}$
	std. dev.	$1.82\cdot10^{1}$	$3.22\cdot10^{-1}$	$1.710\cdot10^{3}$
Rosenbrock	mean	$3.167\cdot10^{2}$	$1.29\cdot10^{2}$	$1.81\cdot10^{3}$
	median	$9.17\cdot10^{1}$	$2.59\cdot10^{1}$	$2.29\cdot10^{2}$
	std. dev.	$6.127\cdot10^{2}$	$2.64\cdot10^{2}$	$3.77\cdot10^{3}$
Zakharov	mean	$7.60\cdot10^{2}$	$2.21\cdot10^{-11}$	$1.76\cdot10^{1}$
	median	$2.19\cdot10^{1}$	$6.88\cdot10^{-17}$	2.86
	std. dev.	$2.11\cdot10^{3}$	$3.62\cdot10^{-10}$	$4.46\cdot10^{1}$
Griewank	mean	$8.06\cdot10^{-1}$	$7.50\cdot10^{-2}$	$8.94\cdot10^{-2}$
	median	$3.32\cdot10^{-1}$	$6.64\cdot10^{-2}$	$7.23\cdot10^{-2}$
	std. dev.	$9.25\cdot10^{-1}$	$4.68\cdot10^{-2}$	$6.46\cdot10^{-2}$
Rastrigin	mean	4.12	2.87	7.39
	median	3.08	1.98	7.27
	std. dev.	3.25	1.88	2.82
Ackley	mean	$2.00\cdot10^{1}$	$2.00\cdot10^{1}$	$2.00\cdot10^{1}$
	median	$2.00\cdot10^{1}$	$2.00\cdot10^{1}$	$2.00\cdot10^{1}$
	std. dev.	$2.00\cdot10^{-2}$	$2.99\cdot10^{-2}$	$5.40\cdot10^{-3}$
Michalewitz	mean	2.05	1.99	$9.94\cdot10^{-1}$
	median	2.06	2.07	$9.65\cdot10^{-1}$
	std. dev.	$4.49\cdot10^{-1}$	$5.32\cdot10^{-1}$	$2.38\cdot10^{-1}$
Perm	mean	$5.50\cdot10^{14}$	$2.08\cdot10^{1}$	$2.46\cdot10^{12}$
	median	$1.43\cdot10^{3}$	4.79	$2.78\cdot10^{3}$
	std. dev.	$2.85\cdot10^{15}$	$9.85\cdot10^{1}$	$1.96\cdot10^{13}$
Spherical	mean	$2.45\cdot10^{-6}$	$4.48\cdot10^{-19}$	$1.22\cdot10^{-1}$
	median	$4.43\cdot10^{-7}$	$2.29\cdot10^{-23}$	$9.85\cdot10^{-2}$
	std. dev.	$1.50\cdot10^{-5}$	$1.39\cdot10^{-17}$	$1.18\cdot10^{-1}$

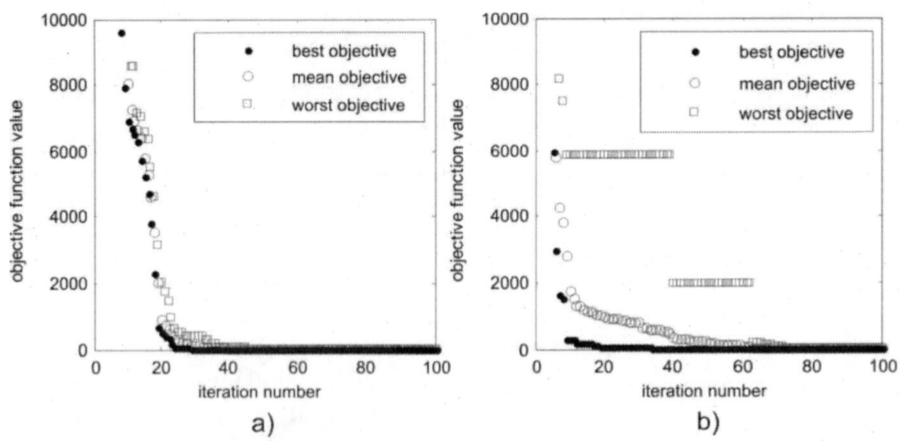

Figure 2. Objective values during 100 iterations. a) PSO swarm; b) GPSO swarm.

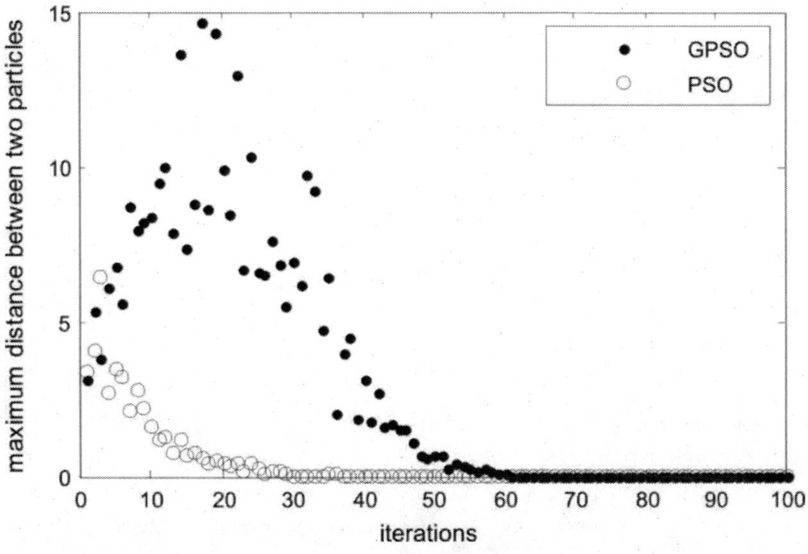

Figure 3. Maximum distance between two particles during the optimization process for GPSO and PSO.

Figure 4 illustrates one of the most important advantages of PSO algorithm and its modifications over other evolution algorithms – its high calculation speed, as a consequence of algorithm simplicity. Calculation of minimum of Ackley's function is depicted on Figure 4 as an example, but the calculation time is similar for all other benchmarks. One can notice that both PSO and GPSO algorithms finish the calculation in less than 0.1 seconds, while GA needs about 1.3 seconds to complete calculations for same swarm size and same number of iterations. This property of PSO is very promising and implies its application in real-time optimization problems.

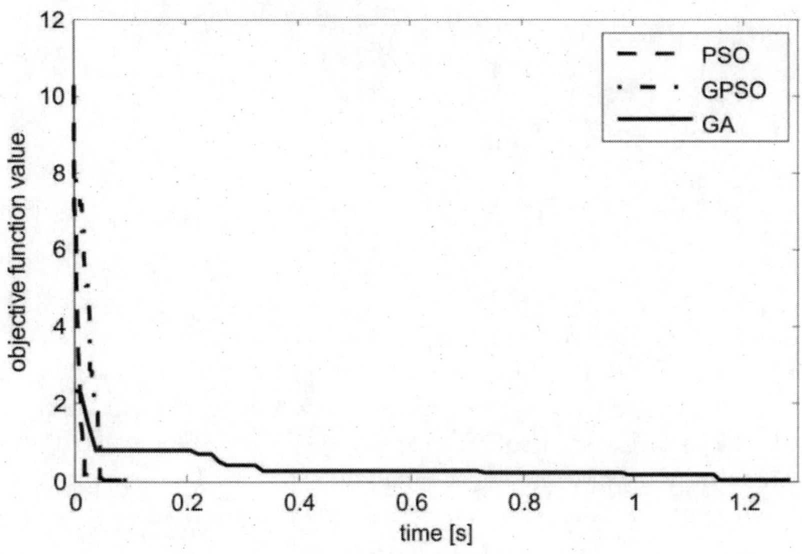

Figure 4. Calculation time for PSO, GPSO and GA.

Two other experiments were also conducted, one with same swarm size and iteration number, but with initial hypercube centered on global optimal point for each benchmark, and the other with the same swarm size but with 1000 iterations. Both these experiments showed results that are consistent with previously described experiment (see (Kanović et al., 2011) for more details).

3. GPSO IN GEAR GEOMETRY OPTIMIZATION

3.1. Essential Terms Definition

In this section, we will present an application of GPSO algorithm in optimal design of high contact ratio gear pairs. High contact ratio (HCR) gear pairs are special gear pairs having at least two pairs of teeth in contact. High contact ratio is obtained with increased addendum height, larger than in standard gearing. Proposed geometry of HCR gearings is complicated due to the fact that there is a possibility of occurring meshing interference, much larger than in standard involute profiles. Also, there is a higher risk of too small thickness of a tooth tip and significantly less favorable values of specific slips into the flanks. A pair of HCR gears with its basic geometry is depicted in Figure 5.

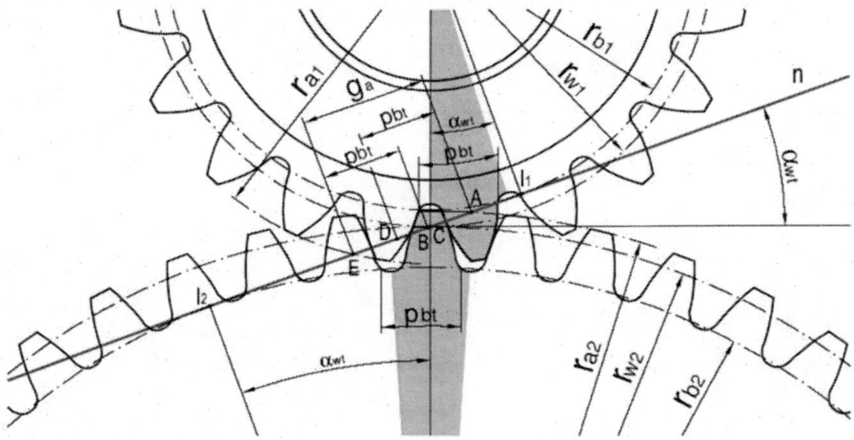

Figure 5. HCR gear geometry.

The distance between two corresponding flanks is called the base pitch. If the value of the base pitch is higher than the value of the path of contact, contact between the two following teeth does not occur, and when the two mating teeth release, the continuity in meshing is not guaranteed. The correct working of the gear will be assured if the value of the path of contact is higher than that of the base pitch (see Figure 5). The ratio between the length of contact g_a, and pitch on the base cylinder p_{bt} is given by the following formula:

$$\varepsilon_\alpha = \frac{\text{length of contact}}{\text{base pitch}} = \frac{g_a}{p_{bt}} = \frac{\sqrt{r_{a1}^2 - r_{b1}^2} + \sqrt{r_{a2}^2 - r_{b2}^2} - a_w \sin\alpha_{wt}}{p_{bt}} \quad (31)$$

and it is called the transverse contact ratio. In (31), a_w is centre distance, α_{wt} is working pressure angle, r_{a1}, r_{b1}, r_{a2}, and r_{b2} are pitch and base radius of pinion and gear wheel, respectively (Dudley, 1995). So, contact ratio can be defined as the average number of teeth in contact at one time.

It can be expected that high-quality HCR gear meshes have lower mesh-induced vibration and noise than normal contact ratio (NCR) gear meshes (Kasuba,1981). It is well known that increasing the average number of teeth in contact leads to excluding or reduction of the vibration amplitude. First, it was established experimentally that dynamic loads decrease with increasing contact ratio in spur gearing (Kasuba, 1981). Moreover, in order to get a further reduction of the vibration, HCR gear profiles can be optimized. Sato et al. (Sato et al., 1983) found that HCR gears are less sensitive with respect to manufacturing errors. In particular, such kind of gears allows larger tolerance in the tip relief length. Moreover, they found that, in the absence of pressure angle error, the best contact ratio should be about 2. Kahraman and Blankenship (Kahraman et al., 1999) published an experimental work on HCR gear vibration; they found that the best behavior is obtained with an integer contact ratio, even though other specific non integer (rational) contact ratios can minimize the amplitude of some specific harmonics of the static transmission error. It is important to note that in (Kahraman et al., 1999) HCR gears were obtained by modifying the outside diameter; the other macro-geometric parameters, e.g. the number of teeth, were left unchanged.

Dynamic loads and noise are reduced by using high contact ratio gears. According to results of different measurements of gear pair, reduction of noise proved to be the best using HCR gearing with the value of contact ratio $\varepsilon_\alpha = 2$. Decrease in noise is caused by $\varepsilon_\alpha = 2$ because there are always two pairs of teeth in contact, which means when one pair of teeth go out from the contact, another pair of teeth is coming in contact and applied force is considerably smaller since it is divided on two pairs of teeth. Therefore, it is a general conclusion that gearing in automotive industry should be done with $\varepsilon_\alpha = 2$ in order to reduce the noise and dynamic forces.

The increase in contact ratio can be implemented in two ways: by decreasing pressure angle α and by increasing tooth height h. Obviously, the use of a standard pressure angle and standard tools is preferable (Podzharov et al., 2003). Therefore, the most favorable solution is obtained by increasing addendum height. However, there are a lot of geometrical and manufacturing constraints that have to be satisfied which limit increasing the contact ratio.

The equation (31) shows that $\varepsilon_\alpha = f\,(g_a, p_{bt})$. Tooth pitch on the base circle of standard involute gearing is equal to base pitch on HCR gearing, and it is considered as constant. This means that achieving the greatest value of the contact ratio ε_α has to be obtained by the greatest possible increase of length of contact g_a, which is defined as:

$$g_a = \sqrt{r_{a1}^2 - r_{b1}^2} + \sqrt{r_{a2}^2 - r_{b2}^2} - a_w \sin \alpha_{wt} \tag{32}$$

where the tip diameters of pinion and gear wheel are as follows:

$$r_{a1} = r_1 + \left(h_{a1} + x_1 \cdot m_n\right) \tag{33}$$

$$r_{a2} = r_2 + \left(h_{a2} + x_2 \cdot m_n\right) \tag{34}$$

In (33) and (34), h_{a1}, h_{a2} are addendum heights, x_1 and x_2 are factors of correction of the pinion and gear wheel and m_n is gear module.

The goal of the geometric parameters optimization of HCR gearing is to achieve the desired value $\varepsilon_\alpha = 2$, as explained earlier. The variables, i. e. optimization parameters are addendum heights of teeth h_{a1}, h_{a2} and factors of correction x_1 and x_2, while other basic parameters of the gear pair are known.

In the sequel, we will define some constraints that must be taken into account during the optimization process. These constraints are related to interference during production, removal of meshing interference and minimum arc thickness of the tooth tip. All relationships defined in the sequel are based on gear geometry, but their derivation will not be explained in details, since it can be found in the literature; we will only show the final form of constraints to clarify all the complexity of the optimization procedure.

3.1.1. Interference during the Production

This interference occurs in the production process of gear forming when the tooth of the rack tool is in collision with a produced transition curve of the gear wheel, resulting in a so-called undercut tooth, as shown in Figure 6.

Interference during the production will not occur if following condition is satisfied (Dudley, 1995):

$$g_{F_{Z1}} = r_{b1} \cdot \tan\alpha_t - \frac{m_n}{\sin\alpha_t}\left(\frac{h_{a1}}{m_n} - x_1\right) \geq 0 \tag{35}$$

where α_t is transverse pressure angle. This condition must also be valid for gear wheel, so we have:

$$g_{F_{Z2}} = r_{b2} \cdot \tan\alpha_t - \frac{m_n}{\sin\alpha_t}\left(\frac{h_{a2}}{m_n} - x_2\right) \geq 0. \tag{36}$$

3.1.2. Meshing Interference

Meshing interference is referred in the case of a collision between curves of teeth profiles as interference between these curves. It means that the meshing interference may occur as a collision of head of gear wheel and the transition curve of pinion, as shown in Figure 7, and / or head of pinion and the transition curve of gear wheel.

According to Figure 7 it is possible to assess the interference between meshed head of gear wheel and transition curve of pinion, where the correct image in this case is only possible if the head wheel comes into engagement with the pinion of the involute part which is bordered with the meshing line and point F_{z1} (last point on the pinion involute). Meshing interference will not occur only if distance g_{FzA} is bigger than distance g_{Fz1}. This condition can be interpreted as (Podzharov et al., 2003):

$$r_{a2} \leq a_w \frac{\sin\alpha_{wt}}{\sin\alpha_{at2}} - r_{b1}\frac{\tan\alpha_t}{\sin\alpha_{at2}} + \frac{m_n}{\sin\alpha_t \cdot \sin\alpha_{at2}}\left(\frac{h_{a1}}{m_n} - x_1\right). \tag{37}$$

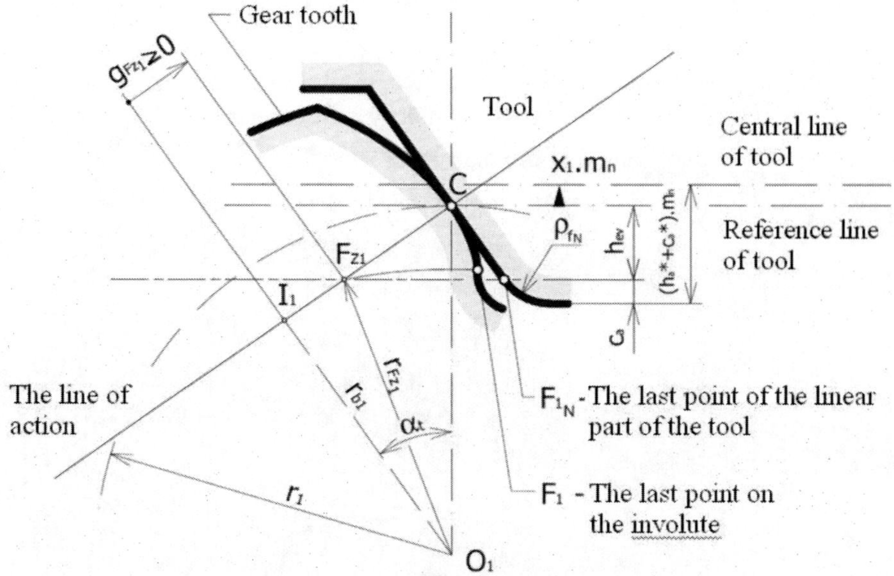

•

Figure 6. Interference during the production with tooth rack tool.

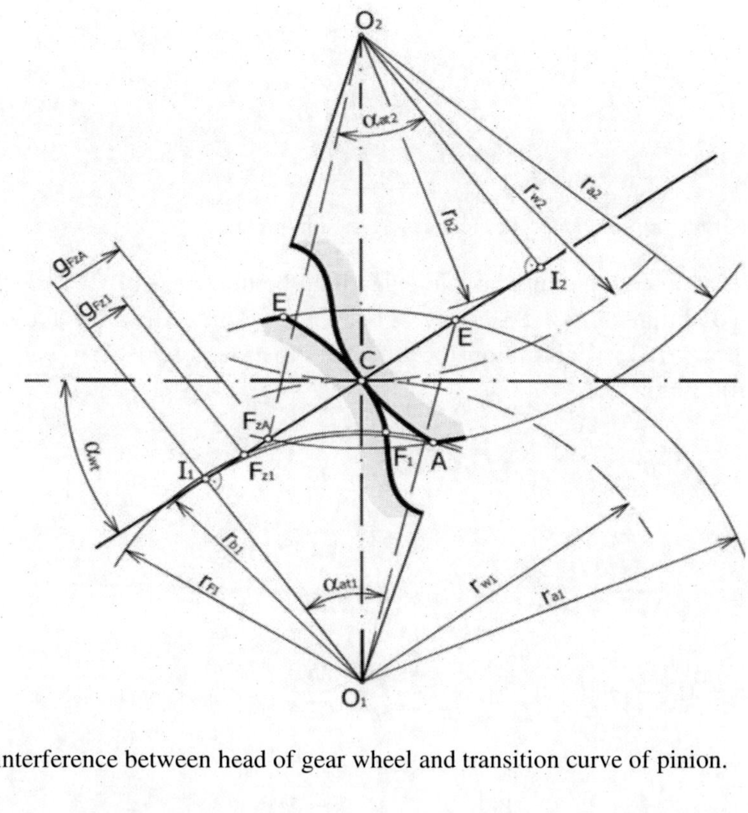

•

Figure 7. Meshing interference between head of gear wheel and transition curve of pinion.

Same condition must be valid for gear wheel, so we have:

$$r_{a1} \le a_w \frac{\sin\alpha_{wt}}{\sin\alpha_{at1}} - r_{b2}\frac{\tan\alpha_t}{\sin\alpha_{at1}} + \frac{m_n}{\sin\alpha_t \cdot \sin\alpha_{at1}}\left(\frac{h_{a2}}{m_n} - x_2\right). \tag{38}$$

3.1.3. Minimum Thickness of the Tooth Head Circle

Changing addendum height h_{a1} will certainly influence the total thickness of the tooth on tip circle s_{a1}, shown in Figure 8. Greater tooth height, as well as a positive correction factor, may affect the thickness of the tooth on tip circle under the permissible value.

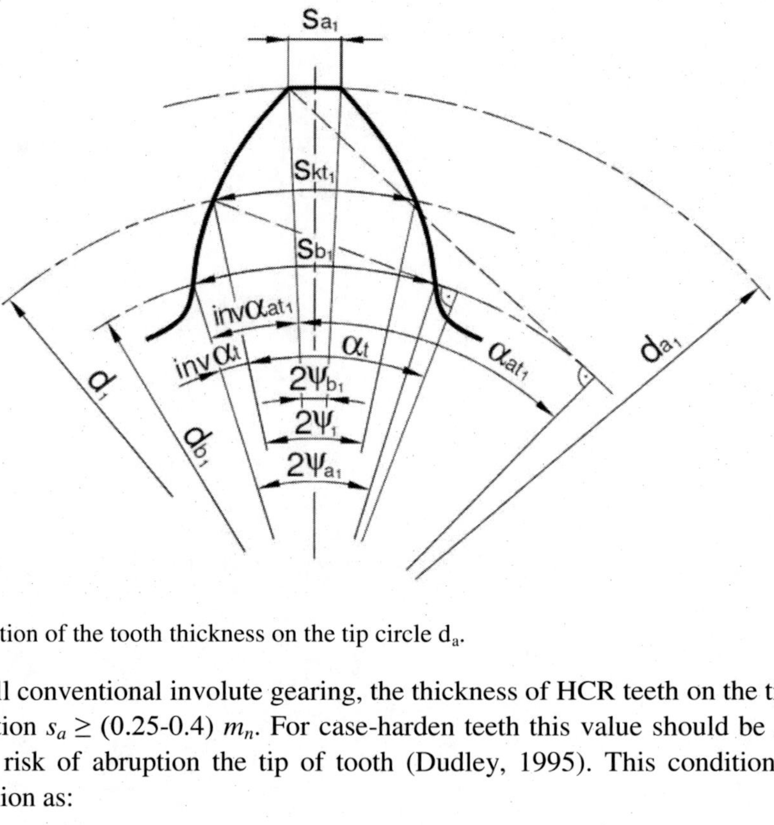

Figure 8. Determination of the tooth thickness on the tip circle d_a.

Same as for all conventional involute gearing, the thickness of HCR teeth on the tip circle is given by condition $s_a \geq$ (0.25-0.4) m_n. For case-harden teeth this value should be $s_a \geq 0.4\ m_n$, due to lower risk of abruption the tip of tooth (Dudley, 1995). This condition can be formulated for pinion as:

$$0.4 \cdot m_n \le \left(d_1 + 2\cdot\left(\frac{h_{a1}}{m_n} + x_1\right)\cdot m_n\right)\cdot\left(\frac{p_t + 4x_1\, m_n \tan\alpha_t}{2d_1} + \text{inv}\alpha_t - \text{inv}\alpha_{at1}\right) \tag{39}$$

and for gear wheel as:

$$0.4 \cdot m_n \le \left(d_2 + 2\cdot\left(\frac{h_{a2}}{m_n} + x_2\right)\cdot m_n\right)\cdot\left(\frac{p_t + 4x_2\, m_n \tan\alpha_t}{2d_2} + \text{inv}\alpha_t - \text{inv}\alpha_{at2}\right). \tag{40}$$

3.2. Optimization Criterion Formulation

The goal of geometry parameters optimization is to achieve that contact ratio equals two ($\varepsilon_\alpha = 2$), so that always two pairs of gears are engaged. With this value of contact ratio it is expected to reduce the vibration and gear noise. At the same time, all previously mentioned constraints, defined by equations (35), (36), (37), (38), (39) and (40) must be valid. Further, there are some other constraints related to search space, i. e. to the permissible values of variables in optimality criterion (Dudley, 1995), such as:

addendum height on pinion teeth h_{a1} and on gear teeth h_{a2} must be in range $[1/m_n, 1.5/m_n]$;

correction factors of pinion x_1 and gear wheel x_2 must be in range [-1, 1].

For optimization of HCR geometry following data is used: teeth number of pinion $z_1 = 21$, teeth number of gear wheel $z_2 = 51$, module $m_n = 4$ mm, modified centre distance $a_w = 144$ mm, and pressure angle $\alpha_n = 20° = \pi/9$ rad. Optimality criterion is defined as:

$$C = \left(\varepsilon_\alpha - 2\right)^2 \qquad (41)$$

whose global minimum results in desired value of contact ratio. Using GPSO algorithm, a very accurate solution (of order 10^{-15}) was obtained, with all constraints satisfied.

This problem, however, does not have only one global minimum, and GPSO algorithm provided more solutions, allowing to designer to choose the most appropriate one for given construction parameters. We must emphasize that in this example only the value of the contact ratio was optimized.

Further research could be conducted in optimizing not only geometry of the gear pair, but also parameters related to potential damage occurrence, such as pitting or scuffing.

This would further improve the efficiency and quality of optimized gear pairs.

4. GPSO in Architecture and Urban Design: Optimal Buildings Layout in Terms of Energy Efficiency

This section presents a new approach to architecture and urban design that allows increasing the energy efficiency of buildings or groups of buildings that are built closely together. The goal is to maximize the effect of sunlight and create shadow minimization on each structure. The problem is solved using the GPSO algorithm with niching (Passaro, 2007).

In recent years, investors have put great emphasis on the development of eco-technologies across almost all sectors. *Eco-technology* tends to *improve energy efficiency and to reduce the amount of harmful waste.* Buildings are the single largest consumers of energy, and this consumption has a *tendency to increase over the years due to population growth.* For that reason, the greatest potential for reducing energy consumption is in the area *of energy efficient buildings.* The main goal of energy efficiency in buildings *is to develop and implement mechanisms to permanently reduce energy consumption in new buildings and to properly reconstruct existing buildings.*

There are many ways to harness solar energy. One possibility is the use of *photovoltaic panels.* Maximum use of natural light and thermal solar energy is promoted as a way to use sunlight to save energy. So in construction, insolation is an important consideration and an essential climate variable for human comfort and energy efficiency. Our main concern is either maximising solar gain in northern latitudes or minimising it in warmer climates while keeping the maximum area. In a full application, the constraints will be much more multidimensional and complex in nature, but to simplify the model, only solar and functional (area) constraints are considered.

In this research, the idea was to use optimization algorithms to organize buildings within a defined area. Likewise, the number of structures and their dimensions were predetermined and the shapes considered simplified rectangular prisms. With these restrictions in mind, there are several aspects to consider.

The search space in GPSO is composed of particles. In this structure, one particle consists of three parts: the number that describes how many buildings are in the area, building information, and scheme which describes how the buildings are distributed. Each building unit has its position length, width and height. For the sake of simplicity, we assume that dimensions are determined by width *R*, length *r*, height *h*, and positions *x*, *y* and *z*. The dimension of search space is determined by dimension of the architecture area - L_1 (width) and L_2 (length). Hence for an individual unit, the phenotype is presented as a sequence in the following format:

$$\xi = (n, \mathrm{L}_1, \mathrm{L}_2, B_1, \ldots, B_n) \tag{42}$$

In Eq.(42), *n* stands for the number of buildings, and B_1;...; *Bn* represent building parameters, each of which has the form:

$$B_k = (x_k, y_k, z_k, h_k, R_k, r_k) \tag{43}$$

Our task is to determine the *x* and *y* coordinates, while the other parameters are known. The value of the *x* and *y* parameters is limited by the dimensions of the architectural area, so parameter *x* will be in the interval [-L_1/2 L_1/2] and parameter *y* will be in the interval [-L_2/2 L_2/2].

The *next step is to define* the value of the optimality criterion of the problem. An energy function is defined to describe our goal and that is the total shadow an individual has.

The length of the shadow cast on the ground depends on largest sunlight angle in middle of the day (solar elevation angle), represented with $v_0 \in (0,1)$. Also, let $vsun(t)$ represent the direction of the sunlight described by the following equation (Liu et al., 2007) (Mingxi et al., 2005):

$$vsun(t) = (v(t)\cos(\pi - t), v(t)\sin(\pi - t), -v_0\sin(\pi - t)) \tag{44}$$

$$v(t) = \sqrt{1 - v_0^2 sin^2(\pi - t)} \tag{45}$$

The sun is continually moving throughout the *day, so the strength of the sunlight changes at different times of the day. This can be described by equation:*

$$\rho(t) = \rho_0 \sin(\pi - t), \rho_0 - \textit{coefficient of insolation} \quad (46)$$

If we have a stick with height h_0, whose coordinate position is $(x_0, y_0, 0)$, then the coordinates of its shadow in the x – y plane at time *t*, which varies from 0 to π, will be measured by the following equation:

$$x = x_0 + \frac{h_0}{v_0} v(t) ctg(\pi - t) \quad (47)$$

$$y = y_0 + \frac{h_0}{v_0} v(t) \quad (48)$$

The Figure 99 below shows how shadows change during the day.

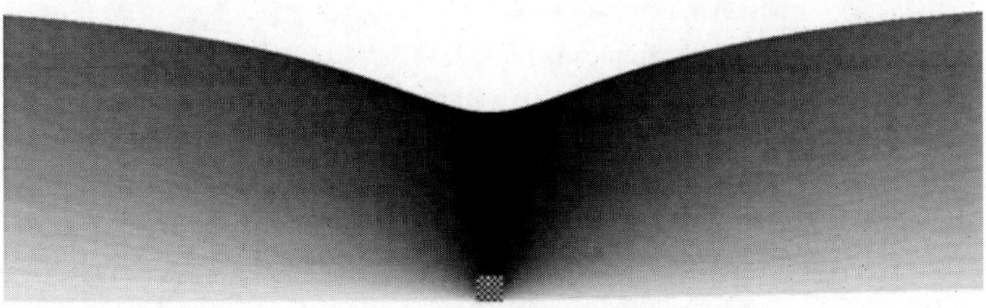

•

Figure 9. Shadow figure.

Now, we have building with central position(x_0, y_0, z_0), height h_0, width R_0, length r_0 and with *fixed* points in *time* t_0. Thus we can define the edge of the building with

$$\Theta_0 = (x_0 - 0.5R_0, x_0 + 0.5R_0) \times (y_0, y_0 + r_0) \quad (49)$$

Then, equations (47) and (48) are implemented on all points along building edge, and in that manner the shadow region χ_{Θ_0} is denoted as a function of Θ_0. If there is a second building ξ with central position $(x_1, y_1, 0)$, height h_1, width R_1, length r_1 we now compute the total sunlight building ξ receives.

Area of the second building which could be under shadow of the first building is defined as:

$$D_1 = (x_1 - 0.5R_1, x_1 + 0.5R_1) \times (y_1, y_1 + h_1) \quad (50)$$

So, the shadow the second building receives is:

$$R_s(1,0) = \iint_{D_1} \chi_{\Theta_0} \rho(t)\, dx\, dy \quad (51)$$

Therefore, the shadows that all buildings make for each other in *fixed time* t_0 is

$$R_s(\xi) = \sum_{\substack{i,j=1 \\ i \neq j}}^{n} \iint_{D_{(i)}} \chi_{\Theta_{(j)}} \rho(t)\, dx\, dy \tag{52}$$

Then the total shadow an individual has during the day is

$$R_u(\xi) = \int_t R_s(\xi) dt \tag{53}$$

```
begin
   initialize population S
      perform GPSO algorithm on S
      niche core number = 0
      put the best solution into niche core set
      for (j=1 to number of niches N)
         niche core number = j
         initialize new population S
         initialize parameters for GPSO
         Perform the GPSO algorithm on S and
            when calculate fitness function:
         for (each particle p in S)
               for (k=0 to niche core number)
               distance=calculate distance between p and solution in
                  k- niche core
                  if (distance < niche radius)
                        fitness value is increased
                  else
                        fitness value is calculated regularly
                  end
               end
         end
         put the best solution into j-niche core
      end
end
```

Figure 10. Niche PSO pseudo-code.

Now we can formulate the optimization problem as calculation of individual ξ in such way that $R_u(\xi)$ is minimized. This problem is solved by the niche GPSO algorithm. The niching particle swarm optimization (Passaro, 2007) algorithm is an improved version of

PSO, which enables more efficient search and decreases the probability of ending in local optima. It has been successfully applied to find multiple solutions to function optimization problems. Niches in the fitness landscape are identified by monitoring changes in the fitness of particles. We define the distance between two individuals and if one individual is closer to the existing optima, then its fitness value is "worsened", i.e. increased in case of minimization. The pseudo-code of the Niche GPSO algorithm applied in this research is given in Figure 10. We define distance between two individuals as Euclidean distance $d(\xi_i, \xi_j)$. So if ξ_j core of niche j and $d(\xi_i, \xi_j) < niche\ radius$, then define fitness value of i to be as follows

$$df(\xi_i) = f(\xi_i)\ \frac{niche\ radius}{d(\xi_i, \xi_j)} \tag{54}$$

In the following example the number of buildings is 21 and architectural area has dimensions of 1000m x 1000m. Algorithm with five niches core demonstrated best results. Niche radius between two different individuals is 250 m. In Table 1 we see some experimental results and Figure 11 shows the value of optimization function during the time in different niches.

Table 1.

Niche No	Best results	Simulation time
0	220	16h 29min 22sec
1	0	
2	112.52	
3	0	
4	43.24	
5	65.15	

In this experiment, two global optima are found. This means that fitness functions in both cases are equal to zero. One of the schemes of an arrangement of buildings is shown in Figure 12.

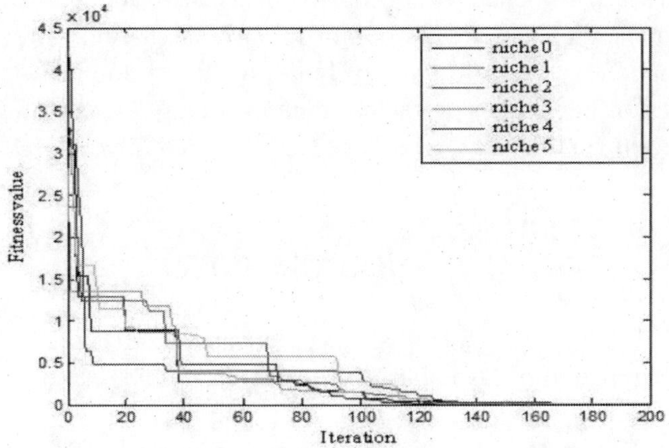

•

Figure 11. Fitness curve.

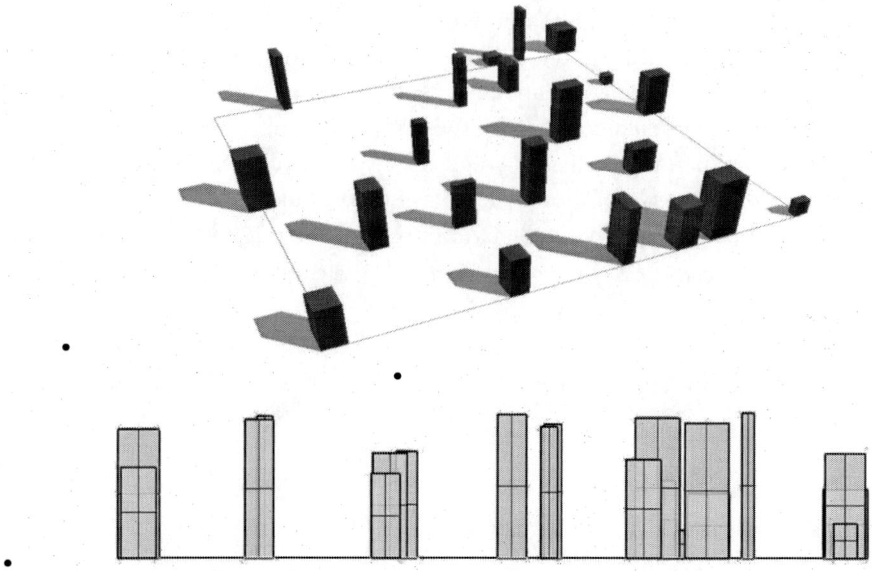

Figure 12. Organization of buildings in the area.

Experimental results presented in this section showed that GPSO algorithm with niching can be successfully applied in solving problems of optimal architectural and urban design and that it can be used as efficient tool to increase energy efficiency of urban environment.

Conclusion

Swarm-based methods proved to be a powerful and robust tool for tackling the global optimization problems in science and engineering. In this chapter, the idea of PSO algorithm modification, named GPSO, is presented. This modification incorporates requirements concerning exploration and exploitation properties in sense of the linear control theory. Convergence analysis of this algorithm has been conducted, and the influence of parameters on particle dynamics and optimizer's performance has been investigated. A broad empirical study based on various benchmark problems was also conducted. Finally, two practical engineering application examples, in optimization of high contact ratio gear pairs geometry and in determination of optimal layout of buildings in a settlement in terms of solar energy effect maximization have been presented, demonstrating the potential of this algorithm for practical engineering problems.

References

Alireza, A. (2011) PSO with Adaptive Mutation and InertiaWeight and Its Application in Parameter Estimation of Dynamic Systems. *Acta Automatica Sinica* Vol. 37, No. 5, 541-548.

Åström, K.J. and Wittenmark, B. (1997), *Computer-Controlled Systems - Theory and Design, 3rd Ed.* New Jersey: Prentice Hall,.

Bouallegue, S., Haggege, J., Ayadi, M., Benrejeb, M. (2012) PID-type fuzzy logic controller tuning based on particle swarm optimization. *Engineering Applications of Artificial Intelligence* 25, 484–493.

Chatterjee, A., Ghoshal, S.P., Mukherjee V. (2011) Craziness-based PSO with wavelet mutation for transient performance augmentation of thermal system connected to grid. *Expert Systems with Applications* 38, 7784–7794.

Chen, M. R., Li, X., Zhang, X. and Lu, Y. Z. (2010). A novel particle swarm optimizer hybridized with extremal optimization. *Applied Soft Computing 10*, 367-373.

Clerc, M. and Kennedy, J. (2002). The Particle Swarm - Explosion, Stability and Convergence in a Multidimensional Complex Space. *IEEE Transactions on Evolutionary Computation 6* (1), 58-73.

Cui, Z., Cai, X. and Zeng, J., Sun, G. (2008). Particle swarm optimization with FUSS and RWS for high dimensional functions. *Applied Mathematics and Computation 205*, 98-108.

DeBao, C. and ChunXia, Z. (2009). Particle swarm optimization with adaptive population size and its application. *Applied Soft Computing 9*, 39–48.

Dimopoulos, G.G. (2007). Mixed-variable engineering optimization based on evolutionary and social metaphors. *Computer methods in applied mathematics and engineering 196*, 803-817.

Dong, et al., (2005). An Application of Swarm Optimization to Nonlinear Programming. *Computers and Mathematics with Applications 49*, 1655-1668.

Dorigo, M. and Blum, C. (2005). Ant colony optimization theory: A survey. *Theoretical Computer Science 344*, 243-278.

Dudley, D.W., (Ed.) (1995) *Gear motor Handbook*, Berlin: Springer.

Eberhart, R. C., and Shi, Y. (2000). Comparing inertia weights and constriction factors in particle swarm optimization. *Proceedings of the Congress on Evolutionary Computation* (84-88), La Jolla, California.

Fang, H., Chen, L., Shen, Z. (2011) Application of an improved PSO algorithm to optimal tuning of PID gains for water turbine governor. *Energy Conversion and Management* 52, 1763–1770.

He, Q. and Wang, L. (2007). An effective co-evolutionary particle swarm optimization for constrained engineering design problems. *Engineering Applications of Artificial Intelligence 20*, 89-99.

Jiang, M., Luo, Y.P. and Yang, S.Y. (2007). Stochastic convergence analysis and parameter selection of standard particle swarm optimization algorithm. *Information Processing Letters 102*, 8-16.

Jiang, Y., Hu, T., Huang, C. C., Wu, X. (2007). An improved particle swarm optimization algorithm. *Applied Mathematics and Computation 193*, 231-239.

Jiang, Y., Liu, C., Huang, C. and Wu, X. (2010). Improved particle swarm algorithm for hydrological parameter optimization. *Applied Mathematics and Computation 217*, 3207-3215.

Jie, J., Zeng, J., Han, C., Wang, Q. (2008) Knowledge-based cooperative particle swarm optimization. *Applied Mathematics and Computation* 205, 861-873.

Kahraman, A., Blankenship, G.W. (1999) Effect of Involute Contact Ratio on Spur Gear Dynamics. *ASME Journal of Mechanical Design 121*, 112-118.

Kanović, Ž., Rapaić, M., Erdeljan, A. (2008). Generalized PSO Algorithm In Optimization Of Water Distribution, *Proceedings of Planning and Management of Water Resources Systems*, (203-210), Novi Sad.

Kanović, Ž., Rapaić, M. and Jeličić, Z. (2011). Generalized Particle Swarm Optimization Algorithm - Theoretical and Empirical Analysis with Application in Fault Detection. *Applied Mathematics and Computation 217*, 10175-10186.

Kasuba,R. (1981) Dynamic Loads in Normal and High Contact Ratio Spur Gearing. *Proc. of International Symposium on Gearing and Power Transmissions*, (49-55) Tokyo.

Kennedy, J. and Eberhart, R.C. (1995). Particle Swarm Optimization. *Proceedings of IEEE International Conference on Neural Networks* (1942-1948), Perth, Australia.

Kirkpatrick, S., Gellat, C.D. and Vecchi, M.P. (1983). Optimization by Simulated Annealing. *Science, 220* (4598), 671-680.

Liang, J. J., Suganthan, P.N. (2005). Dynamic multi-swarm particle swarm optimizer. *Proceedings of IEEE Swarm Intelligence Symposium* (124-129), Pasadena, California.

Liu, H., Duan, H., Liu, X. (2007) Particle swarm optimization based on dynamic niche technologz with applications to conceptual design. *Shandong: Advance in Engineering Software 38*, 668-676.

Majhi, B., Panda, G. (2011) Robust identification of nonlinear complex systems using low complexity ANN and particle swarm optimization technique. *Expert Systems with Applications* 38, 321–333.

Mallipeddi, R., Jeyadevi, S., Suganthan , P.N., Baskar, S. (In press) Efficient constraint handling for optimal reactive power dispatch problems. *Swarm and Evolutionary Computation.*

Menhas, M.I., Wang, L., Fei, M., Pan, H. (2012) Comparative performance analysis of various binary coded PSO algorithms in multivariable PID controller design. *Expert Systems with Applications* 39, 4390–4401.

Michalewicz, Z. (1999). *Genetic Algorithms + Data Structures = Evolution Programming, 3rd Ed.,* Berlin: Springer.

Mingxi T., Hamilton J., Liu, F. X. (2005) An eco-conscious housing design model based on co-evolution. *Shandong: Advance in Engineering Software 36,* 115-125

Niu, B., Zhu, Y., He, X. and Wu, H. (2006). MCPSO: A multi-swarm cooperative particle swarm optimizer. *Applied Mathematics and Computation 185*, 1050-1062.

Olamaei, Y., Niknam, T. and Gharehpetian, G. (2008). Application of particle swarm optimization for distribution feeder reconfiguration considering distributed generators. *Applied Mathematics and Computation 201*, 575-586.

Ozcan, E. and Mohan, C.K. (1998). Analysis of Simple Particle Swarm Optimization System. *Intelligent Engineering Systems Through Artificial Neural Networks* ***8,*** 253-258.

Papoulis, A. (1965). *Probability, Random Variables and Stochastic Processes.* New York: McGraw-Hill.

Passaro, A. (2007) *Niching in Particle Swarm Optimization*. Universit`a di Pisa.

Podzharov, E., Mozuras, A., Sanchez, J. A. A. (2003) Design of High Contact Ratio Spur Gears to Reduce Static and Dynamic Transmission Error. *Ingeneria Mecanica, Septiembre 2003, Vol.1*, 85-90.

Ramesh, L., Chakraborthy, N., Chowdhury, S.P., Chowdhury, S. (2012) Intelligent DE algorithm for measurement location and PSO for bus voltage estimation in power distribution system. *Electrical Power and Energy Systems* 39, 1–8.

Rapaić, M.R. and Kanović, Ž. (2009). Time-varying PSO - convergence analysis, convergence related parameterization and new parameter adjustment schemes. *Information Processing Letters 109*, 548-552.

Ratnaweera, A., Saman, K.H. and Watson, H.C. (2004). Self-Organizing Hierarchical Particle Swarm Optimizer With Time-Varying Acceleration Coefficients. *IEEE Transactions on Evolutionary Computation 8* (3), 240-255.

Shahzad, F., Rauf Baig, A., Masood, S., Kamran, M., Naveed, N. (2009) Opposition-Based Particle Swarm Optimization wit Velocity Clamping (OVCPSO). In: Yu, W., Sanchez, E. N., (Ed.). *Advances in Computational Intelligence*. (pp. 339-348)Berlin Heidelberg: Springer.

Samanta, B. and Nataraj, C. (2009). Use of particle swarm optimization for machinery fault detection. *Engineering Applications of Artifficial Intelligence 22*, 308-316.

Sato, T., Umezawa, K., Ishikawa, J. (1983) Effect of Contact Ratio and Profile Correction of Spur Gears on the Rotational Vibrations. *Bulletin of the JSME 26* (221) 2010-2016.

Schutte, J.C. and Groenwold, A.A. (2005). A Study of Global Optimization Using Particle Swarms. *Journal of Global Optimization 31*, 93-108.

Senthil Arumugam, M. and Rao, M.V.C. (2007). On the improved performances of the particle swarm optimization algorithms with adaptive parameters, cross-over operators and root mean square (RMS) variants for computing optimal control of a class of hybrid systems. *Applied Soft Computing 8*, 324-336.

Senthil Arumugam, M., Rao, M.V.C., Tan, A.W.C. (2009). A novel and effective particle swarm optimization like algorithm with extrapolation technique. *Applied Soft Computing*, Volume 9, Issue 1, 308–320.

Stepanić, P., Latinović, I. and Đurović, Ž. (2009). A new approach to detection of defects in rolling element bearings based on statistical pattern recognition. *The International Journal of Advanced Manufacturing Technology 45*, 91-100.

Shi, Y. and Eberhart, R.C. (1999). Empirical study of particle swarm optimization. *Proceedings of IEEE International Congress on Evolutionary Computation 3*, (101-106) Washington DC, USA.

Thangaraj, R., Pant, M., Abraham, A., Bouvry, P. (2011) Particle swarm optimization: Hybridization perspectives and experimental illustrations. *Applied Mathematics and Computation 217*, 5208-5226.

Van der Bergh, F. (2001), *An Analysis of Particle Swarm Optimizers.* PhD Thesis, University of Pretoria, Pretoria.

In: Self-organization: Theories and Methods
Editor: WenJun Zhang
ISBN: 978-1-62618-865-5

Chapter 5

WEIGHTS AND STRUCTURE DETERMINATION OF ARTIFICIAL NEURONETS

Yunong Zhang*,* Xiaotian Yu, Lin Xiao, Weibing Li and Zhengping Fan
School of Information Science and Technology, Sun Yat-sen University,
Guangzhou, Guangdong, China

Abstract

Artificial neuronets (AN), especially with error back-propagation (BP) training algorithms, have been widely investigated and applied in various science and engineering fields. However, BP-type neuronets, which are self-adaptive systems, have shown some inherent weaknesses, such as, the possibility of being trapped into local minima, the difficulty in choosing appropriate learning rate, and most importantly, the inability to determine the optimal neuronet structure. To solve the inherent weaknesses of AN, lots of improvements for BP-type algorithms have been investigated. However, as researchers (including the authors) realize and experience quite frequently, the inherent weaknesses of BP-type neuronets still exist. In this chapter, differing from others' algorithmic improvements on the training procedure, our way about the problem solving exploits some elegant structure-design, parameter-setting, pseudoinverse and numerical optimization techniques. In other words, a new type of AN using linearly-independent or orthogonal polynomials as activation functions, is presented and analyzed by us (the authors). These finally lead us to propose a weights and structure determination (WASD) method, which is based on a weights-direct-determination (WDD) method, for our presented feedforward AN. Based on the authors' previous work, single-input neuronets equipped with the WASD method have successfully overcome the above weaknesses. To investigate and verify two- or multiple-input neuronets equipped with this method, the authors firstly put forward various novel neuronets based on different activation functions. Then, corresponding WASD algorithms are proposed for the presented neuronets. For better performance (e.g., more efficiency and conciseness in self-organizing systems), the authors further propose pruning techniques in the neuronet structure determination. Finally, based on various target functions, numerical results further substantiate the efficacy of the proposed neuronets

*E-mail addresses: zhynong@mail.sysu.edu.cn; jallonzyn@sina.com

equipped with the corresponding WASD algorithms, which shows the better performance in terms of training (or say, approximation or learning), generalization (or say, testing or validation) and prediction.

Keywords: Artificial neuronets (AN); Back-propagation (BP); Weights and structure determination (WASD) method; Weights-direct-determination (WDD) method.

1. Introduction

With parallel-processing nature, distributed storage, self-adaptive and self-learning abilities, artificial neuronets (AN) (i.e., artificial neural networks), especially with error back-propagation (BP) training algorithms, have been investigated and applied widely in many scientific, engineering and practical fields (Ding et al., 2011; Sadeghi, 2000; Narendra and Mukhopadhyay, 1993; Zhang et al., 2008a; Wang and Sui, 2007). However, BP algorithms are essentially gradient-based iterative methods, which adjust the AN weights to bring the network input/output behavior into a desired mapping by taking a gradient-based descent direction (Yu and Chen, 1995; Zhang et al., 2008b; Wilson and Martinez, 2001). As self-organizing neural networks, BP-type neuronets have some inherent weaknesses (Yu and Chen, 1995; Wilson and Martinez, 2001; Zhang et al., 2008a; Zhang and Ruan, 2009; Zhang et al., 2009a), e.g., local minima, slow convergence and uncertainties about the optimal number of hidden-layer neurons for three-layer neuronets (which include the input layer, the hidden layer and the output layer). Because of these inherent weaknesses, many improved BP-type algorithms have been proposed and investigated (Wilson and Martinez, 2001; Jenkins, 2006). Generally speaking, there are two general types of improvements. On one hand, BP-type algorithms could be improved based on the standard gradient-descent method. On the other hand, numerical optimization techniques could be employed for the network training. It is worth pointing out that people usually pay more attention to the learning-algorithm itself to ameliorate the performance of BP-type neuronets (Jenkins, 2006; Zhang et al., 2009a). However, almost all of improved BP-type algorithms have not overcome the above inherent weaknesses.

In order to thoroughly overcome the above weaknesses of BP-type neuronets and improve their performance (e.g., in terms of the training efficiency, the generalization accuracy and the conciseness of neuronets), the authors have focused on the usage of different activation functions and the determination of the optimal neuronet structure in modeling the self-organizing neuronets. Differing from algorithmic improvements about the BP iterative-training procedure, the method of using different activation functions and improving the conciseness of the neuronet structure to achieve better performance has been substantiated effectively in the authors' previous work (Zhang et al., 2008a; Zhang and Ruan, 2009; Zhang et al., 2009a; Zhang et al., 2010a; Zhang et al., 2012a). Note that feedforward neuronets investigated in the previous work are mostly the single-input or two-input neuronets. In view of the fact that multiple-input systems are the most frequently encountered systems (Fu and Chai, 2007), it is worth further investigating modeling multiple-input neuronets in this way. In this chapter, artificial neuronets with two, three or more inputs and using linearly-independent or orthogonal polynomials as activation functions are proposed and analyzed by the authors.

From the previous work, the authors have proposed a weights-direct-determination (WDD) method to determine the weights between the hidden layer and the output layer for three-layer neuronets (Zhang et al., 2009a; Zhang et al., 2010a; Zhang et al., 2012a; Zhang et al., 2012b). For three-layer neuronets, the number of hidden-layer neurons can influence greatly the overall performance of neuronets. Specifically speaking, with too few hidden-layer neurons, the neuronet may fail to achieve the expected learning accuracy, while excess hidden-layer neurons may result in over-fitting phenomenon and higher computational complexity (Zhang et al., 2010a; Zhang et al., 2012a; Zhang et al., 2012b). In light of the above analysis, it is meaningful and important to obtain the optimal number of hidden-layer neurons (or say, the optimal structure of neuronets). Thus, based on the WDD method, the authors propose a corresponding weights and structure determination (WASD) algorithm for each presented neuronet. In addition, for better performance (e.g., more efficiency and conciseness in self-organizing systems), the authors further propose pruning techniques in the neuronet structure determination. Finally, based on various target functions, numerical results further substantiate the efficacy of the proposed neuronets equipped with the corresponding WASD algorithms, which shows the better performance in terms of training (or say, approximation or learning), generalization (or say, testing or validation) and prediction.

The remainder of this chapter is thus organized as follows. Section 2 proposes three neuronets with two inputs (i.e., the 2-input neuronet activated by Legendre polynomials, the 2-input neuronet activated by Chebyshev polynomials of Class 1 and the 2-input neuronet activated by Chebyshev polynomials of Class 2). For each 2-input neuronet, a corresponding WASD algorithm is proposed. Then, numerical results are shown to verify the efficacy of each neuronet equipped with the corresponding WASD algorithm. In Section 3, the authors propose and investigate two neuronets with three inputs (i.e., neuronets with Euler-polynomial and power activation functions). The corresponding WASD algorithms without/with pruning techniques are presented and verified. Besides, the authors discuss and propose multiple-input neuronets with different activation functions (i.e., Euler, Bernoulli, Chebyshev and Hermit polynomials) and the corresponding WASD algorithms in Section 4. Numerical results further verify the efficacy of the proposed multiple-input neuronets equipped with the WASD algorithms. Final remarks are given in Section 5.

2. WASD of AN with Two Inputs

To lay a solid basis for future solution of multiple-input neuronets, 2-input feedforward neuronets activated by different polynomials (i.e., Legendre polynomials, Chebyshev polynomials of Class 1 and Chebyshev polynomials of Class 2) are investigated in this section. Then the corresponding WASD algorithms are investigated in detail. Numerical results show the efficacy of the proposed neuronets equipped with the WASD algorithms.

2.1. The First Theoretical Basis and Analysis

Before constructing 2-input neuronets, the related theoretical basis and analysis are presented as below. Based on the theory of the multivariate function approximation (Liang and Li, 2005; Xu and Yang, 1987), we have the following proposition.

Proposition 1. For two continuous independent variables x_1 and x_2, let $f(x_1,x_2)$ denote a given continuous function. Then, there exist polynomials $g_k(x_1)$ and $h_k(x_2)$ (with $k = 1,2,3,\cdots$) to formulate $f(x_1,x_2)$; i.e.,

$$f(x_1,x_2) = \sum_{k=1}^{\infty} g_k(x_1)h_k(x_2). \tag{1}$$

With out loss of generality, let $\{\varphi_i(x)$ (with $i = 1,2,\cdots)\}$ denote a group of linearly-independent orthogonal polynomials. According to the above proposition, the weighted orthogonal polynomials can be used to formulate $f(x_1,x_2)$. That is, let $g_k(x_1) = a_m\varphi_m(x_1)$ and $h_k(x_2) = b_n\varphi_n(x_2)$, where a_m and b_n denote the weights for φ_m and φ_n, respectively (with $m,n = 1,2,3,\cdots$). Then (1) can be reformulated as

$$f(x_1,x_2) = \sum_{m=1}^{\infty}\sum_{n=1}^{\infty} a_m\varphi_m(x_1)b_n\varphi_n(x_2) = \sum_{m=1}^{\infty}\sum_{n=1}^{\infty} \omega_{m,n}\varphi_m(x_1)\varphi_n(x_2), \tag{2}$$

where $\omega_{m,n} = a_m b_n$ denotes the weight for $\varphi_m(x_1)\varphi_n(x_2)$.

In order to further discuss about the approximation of target function $f(x_1,x_2)$ and construct neuronets, several kinds of orthogonal polynomials are defined as blow.

Definition 1. For the variable $x \in [-1,1]$, Legendre orthogonal polynomials can be defined as below (Zhang et al., 2009b):

$$\varphi_{i+2}^L(x) = \frac{2i+1}{i+1}x\varphi_{i+1}^L(x) - \frac{i}{i+1}\varphi_i^L(x), \text{with } \varphi_1^L(x) = 1 \text{ and } \varphi_2^L(x) = x,$$

where $\varphi_i^L(x)$ denotes the Legendre orthogonal polynomial of degree i (with $i = 1,2,\cdots$).

Definition 2. For the variable $x \in [-1,1]$, Chebyshev polynomials of Class 1 can be defined as below (Zhang et al., 2012c):

$$\varphi_{i+2}^{C_1}(x) = 2x\varphi_{i+1}^{C_1}(x) - \varphi_i^{C_1}(x), \text{with } \varphi_1^{C_1}(x) = 1 \text{ and } \varphi_2^{C_1}(x) = x,$$

where $\varphi_i^{C_1}(x)$ denotes the Chebyshev polynomial of Class 1 of degree i (with $i = 1,2,\cdots$).

Definition 3. For the variable $x \in [-1,1]$, Chebyshev polynomials of Class 2 can be defined as below (Zhang et al., 2012a):

$$\varphi_{i+2}^{C_2}(x) = 2x\varphi_{i+1}^{C_2}(x) - \varphi_i^{C_2}(x), \text{with } \varphi_1^{C_2}(x) = 1 \text{ and } \varphi_2^{C_2}(x) = 2x,$$

where $\varphi_i^{C_2}(x)$ denotes the Chebyshev polynomial of Class 2 of degree i (with $i = 1,2,\cdots$).

2.2. Neuronet of Legendre Polynomials

In this subsection, Legendre orthogonal polynomials are used to construct a neuronet, i.e., $\varphi_i(x)$ is replaced with $\varphi_i^L(x)$ in (2). Then two weights and structure determination (WASD) algorithms with different growing speeds are built up to determine the optimal weights and structure of the proposed 2-input Legendre orthogonal polynomial neuronet (2ILOPN). Numerical studies further verify the efficacy of the proposed 2ILOPN equipped with the two aforementioned WASD algorithms.

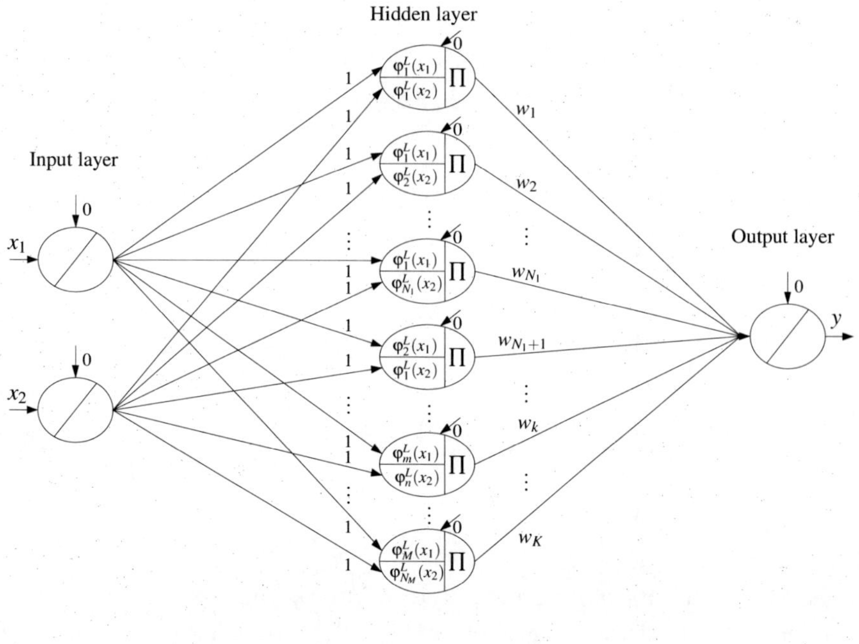

Figure 1. Model structure of the 2ILOPN.

2.2.1. Model Structure of 2ILOPN

The target function $f(x_1,x_2)$ can be approximated via a finite number of basis functions $\{\varphi_m^L(x_1)\varphi_n^L(x_2)\}$ with the corresponding optimal weights $\{\omega_{m,n}\}$ based on formula (2). Thus we have the following formula:

$$f(x_1,x_2) \approx \sum_{m=1}^{M}\sum_{n=1}^{N_m} \omega_{m,n}\varphi_m^L(x_1)\varphi_n^L(x_2).$$

Note that the basis functions $\{\varphi_m^L(x_1)\varphi_n^L(x_2)\}$ can be sequenced in different orders (Dunkl and Xu, 2001). This implies that N_m (i.e., the maximum of n corresponding to a specific value of m) can be different for different m, with $m = 1,2,3,\cdots,M$. Therefore, the target function $f(x_1,x_2)$ can be approximated via the following form:

$$f(x_1,x_2) \approx \sum_{m=1}^{M}\varphi_m^L(x_1)\left(\sum_{n=1}^{N_m}\omega_{m,n}\varphi_n^L(x_2)\right) = \sum_{k=1}^{K} w_k\phi_k^L(x_1,x_2),$$

where $K = \sum_{m=1}^{M} N_m$ denotes the total number of basis functions used to approximate $f(x_1,x_2)$. In addition, $\{w_k = \omega_{m,n},$ with $k = 1,2,3,\cdots,K\}$ denotes the set of weights for $\{\phi_k^L(x_1,x_2)\}$, and $\phi_k^L(x_1,x_2) = \phi_k^L([x_1,x_2]^{\mathrm{T}}) = \varphi_m^L(x_1)\varphi_n^L(x_2)$ with superscript ^{T} denoting transpose operator. Besides, with $k = \sum_{d=1}^{m} N_d - N_m + n$, we detail $\{\phi_k^L(x_1,x_2)\}$ shown as below:

$$\begin{cases} \phi_1^L(x_1,x_2) = \varphi_1^L(x_1)\varphi_1^L(x_2), \\ \phi_2^L(x_1,x_2) = \varphi_1^L(x_1)\varphi_2^L(x_2), \\ \vdots \\ \phi_{N_1}^L(x_1,x_2) = \varphi_1^L(x_1)\varphi_{N_1}^L(x_2), \\ \phi_{N_1+1}^L(x_1,x_2) = \varphi_2^L(x_1)\varphi_1^L(x_2), \\ \vdots \\ \phi_{N_1+N_2}^L(x_1,x_2) = \varphi_2^L(x_1)\varphi_{N_2}^L(x_2), \\ \vdots \\ \phi_k^L(x_1,x_2) = \varphi_m^L(x_1)\varphi_n^L(x_2), \\ \vdots \\ \phi_K^L(x_1,x_2) = \varphi_M^L(x_1)\varphi_{N_M}^L(x_2). \end{cases}$$

As mentioned above, the basis functions used to approximate $f(x_1,x_2)$ can be generated in a given order. For better understanding, two typical orders are presented, which can be expressed as the following limitations via a given positive integer i larger than 1 (i.e., $i > 1$):
Limitation I. $M = i$, and $N_m = i$ with $m = 1,2,\cdots,M$;
Limitation II. $M = i-1$, and $N_m = i-m$ with $m = 1,2,3,\cdots,M$.
So, the total number of basis functions K can be determined by $K = i^2$ corresponding to Limitation I or $K = i(i-1)/2$ corresponding to Limitation II.

2.2.2. WASD Algorithm for 2ILOPN

In light of the previous work (Zhang and Ruan, 2009; Zhang et al., 2009a; Zhang et al., 2010a), the number of hidden-layer neurons can affect greatly the performance of neuronets. Thus, to obtain the optimal number of hidden-layer neurons (as well as the final optimal weights) of the proposed 2ILOPN, two weights and structure determination (WASD) algorithms of growing type are then comparatively developed and investigated.

For the neuronet shown in Figure 1, the weights can be determined directly by using a pseudoinverse based method, i.e., the so-called weights-direct-determination (WDD) method. Specifically, to present the WASD algorithm (including the WDD method) more clearly, we give $(X_p,\gamma_p)|_{p=1}^{Q}$ as the sample pairs for training, where $X_p \in R^2$ denotes the pth input vector $[x_{p1},x_{p2}]^{\mathrm{T}}$, and γ_p denotes the pth desired output (or say, target output). Then, the mean square error (MSE) function is defined as

$$E = \frac{1}{2}\sum_{p=1}^{Q}\left(\gamma_p - \sum_{k=1}^{K} w_k\phi_k(X_p)\right)^2 /Q, \tag{3}$$

where K denotes the total number of the hidden-layer neurons, and $\phi_k(\cdot)$ denotes the activation function of the kth hidden-layer neuron of the proposed neuronet [e.g., for the 2ILOPN, $\phi_k(\cdot) = \phi_k^L(\cdot)$]. It is worth mentioning that, if the optimal linking weights are determined, the MSE function (3) achieves its minimal value. Then, from our previous work (Zhang

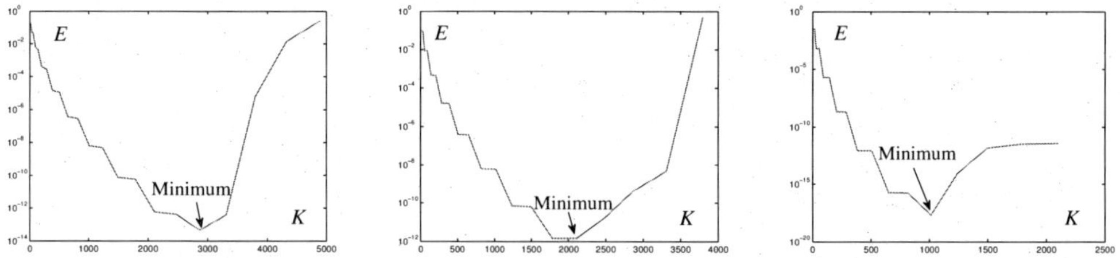

(a) For target function (5) (b) For target function (6) (c) For target function (7)

Figure 2. Relationship between the MSE (i.e., E with $Q = 1156$) and the number of hidden-layer neurons (i.e., K) with Limitation I.

and Ruan, 2009; Zhang et al., 2009a; Zhang et al., 2010a), the optimal linking weights can be determined directly by using the following WDD method:

$$w = (\Phi^{\mathrm{T}}\Phi)^{-1}\Phi^{\mathrm{T}}\gamma, \tag{4}$$

where w is the vector of the linking weights, γ is the vector of the desired outputs, and Φ is the input-activation matrix, which are defined respectively as

$$w = \begin{bmatrix} w_1 \\ w_2 \\ \vdots \\ w_K \end{bmatrix} \in R^K, \ \gamma = \begin{bmatrix} \gamma_1 \\ \gamma_2 \\ \vdots \\ \gamma_Q \end{bmatrix} \in R^Q, \ \Phi = \begin{bmatrix} \phi_1(X_1) & \phi_2(X_1) & \dots & \phi_K(X_1) \\ \phi_1(X_2) & \phi_2(X_2) & \dots & \phi_K(X_2) \\ \vdots & \vdots & \ddots & \vdots \\ \phi_1(X_Q) & \phi_2(X_Q) & \dots & \phi_K(X_Q) \end{bmatrix} \in R^{Q\times K}.$$

It is worth pointing out that w can also be written as $w = \mathrm{pinv}(\Phi)\gamma$, where $\mathrm{pinv}(\Phi)$ denotes the pseudoinverse of input-activation matrix Φ and can be evaluated directly via MATLAB routine "pinv" (Su and Ruan, 2008; Mathews and Fink, 2007).

Before presenting such WASD algorithms, a variety of target functions have been tested to investigate the relationship between the MSE function (3) and the number of hidden-layer neurons, such as the following three target functions:

$$f(x_1,x_2) = 10\sin(x_1)e^{-(2x_1)^2-(2x_2)^2}, \tag{5}$$
$$f(x_1,x_2) = 4e^{-x_1^2-(2x_2)^2}+10, \tag{6}$$
$$f(x_1,x_2) = \sin(\pi x_1 x_2)+20. \tag{7}$$

Figures 2 and 3 show the so-called UV plots of the above three target functions (5)-(7) with Limitations I and II, respectively, which indicate the relationship between the MSE (i.e., E with the data number $Q = 1156$) and the number of hidden-layer neurons of the 2ILOPN (i.e., K). As seen from Figures 2 and 3, there is a minimum of MSE for each target function. Note that such a phenomenon (i.e., the existence of the minimum of E) appears for many other target functions (with related results omitted here due to similarity). Therefore, two WASD algorithms with different growing speeds are finally developed to determine the optimal number of hidden-layer neurons (or say, the optimal model structure) of the 2ILOPN by searching the minimal MSE. For better understanding, the flowchart

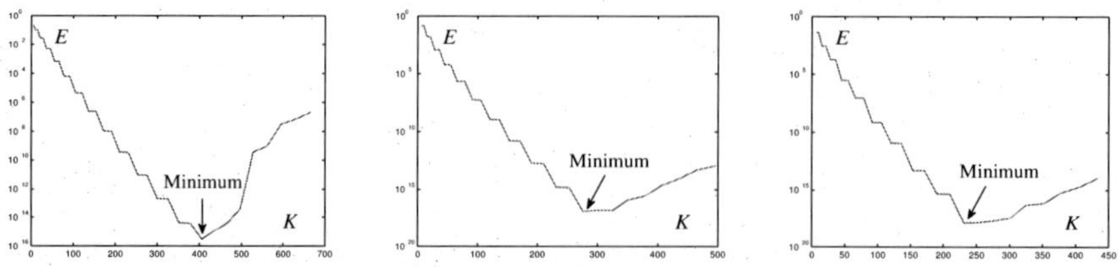

(a) For target function (5) (b) For target function (6) (c) For target function (7)

Figure 3. Relationship between the MSE (i.e., E with $Q = 1156$) and the number of hidden-layer neurons (i.e., K) with Limitation II.

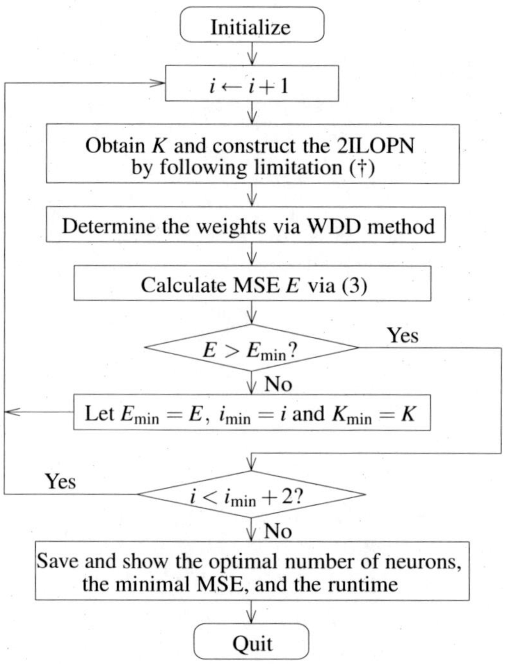

Figure 4. Flowchart of the WASD algorithm for the 2ILOPN. *Reproduced from Y. Zhang, J. Chen et al., Growing-type weights and structure determination of 2-input Legendre orthogonal polynomial neuronet, Figure 4, Proceedings of the 2012 IEEE International Symposium on Industrial Electronics, pp. 852-857. ©[2012] IEEE. Reprinted, with permission.*

of the WASD algorithms is shown in Figure 4. Note that the only difference between the two WASD algorithms developed in this subsection is the limitation exploited in the algorithms. That is, the limitation marked with (†) in Figure 4 can be chosen as Limitation I or Limitation II, which will lead to different growing speeds of the number of hidden-layer neurons. Besides, the procedure is detailed as below.

a) The procedure has the following variables.

- E denotes the MSE of the current 2ILOPN;
- E_{min} denotes the minimal MSE found;

Table 1. Approximation and testing results of the 2ILOPN on target functions (5)-(7)

		$K_{\min}$	E_{app}	E_{test}	runtime (s)
Function (5)	Limitation I	2869	4.620×10^{-14}	3.303×10^{-13}	410.941
	Limitation II	406	2.969×10^{-16}	4.964×10^{-16}	17.642
Function (6)	Limitation I	2108	1.467×10^{-12}	3.363×10^{-12}	251.953
	Limitation II	276	1.227×10^{-17}	4.357×10^{-16}	7.102
Function (7)	Limitation I	1014	2.097×10^{-18}	1.202×10^{-18}	89.555
	Limitation II	231	1.323×10^{-18}	1.463×10^{-18}	4.688

• i denotes the current positive integer used to construct the 2ILOPN with Limitation I or Limitation II;

• $i_{\min}$ denotes the positive integer corresponding to $E_{\min}$;

• K denotes the current number of hidden-layer neurons;

• $K_{\min}$ denotes the optimal number of hidden-layer neurons.

b) The procedure has the following steps.

• **Step 1.** Obtain the set of training data-pairs $\{X_p, \gamma_p\}_{p=1}^{Q}$. Initialize the structure of the 2ILOPN with $i = 2$, and set $i_{\min} = 2$. Besides, $E_{\min}$ is set initially large enough (e.g., 10).

• **Step 2.** Let $i \leftarrow i+1$. If the limitation marked with (†) is chosen as Limitation I, then $K = i^2$; otherwise [if the limitation marked with (†) is chosen as Limitation II], $K = i(i-1)/2$. Then construct the 2ILOPN.

• **Step 3.** Calculate the weights between hidden-layer and output-layer neurons via WDD (4) and the MSE E via (3).

• **Step 4.** If $E > E_{\min}$, proceed to Step 5; otherwise, let $E_{\min} = E$, $i_{\min} = i$ and $K_{\min} = K$, and proceed to Step 2.

• **Step 5.** If $i < i_{\min} + 2$, then proceed to Step 2; otherwise, proceed to Step 6.

• **Step 6.** Output and save the optimal number of hidden-layer neurons $K_{\min}$, the minimal MSE $E_{\min}$ and the runtime. Then terminate the procedure.

2.2.3. Numerical Studies of 2ILOPN

In this subsubsection, numerical studies are performed to verify the efficacy of the proposed 2ILOPN equipped with one of the aforementioned two WASD algorithms. Note that, due to results' similarity, the numerical results are shown only for the aforementioned target functions (5)-(7).

Table 1 shows the numerical results of the 2ILOPN equipped with the WASD algorithms. For each target function, we sample uniformly over region $[-1,1]^2$ with gap-size 0.06 to generate the training data set $\{X_p, \gamma_p\}_{p=1}^{1156}$ (i.e., $Q = 1156$), which is used for approximation (or say, learning). In addition, for testing, the data set $\{X_p, \gamma_p\}_{p=1}^{4761}$ is generated by sampling uniformly over region $[-1,1]^2$ with gap-size 0.029. As shown in Table 1, the optimal number of hidden-layer neurons (corresponding to the minimal MSE) for each target function (i.e., $K_{\min}$) is determined successfully via the WASD algorithms. Besides, E_{app} and E_{test} shown in Table 1 denote the MSEs of approximation and testing, respectively. From Table 1, we see that the 2ILOPN equipped with either of the two algorithms can approximate accurately the target functions with tiny E_{app} (i.e., of order $10^{-12} \sim 10^{-18}$).

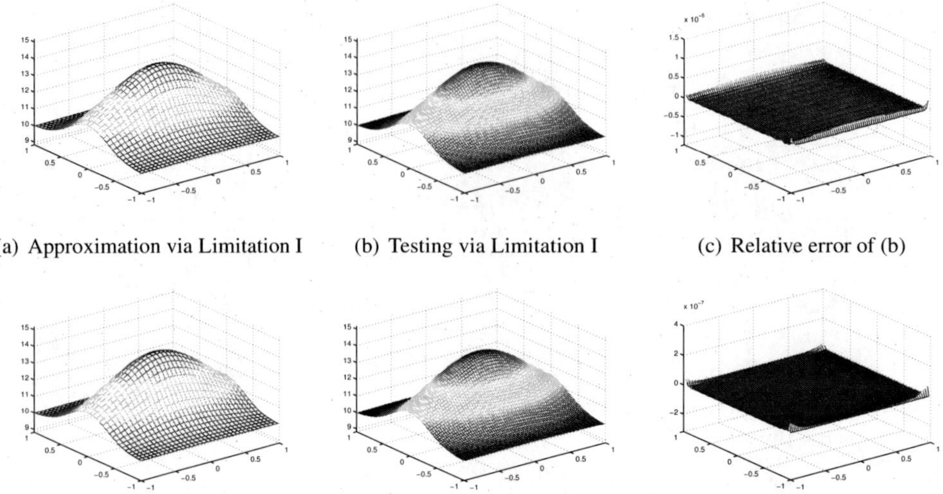

(a) Approximation via Limitation I (b) Testing via Limitation I (c) Relative error of (b)

Figure 5. Approximation and testing results of the 2ILOPN with either Limitation I or Limitation II about target function (6). *Reproduced from Y. Zhang, J. Chen et al., Growing-type weights and structure determination of 2-input Legendre orthogonal polynomial neuronet, Figure 5, Proceedings of the 2012 IEEE International Symposium on Industrial Electronics, pp. 852-857. ©[2012] IEEE. Reprinted, with permission.*

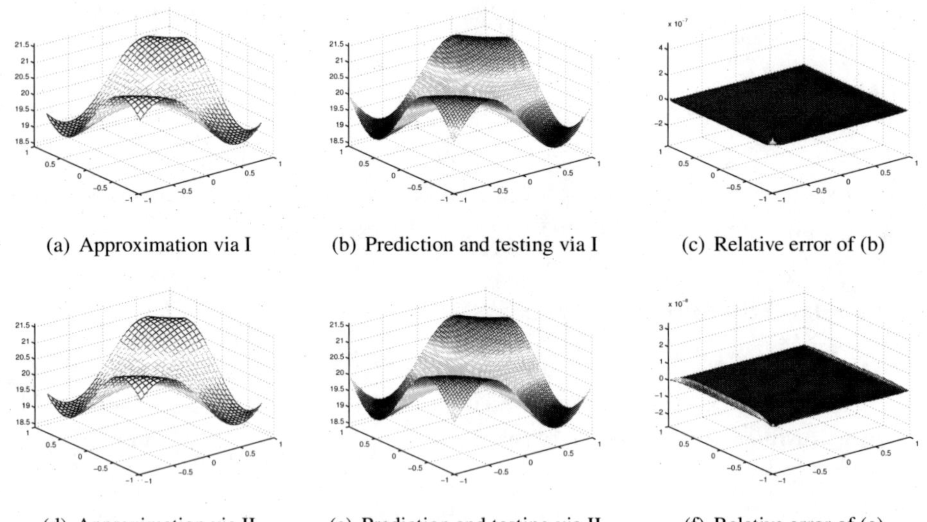

(a) Approximation via I (b) Prediction and testing via I (c) Relative error of (b)

(d) Approximation via II (e) Prediction and testing via II (f) Relative error of (e)

Figure 6. Approximation, prediction and testing results of the 2ILOPN with either Limitation I or Limitation II about target function (7). *Reproduced from Y. Zhang, J. Chen et al., Growing-type weights and structure determination of 2-input Legendre orthogonal polynomial neuronet, Figure 6, Proceedings of the 2012 IEEE International Symposium on Industrial Electronics, pp. 852-857. ©[2012] IEEE. Reprinted, with permission.*

Moreover, as shown in Table 1, the proposed 2ILOPN can achieve superior performance on testing. These results substantiate the abilities of approximation and generalization of the 2ILOPN. Note that, for the above three target functions (5)-(7), the WASD algorithm with Limitation II takes little time to determine the corresponding optimal numbers of hidden-layer neurons, which shows the efficacy. For better understanding, Figure 5 illustrates the numerical results of the 2ILOPN working on target function (6) [with the numerical results on (5) and (7) omitted due to results' similarity], which intuitively and visually show the good performance of the 2ILOPN equipped with either of the two WASD algorithms (in terms of approximation and testing).

For further investigation, we sample the target function (7) over region $[-0.9,0.9]^2$ with gap-size 0.06 to generate the training data set $\{X_p,\gamma_p\}_{p=1}^{961}$ (i.e., $Q=961$). Based on such a data set, the 2ILOPN is used to predict the target function (7) over region $\{[-1,1]^2-[-0.9,0.9]^2\}$, and the corresponding numerical results are illustrated in Figure 6. As seen from the figure, the proposed 2ILOPN equipped with either of the two WASD algorithms can achieve excellent performance in terms of prediction and testing, which demonstrates the excellent abilities of the 2ILOPN.

In summary, we can conclude from Table 1 and Figures 5 and 6 that, with superior approximation, generalization and prediction performance, less runtime, and simpler structure of the 2ILOPN, the WASD algorithm with Limitation II is much better than the one with Limitation I (though both algorithms work excellently).

2.3. Neuronet of Chebyshev Polynomials of Class 1

In this subsection, a new type of feedforward 2-input neuronet using Chebyshev polynomials of Class 1 (2INCP1) is constructed and investigated. In addition, a new WASD algorithm is proposed to determine the optimal number of hidden-layer neurons of the 2INCP1, which includes a procedure of pruning the proposed neuronet (after the neuronet grows up). Numerical results further substantiate the efficacy of the 2INCP1 equipped with the so-called WASD algorithm.

2.3.1. Model Structure of 2INCP1

Based on the definition of the Chebyshev polynomials of Class 1, we can obtain a new type of 2-input neuronet by using $\{\varphi_i^{C_1}(x)\}$ in (2). In light of the graded lexicographic order (Dunkl and Xu, 2001) and the theoretical analysis, basis functions $\{\varphi_i^{C_1}(x_1)\varphi_j^{C_1}(x_2)(\text{with } i,j=1,2,\cdots)\}$ with graded lexicographic order can be used to best estimate the target function $f(x_1,x_2)$. In detail, let t be the total degree of $\varphi_i^{C_1}(x_1)$ and $\varphi_j^{C_1}(x_2)$, i.e., $t=i+j$. Then, a maximum value of t defined as T is given to obtain the corresponding basis functions used to approximate the target function. In other words, t is constrained by T (i.e., $t=0,1,\cdots,T$). Thus, we can obtain the order of

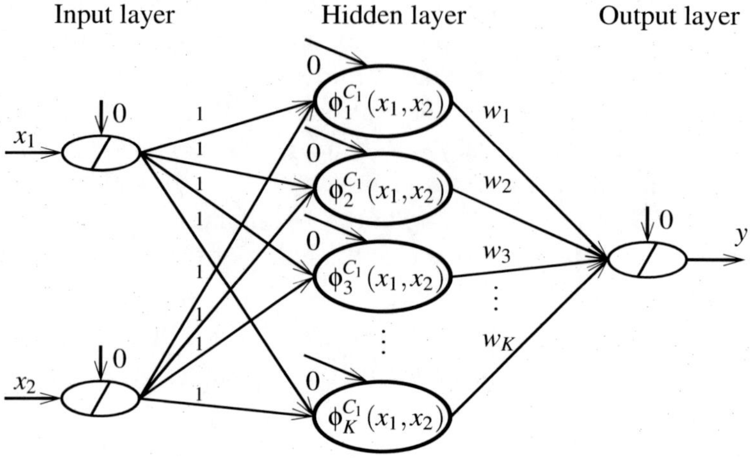

Figure 7. Model structure of the 2INCP1.

$\{\phi_k^{C_1}(x_1,x_2)=\varphi_i^{C_1}(x_1)\varphi_j^{C_1}(x_2), k=(i+j)(i+j+1)/2+i+1\}$ (with $k=1,2,\cdots,K$) as

$$\begin{cases} \phi_1^{C_1}(x_1,x_2)=\varphi_0^{C_1}(x_1)\varphi_0^{C_1}(x_2), \\ \phi_2^{C_1}(x_1,x_2)=\varphi_0^{C_1}(x_1)\varphi_1^{C_1}(x_2), \\ \phi_3^{C_1}(x_1,x_2)=\varphi_1^{C_1}(x_1)\varphi_0^{C_1}(x_2), \\ \quad\vdots \\ \phi_{t(t+1)/2+1}^{C_1}(x_1,x_2)=\varphi_0^{C_1}(x_1)\varphi_t^{C_1}(x_2), \\ \phi_{t(t+1)/2+2}^{C_1}(x_1,x_2)=\varphi_1^{C_1}(x_1)\varphi_{t-1}^{C_1}(x_2), \\ \quad\vdots \\ \phi_{t(t+1)/2+t+1}^{C_1}(x_1,x_2)=\varphi_t^{C_1}(x_1)\varphi_0^{C_1}(x_2), \\ \quad\vdots \\ \phi_K^{C_1}(x_1,x_2)=\varphi_T^{C_1}(x_1)\varphi_0^{C_1}(x_2). \end{cases}$$

On the above-presented theoretical basis, the 2INCP1 is constructed with its model structure shown in Figure 7. The linking weights between the input-layer and hidden-layer neurons are fixed to be 1. In addition, all neuronal thresholds are fixed to be 0.

2.3.2. WASD Algorithm for 2INCP1

The performance of neuronets relates closely to the number of hidden-layer neurons (NOHLN), which is also true in the case of the 2INCP1. Therefore, the main purpose of the WASD algorithm based on the above-presented WDD method in (4) is to determine the optimal NOHLN of the 2INCP1. For better understanding, the flowchart of such a pruning-included WASD algorithm is shown in Figure 8, which consists of two major procedures (i.e., Procedures I and II). It is worth pointing out that "eps" shown in Procedure II of Figure 8 equals 2^{-52} (Su and Ruan, 2008) and is used to avoid the situation of u being zero.

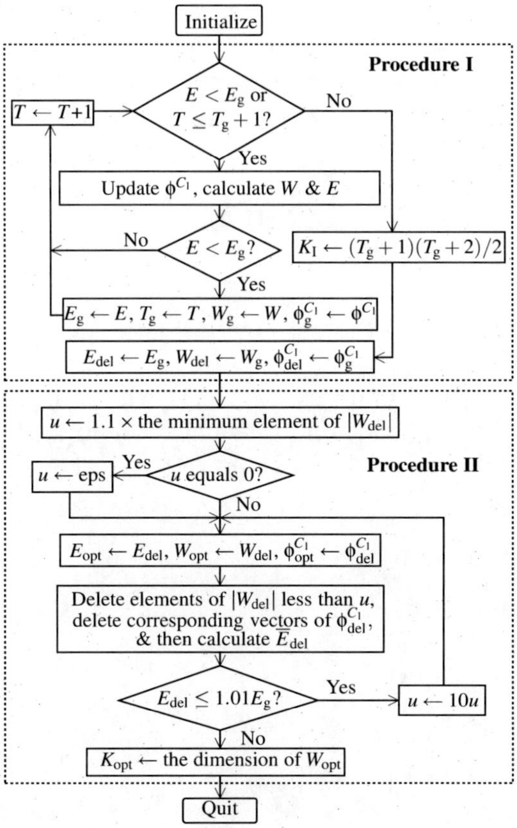

Figure 8. Flowchart of the pruning-included WASD algorithm of the 2INCP1. *Reproduced from Y. Zhang, Y. Yin et al., Pruning-included weights and structure determination of 2-input neuronet using Chebyshev polynomials of Class 1, Figure 2, Proceedings of the 10th World Congress on Intelligent Control and Automation, pp. 700-705. ©[2012] IEEE. Reprinted, with permission.*

Actually, for the proposed WASD algorithm, a variety of target functions have been investigated, such as the typical ones:

$$f(x_1,x_2) = \frac{x_2\cos\left(\sqrt{100x_1^2+100x_2^2+1}\right)}{e^{-2x_1^2-2x_2^2}+2} + 20, \tag{8}$$

$$f(x_1,x_2) = 4x_1e^{-4x_1^2-4x_2^2} + x_2^2 + 20, \tag{9}$$

$$f(x_1,x_2) = \frac{x_2\sin(\pi x_1x_2)}{e^{x_1^2}} + 20. \tag{10}$$

Procedure I of the WASD algorithm is designed to obtain the NOHLN of the grown 2INCP1 with the minimum MSE in (4) [i.e., E_g corresponding to $K_I = (T_g+1)(T_g+2)/2$]. Besides, Figure 9 shows the magnitude distribution of $|w_k|$ (with $k = 1, 2, \cdots, K_I$). As seen from Figure 9, many weights are tiny, which implies that their corresponding hidden-layer

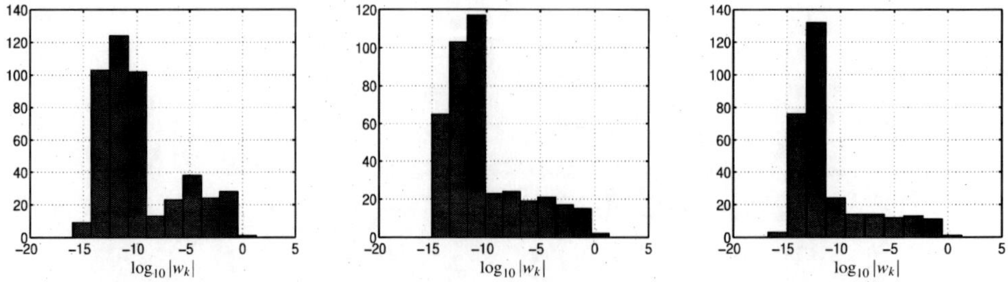

(a) For target function (8) (b) For target function (9) (c) For target function (10)

Figure 9. Histogram of $\log_{10}|w_k|$ with $k = 1, 2, \cdots, K_{\mathrm{I}}$ when the 2INCP1 grows up. *Reproduced from Y. Zhang, Y. Yin et al., Pruning-included weights and structure determination of 2-input neuronet using Chebyshev polynomials of Class 1, Figure 5, Proceedings of the 10th World Congress on Intelligent Control and Automation, pp. 700-705. ©[2012] IEEE. Reprinted, with permission.*

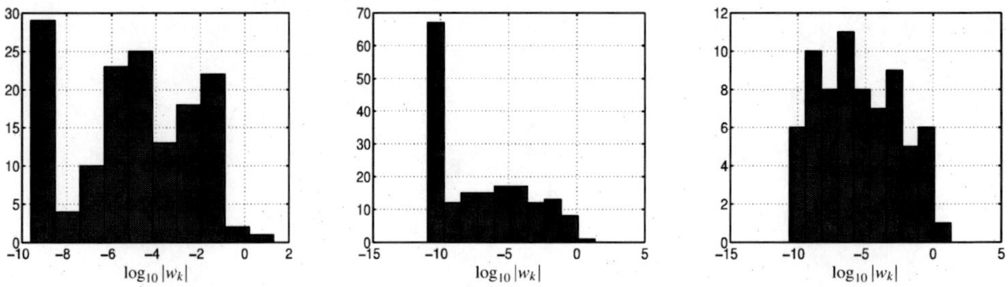

(a) For target function (8) (b) For target function (9) (c) For target function (10)

Figure 10. Histogram of $\log_{10}|w_k|$, with $k = 1, 2, \cdots, K_{\mathrm{opt}}$ after the 2INCP1 is pruned. *Reproduced from Y. Zhang, Y. Yin et al., Pruning-included weights and structure determination of 2-input neuronet using Chebyshev polynomials of Class 1, Figure 6, Proceedings of the 10th World Congress on Intelligent Control and Automation, pp. 700-705. ©[2012] IEEE. Reprinted, with permission.*

neurons almost have no important influence on the performance of the 2INCP1, and thus the structure of the 2INCP1 can be pruned appropriately (by deleting the very much less important hidden-layer neurons). In view of the similar characteristics of many other target functions, Procedure II of the WASD algorithm is developed to obtain the optimal NOHLN (i.e., K_{opt}) of the 2INCP1, with the efficacy of Procedure II demonstrated in Figure 10.

2.3.3. Numerical Studies of 2INCP1

To substantiate the efficacy of the 2INCP1 equipped with the WASD algorithm (i.e., in terms of approximation, validation and prediction), numerical experiments about a variety of target functions are performed. For numerical verification, the set of sample pairs $\{(X_p, \gamma_p), p = 1, 2, \cdots, 2116\}$ is used for approximation. In addition, the set of sample pairs

Table 2. Numerical results via procedure I of the WASD algorithm. *Reproduced from Y. Zhang, Y. Yin et al., Pruning-included weights and structure determination of 2-input neuronet using Chebyshev polynomials of Class 1, Table I, Proceedings of the 10th World Congress on Intelligent Control and Automation, pp. 700-705. ©[2012] IEEE. Reprinted, with permission.*

	Function (8)	Function (9)	Function (10)
K_{I}	465	406	300
E_{app}	2.85×10^{-17}	2.31×10^{-19}	1.11×10^{-21}
E_{pre}	5.25×10^{-6}	9.56×10^{-10}	1.03×10^{-15}
E_{val}	5.81×10^{-17}	2.63×10^{-19}	1.23×10^{-21}
Validation time (s)	6.08	4.69	2.57

Table 3. Numerical results via procedure II of the WASD algorithm. *Reproduced from Y. Zhang, Y. Yin et al., Pruning-included weights and structure determination of 2-input neuronet using Chebyshev polynomials of Class 1, Table II, Proceedings of the 10th World Congress on Intelligent Control and Automation, pp. 700-705. ©[2012] IEEE. Reprinted, with permission.*

	Function (8)	Function (9)	Function (10)
K_{opt}	147	177	71
E_{app}	2.86×10^{-17}	2.31×10^{-19}	1.05×10^{-21}
E_{pre}	5.25×10^{-6}	9.56×10^{-10}	1.03×10^{-15}
E_{val}	5.81×10^{-17}	2.63×10^{-19}	1.17×10^{-21}
Validation time (s)	0.65	0.94	0.17

$\{(X_p, \gamma_p), p = 1, 2, \cdots, 3721\}$ (i.e., $Q = 3721$) for validation is generated by sampling uniformly over region $[-0.9, 0.9]^2$ with new gap-size 0.03. Furthermore, the set of sample pairs $\{(X_p, \gamma_p), p = 1, 2, \cdots, 1920\}$ (i.e., $Q = 1920$) for prediction is generated by sampling uniformly over region $\{[-1.0, 1.0]^2 - [-0.9, 0.9]^2\}$ with new gap-size 0.02.

Tables 2 and 3 show the numerical-verification results of the 2INCP1 equipped with the WASD algorithm. As seen from Table 3, the optimal NOHLN of the pruned 2INCP1 for each target function (i.e., K_{opt}) is determined via Procedure II of the WASD algorithm. Comparing Tables 2 and 3, we can see that more than 56.40% of the hidden-layer neurons have been deleted via Procedure II of the WASD algorithm while the MSEs of approximation, prediction and validation (i.e., E_{app}, E_{pre} and E_{val}) increase by less than 0.36%. In addition, the 2INCP1 with K_{opt} has much shorter validation time than the one with K_{I}, which, together with the above analysis, show the effectiveness/necessity of Procedure II of the WASD algorithm. Thus, by exploiting the proposed WASD algorithm, the optimal NOHLN is automatically determined to achieve the best performance of the 2INCP1 (without prescribing the precision parameter). Note that, for the above three target functions (8)-(10), it takes 33.73, 21.88 and 13.02 seconds, respectively, to determine the corresponding optimal numbers of hidden-layer neurons, which, in some context, shows the efficacy of the proposed WASD algorithm.

With the optimal NOHLN of the 2INCP1 determined, E_{app} and E_{val} illustrated in Table

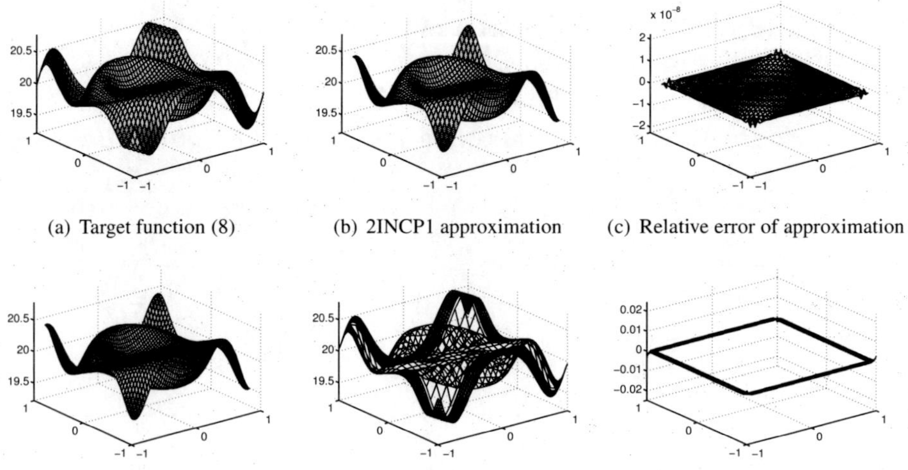

(a) Target function (8) (b) 2INCP1 approximation (c) Relative error of approximation

(d) 2INCP1 validation (e) 2INCP1 prediction (f) Relative error of prediction

Figure 11. Numerical-verification results of the 2INCP1 equipped with the optimal NOHLN about target function (8). *Reproduced from Y. Zhang, Y. Yin et al., Pruning-included weights and structure determination of 2-input neuronet using Chebyshev polynomials of Class 1, Figure 7, Proceedings of the 10th World Congress on Intelligent Control and Automation, pp. 700-705. ©[2012] IEEE. Reprinted, with permission.*

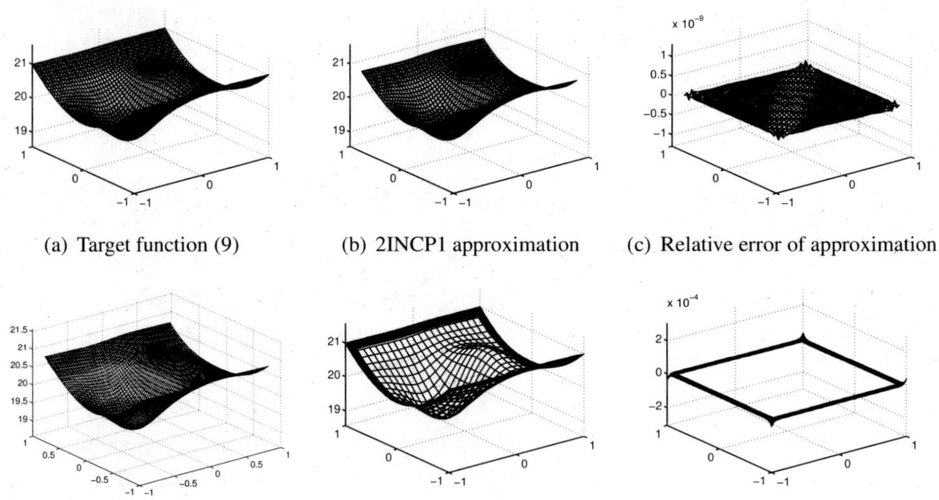

(a) Target function (9) (b) 2INCP1 approximation (c) Relative error of approximation

(d) 2INCP1 validation (e) 2INCP1 prediction (f) Relative error of prediction

Figure 12. Numerical-verification results of the 2INCP1 equipped with the optimal NOHLN about target function (9). *Reproduced from Y. Zhang, Y. Yin et al., Pruning-included weights and structure determination of 2-input neuronet using Chebyshev polynomials of Class 1, Figure 8, Proceedings of the 10th World Congress on Intelligent Control and Automation, pp. 700-705. ©[2012] IEEE. Reprinted, with permission.*

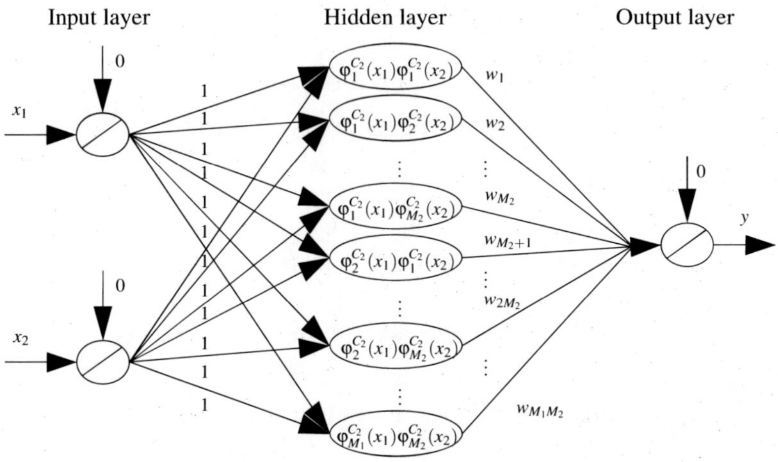

Figure 13. Model structure of the 2INCP2.

3 are very tiny (i.e., of order $10^{-17} \sim 10^{-21}$). For better understanding, the corresponding graphical results are shown in Figures 11 and 12, which substantiate well the superior performance of the 2INCP1 equipped with the WASD algorithm on approximation and validation. In addition, for further investigation, the numerical results of the 2INCP1 on prediction are given in Figure 11(e) and (f) and Figure 12(e) and (f). As seen from these subfigures as well as Table 3, E_{pre} of the 2INCP1 is tiny enough (i.e., of order $10^{-6} \sim 10^{-15}$), which demonstrates the superior performance of the 2INCP1 equipped with the WASD algorithm on prediction.

2.4. Neuronet of Chebyshev Polynomials of Class 2

In this subsection, a new feedforward 2-input neuronet activated by a group of Chebyshev polynomials of Class 2 (i.e., 2INCP2) is constructed and investigated. According to the definition of the Chebyshev polynomials of Class 2, $\varphi_i^{C_2}(x)$ is adopted to replace $\varphi_i(x)$ in (2).

Then, based on (2), we have

$$f(x_1,x_2) \approx \sum_{i=1}^{M_1}\sum_{j=1}^{M_2} w_{ij}\varphi_i^{C_2}(x_1)\varphi_j^{C_2}(x_2) = \sum_{k=1}^{K} w_k\phi_k^{C_2}(x_1,x_2),$$

where $K = M_1M_2$ denotes the total number of Chebyshev polynomials of Class 2 used to approximate $f(x_1,x_2)$, the resultant weight $w_k = w_{ij}$ (with $i = 1,2,\cdots,M_1$ and $j = 1,2,\cdots,M_2$) denotes the weight for $\phi_k^{C_2}(x_1,x_2)$, and $\phi_k^{C_2}([x_1,x_2]^{\mathrm{T}}) = \phi_k^{C_2}(x_1,x_2) =$

$\varphi_i^{C_2}(x_1)\varphi_j^{C_2}(x_2)$. Besides, $k=(i-1)M_2+j$ and

$$\begin{cases} \phi_1^{C_2}(x_1,x_2)=\varphi_1^{C_2}(x_1)\varphi_1^{C_2}(x_2), \\ \phi_2^{C_2}(x_1,x_2)=\varphi_1^{C_2}(x_1)\varphi_2^{C_2}(x_2), \\ \quad\vdots \\ \phi_{M_2}^{C_2}(x_1,x_2)=\varphi_1^{C_2}(x_1)\varphi_{M_2}^{C_2}(x_2), \\ \phi_{M_2+1}^{C_2}(x_1,x_2)=\varphi_2^{C_2}(x_1)\varphi_1^{C_2}(x_2), \\ \quad\vdots \end{cases}$$

Therefore, the target function $f(x_1,x_2)$ can be best estimated by obtaining the optimal weights $\{w_k\}$ for basis functions $\{\phi_k^{C_2}(x_1,x_2)\}$ (with $k=1,2,\cdots,K$).

2.4.1. Model Structure of 2INCP2

According to the above analysis, we propose a new type of 2-input feedforward neuronet activated by Chebyshev polynomials of Class 2. The corresponding model structure of the 2INCP2 is shown in Figure 13. To simplify the neural network model and its computation, the linking weights from input-layer to hidden-layer are all fixed to 1, and all neuronal thresholds are fixed to 0.

2.4.2. WASD Algorithm for 2INCP2

For the neuronet shown in Figure 13, the weights between hidden layer and output layer can be determined directly by using the weights-direct-determination (WDD) method, which is shown in (4). Based on the WDD method, a WASD algorithm is developed to determine the optimal number of hidden-layer neurons, K_{opt}. Before presenting such a WASD algorithm [of which the flowchart is omitted due to the similarity with Figure (4)], a variety of target functions have been tested to investigate the relationship between the MSE function [i.e., (3)] and the number of hidden-layer neurons, such as the following functions:

$$f(x_1,x_2)=\frac{50x_1x_2\sin(\sqrt{10x_1^2+10x_2^2})}{x_1^2+x_2^2}+20, \tag{11}$$

$$f(x_1,x_2)=(2+x_2)e^{-\sin(x_1x_2)}, \tag{12}$$

$$f(x_1,x_2)=(1-x_1)^3\sin(x_2)e^{-x_1^2-x_2^2}+1. \tag{13}$$

Figure 14 illustrates the MSE function in (3) (i.e., E) of the above three target functions versus the number of hidden-layer neurons (i.e., K). As seen from the figure, the curves all look like the character “V” or “U”. From the similar curves and also from other testing results, we can then design the WASD algorithm, which can obtain the minimal value of the MSE (i.e., E_{opt}) and the optimal number of hidden-layer neurons (i.e., K_{opt}).

2.4.3. Numerical Studies of 2INCP2

In this subsubsection, numerical results are presented to substantiate the efficacy and superior abilities of the proposed 2INCP2 in terms of approximation (or say, learning),

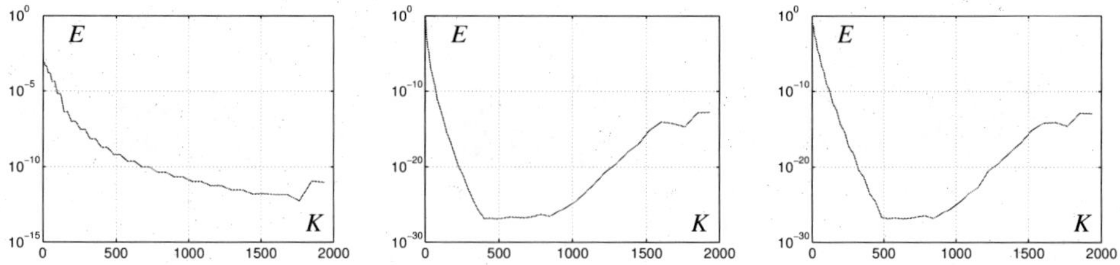

(a) For target function (11) (b) For target function (12) (c) For target function (13)

Figure 14. Relationship between the average MSE and the number of hidden-layer neurons. *Reproduced from Y. Zhang, X. Yu et al., Weights and structure determination of feedforward two-input neural network activated by Chebyshev polynomials of Class 2, Figure 3, Proceedings of the 24th Chinese Control and Decision Conference, pp. 1100-1105. ©[2012] IEEE. Reprinted, with permission.*

Table 4. Approximation, testing and denoising results of the 2INCP2 with $K = 400$. *Reproduced from Y. Zhang, X. Yu et al., Weights and structure determination of feedforward two-input neural network activated by Chebyshev polynomials of Class 2, Table I, Proceedings of the 24th Chinese Control and Decision Conference, pp. 1100-1105. ©[2012] IEEE. Reprinted, with permission.*

	Function (11)	Function (12)	Function (13)
WDD runtime (s)	2.46	2.80	2.43
E_{app}	6.62×10^{-10}	7.17×10^{-25}	2.22×10^{-25}
E_{test}	1.04×10^{-4}	7.82×10^{-25}	2.46×10^{-25}
MRE_{app}	1.41×10^{-5}	3.84×10^{-12}	4.69×10^{-10}
MRE_{test}	1.80×10^{-3}	3.75×10^{-12}	2.59×10^{-10}
E_{den}	6.21×10^{-6}	1.00×10^{-2}	2.00×10^{-3}

Table 5. Numerical results of the 2INCP2 with the WASD algorithm. *Reproduced from Y. Zhang, X. Yu et al., Weights and structure determination of feedforward two-input neural network activated by Chebyshev polynomials of Class 2, Table II, Proceedings of the 24th Chinese Control and Decision Conference, pp. 1100-1105. ©[2012] IEEE. Reprinted, with permission.*

Target function	K_{opt}	WASD runtime (s)	E_{opt}
(11)	1806	1259.7	2.58×10^{-13}
(12)	440	32.90	3.89×10^{-28}
(13)	600	89.95	3.73×10^{-28}

testing (or say, validation), denoising and prediction, with the aid of the WASD algorithm which obtains the optimal number of hidden-layer neurons of the 2INCP2. Note that, due to space limitation and results similarity, the numerical results are shown only for the aforementioned three target functions, i.e., (11)-(13).

Firstly, to verify the abilities of the 2INCP2 in terms of approximation, testing, denois-

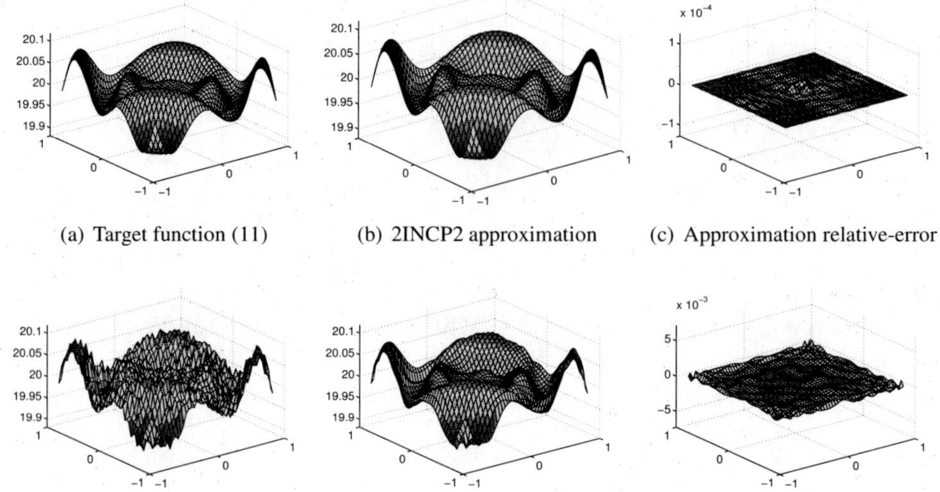

(a) Target function (11)   (b) 2INCP2 approximation   (c) Approximation relative-error

(d) Function with random noise   (e) 2INCP2 denoising   (f) Denoising relative-error

Figure 15. Approximation and denoising results of 2INCP2 on function (11) with $K = 400$. *Reproduced from Y. Zhang, X. Yu et al., Weights and structure determination of feedforward two-input neural network activated by Chebyshev polynomials of Class 2, Figure 4, Proceedings of the 24th Chinese Control and Decision Conference, pp. 1100-1105. ©[2012] IEEE. Reprinted, with permission.*

ing and prediction, the number of the hidden-layer neurons is set to be 400 as an example. We sample uniformly over region $[-0.9, 0.9]^2$ with gap-size 0.04 to generate a square matrix called the input matrix of the training data-set $\{(X_p, \gamma_p), p = 1, 2, \cdots, 2116\}$ (i.e., $Q = 2116$). Besides, we choose the intervals $[-1, -0.9]$ and $[0.9, 1]$ as the unlearned intervals to test the prediction ability of the 2INCP2. The corresponding numerical results are shown in Table 4 and Figures 15 and 16. As seen from Table 4, the training time (or say, the computing time of WDD) is very short, and the MSEs of approximation and testing are quite small. These demonstrate well the efficacy and superior approximation ability of the 2INCP2. Note that MRE_{app} and MRE_{test} represent the maximal relative-error of approximation and testing, respectively. In addition, the MSE of denoising (i.e., E_{den}) shown in Table 4 is small, which substantiates the denoising ability of the 2INCP2. For better understanding, Figures 15 and 16 illustrate the numerical results of the 2INCP2 working on target functions (11) and (13) [with the numerical results on (12) omitted due to results similarity], which intuitively and visually show the good performance of the 2INCP2 (in terms of approximation and denoising). Note that approximation and denoising could be two contradicting abilities, as a "super" neuronet like the 2INCP2 could even learn well the noisy data or the noise; see Figure 14 and Table 4, where the approximation errors are even of order $10^{-10} \sim 10^{-30}$. Furthermore, with the aim of verifying the prediction ability of the well-trained 2INCP2, Figure 16(d) illustrates the graphical results about target function (13). Comparing Figure 16(d) with Figure 16(a) (and also comparing the related numerical results on other target functions), we have the observation that the 2INCP2 equipped with

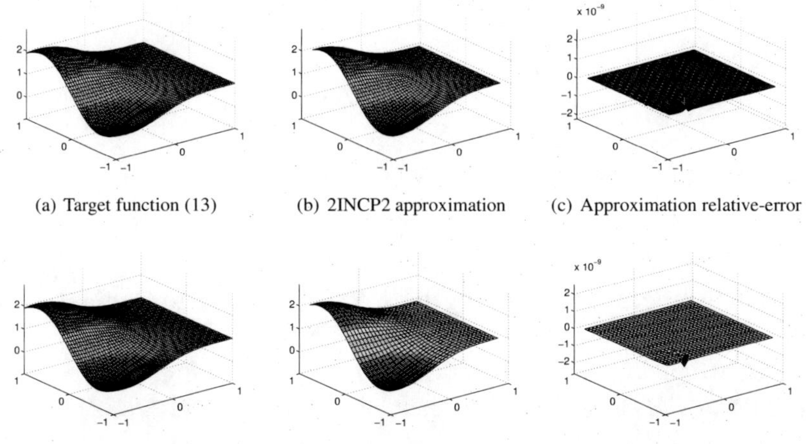

(a) Target function (13) (b) 2INCP2 approximation (c) Approximation relative-error

(d) 2INCP2 prediction (e) 2INCP2 testing (f) Testing relative-error

Figure 16. Approximation, prediction and testing results of 2INCP2 on (13) with $K = 400$. *Reproduced from Y. Zhang, X. Yu et al., Weights and structure determination of feedforward two-input neural network activated by Chebyshev polynomials of Class 2, Figure 5, Proceedings of the 24th Chinese Control and Decision Conference, pp. 1100-1105. ©[2012] IEEE. Reprinted, with permission.*

the WASD algorithm has relatively good or even very good prediction performance.

Secondly, for further investigation, Figure 17 illustrates the results about the searching process of the optimal number of hidden-layer neurons by exploiting the presented WASD algorithm for the 2INCP2. As seen from the figure, which is compared with Figure 14, the optimal number K_{opt} of hidden-layer neurons of the 2INCP2 is obtained efficiently and accurately via the WASD algorithm. In addition, the numerical results shown in Table 5 demonstrate the accuracy and efficacy of the 2INCP2 equipped with the WASD algorithm.

In summary, the above numerical results (i.e., Tables 4 and 5 and Figures 15-17 all demonstrate the efficacy and superior performance of the proposed 2INCP2 equipped with the WASD algorithm.

3. WASD of AN with Three Inputs

In this section, 3-input feedforward neuronets with Euler-polynomial and power activation functions are investigated. The corresponding WASD algorithms are presented for the proposed neuronets, which can be divided into two types: the growing type and the pruning-while-growing type. The pruning technique is adequately discussed in this section. Numerical results substantiate the superior performance of the proposed three-input neuronets equipped with the WASD algorithms in terms of training, validation and prediction.

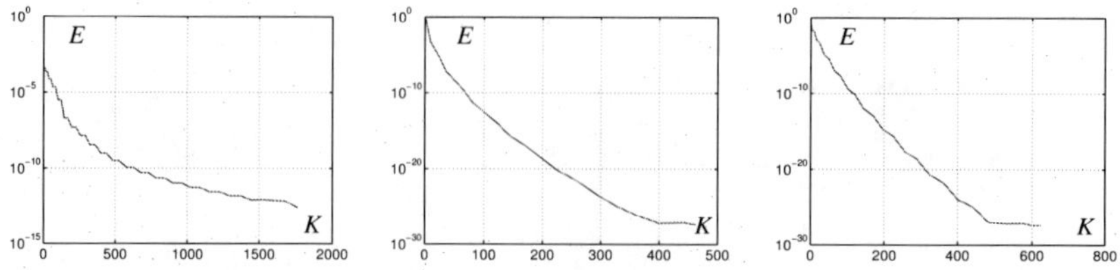

(a) For target function (11) (b) For target function (12) (c) For target function (13)

Figure 17. Searching process by exploiting the WASD algorithm to obtain K_{opt}. *Reproduced from Y. Zhang, X. Yu et al., Weights and structure determination of feedforward two-input neural network activated by Chebyshev polynomials of Class 2, Figure 6, Proceedings of the 24th Chinese Control and Decision Conference, pp. 1100-1105. ©[2012] IEEE. Reprinted, with permission.*

3.1. AN with Growing-type WASD

Based on the function approximation theory and the Weierstrass approximation theorem, a novel 3-input Euler polynomial neuronet (3IEPN) is established in this subsection to fit the 4-dimensional data (i.e., data in the form of three inputs and one output). In order to achieve satisfactory performance of the 3IEPN, a weights and structure determination (WASD) algorithm, which is based on the weights-direct-determination (WDD) method, is built up for the established 3IEPN. As a matter of fact, in the investigation of the 2-input neuronets, WASD algorithms for the 2ILOPN and 2INCP2 are also of the growing type.

3.1.1. The Second Theoretical Basis and Analysis

Theoretically, AN can approximate any continuous target function with any degree of accuracy (Xu and Yang, 1987; Liang and Li, 2005). For better understanding, the detailed descriptions of related essential definition and theorems are given in this subsubsection. Moreover, based on the presented theorems and the authors' previous successful experience (Zhang et al., 2010a; Zhang et al., 2012a; Zhang et al., 2012b; Zhang et al., 2012c), a proposition is put forward to lay a basis for establishing the 3IEPN.

Definition 4. Euler polynomials can be defined as (Zhang et al., 2009a)

$$\varphi_j^E(x) = x^j - \frac{1}{2}\sum_{i=0}^{j-1}\left(\frac{j!}{i!(j-i)!}\varphi_i^E(x)\right), \text{with } \varphi_0^E(x) = 1,$$

where $\varphi_j^E(x)$ denotes a Euler polynomial of degree j (with $j = 1, 2, 3, \cdots$).

Besides, the following theorem is the Weierstrass approximation theorem (Liang and Li, 2005; Liu et al., 2009; Peng and Li, 2009).

Theorem 1. *Let $g(x)$ be a continuous real-valued function defined on $[a,b]$, then for any $\varepsilon > 0$, there must be a polynomial $\rho(x)$ such that, for all $x \in [a,b]$,*

$$|g(x) - \rho(x)| < \varepsilon.$$

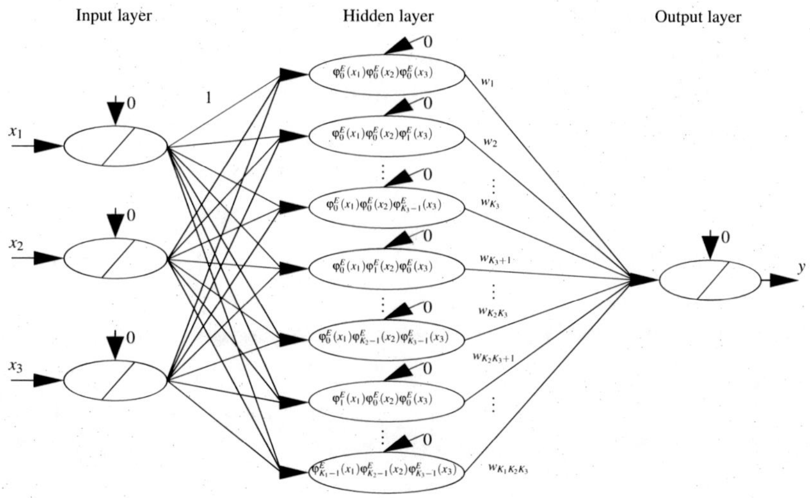

Figure 18. Model structure of the 3IEPN.

It means that any continuous function defined on a closed and bounded interval can be uniformly approximated by a polynomial with any degree of accuracy.

Theorem 2. *For a continuous real-valued function $f(x_1,x_2,\cdots,x_D)$ with D variables, which is defined over $S_D=\{(x_1,x_2,\cdots,x_D)\in R^D|0\le x_q\le 1,q=1,2,\cdots,D\}$, Weierstrass polynomial (i.e., Bernstein polynomial) can be constructed as*

$$B^f_{n_1,n_2,\cdots,n_D}(x_1,x_2,\cdots,x_D) = \sum_{v_1=0}^{n_1}\cdots\sum_{v_D=0}^{n_D} f\left(\frac{v_1}{n_1},\cdots,\frac{v_D}{n_D}\right) p_{n_1,v_1}(x_1)\cdots p_{n_D,v_D}(x_D), \tag{14}$$

where

$$p_{n_q,v_q}(x_q)=C^{v_q}_{n_q}x_q^{v_q}(1-x_q)^{n_q-v_q} \quad (\textit{with } q=1,2,\cdots,D),$$

and $C^{v_q}_{n_q}$ denotes a binomial coefficient; and then

$$\lim_{n_1,n_2,\cdots,n_D\to\infty} B^f_{n_1,n_2,\cdots,n_D}(x_1,x_2,\cdots,x_D)=f(x_1,x_2,\cdots,x_D).$$

That is to say, $f(x_1,x_2,\cdots,x_D)$ can be uniformly approximated by $B^f_{n_1,n_2,\cdots,n_D}(x_1,x_2,\cdots,x_D)$ over S_D. It is worth pointing out that the above theorems were investigated and proved in (Liang and Li, 2005; Liu et al., 2009; Peng and Li, 2009).

Proposition 2. According to the above theorems, $\{p_{n_q,v_q}(x_q),\ v_q=0,1,\cdots,n_q,\ q=1,2,\cdots,D\}$ can be replaced with D groups of Euler polynomials to fit the data of $f(x_1,x_2,\cdots,x_D)$. That is, for the given values of q and v_q, let the v_qth degree of the qth group of Euler polynomials [i.e., $\varphi_{v_q}(x_q)$] replace $p_{n_q,v_q}(x_q)$, where $x_q\in[a,b]$ (with $a,b\in R$ and $a\le b$).

Without loss of generality, the continuous real-valued target function of three variables $f(x_1,x_2,x_3)$ (e.g., with $a=-1.6$, $b=1.6$ and $n_q=K_q-1$, $q=1,2,3$) is investigated in this subsubsection. According to the above proposition, we have

$$f(x_1,x_2,x_3)\approx\sum_{v_1=0}^{K_1-1}\sum_{v_2=0}^{K_2-1}\sum_{v_3=0}^{K_3-1}w_{v_1,v_2,v_3}\varphi^E_{v_1}(x_1)\varphi^E_{v_2}(x_2)\varphi^E_{v_3}(x_3)=\sum_{k=1}^{K}w_k\phi^E_k(x_1,x_2,x_3),$$

where w_{v_1,v_2,v_3} corresponds to $f(v_1/n_1,v_2/n_2,v_3/n_3)$ in (14), K_1, K_2 and K_3 denote the numbers of Euler polynomials used to replace $\{p_{n_1,v_1}(x_1),\ v_1=0,1,\cdots,n_1\}$, $\{p_{n_2,v_2}(x_2),\ v_2=0,1,\cdots,n_2\}$ and $\{p_{n_3,v_3}(x_3),\ v_3=0,1,\cdots,n_3\}$, respectively, and $K=K_1K_2K_3$ denotes the total number of products of Euler polynomials used to fit the data of $f(x_1,x_2,x_3)$. Besides, the weight $w_k=w_{v_1,v_2,v_3}$ denotes the weight for $\phi^E_k(x_1,x_2,x_3)$, and $\phi^E_k(x_1,x_2,x_3)=\varphi^E_{v_1}(x_1)\varphi^E_{v_2}(x_2)\varphi^E_{v_3}(x_3)$, in which, $k=v_1K_2K_3+v_2K_3+v_3$. Then, $\{\phi^E_k(x_1,x_2,x_3)\}$ can be expressed as below:

$$\begin{cases}\phi^E_1(x_1,x_2,x_3)=\varphi^E_0(x_1)\varphi^E_0(x_2)\varphi^E_0(x_3),\\ \phi^E_2(x_1,x_2,x_3)=\varphi^E_0(x_1)\varphi^E_0(x_2)\varphi^E_1(x_3),\\ \qquad\vdots\\ \phi^E_{K_3}(x_1,x_2,x_3)=\varphi^E_0(x_1)\varphi^E_0(x_2)\varphi^E_{K_3-1}(x_3),\\ \phi^E_{K_3+1}(x_1,x_2,x_3)=\varphi^E_0(x_1)\varphi^E_1(x_2)\varphi^E_0(x_3),\\ \qquad\vdots\\ \phi^E_{K_2K_3}(x_1,x_2,x_3)=\varphi^E_0(x_1)\varphi^E_{K_2-1}(x_2)\varphi^E_{K_3-1}(x_3),\\ \phi^E_{K_2K_3+1}(x_1,x_2,x_3)=\varphi^E_1(x_1)\varphi^E_0(x_2)\varphi^E_0(x_3),\\ \qquad\vdots\\ \phi^E_{K_1K_2K_3}(x_1,x_2,x_3)=\varphi^E_{K_1-1}(x_1)\varphi^E_{K_2-1}(x_2)\varphi^E_{K_3-1}(x_3).\end{cases}$$

Therefore, the target function $f(x_1,x_2,x_3)$ can be uniformly approximated by obtaining the optimal weights $\{w_k\}$ for basis functions $\{\phi^E_k(x_1,x_2,x_3)=\phi^E_k(X)\}$ and obtaining the proper number of basis functions (where X denotes the vector of inputs $[x_1,x_2,x_3]^{\mathrm{T}}$).

3.1.2. 3IEPN Model and WASD Algorithm

In this subsubsection, based on the above analysis, a 3IEPN is established. Furthermore, a WASD algorithm is applied to the established neuronet so as to obtain the optimal performance (i.e., training, validation and prediction performance) of the proposed neuronet.

As shown in Figure 18, the 3IEPN adopts a three-layer structure (i.e., the input layer, the hidden layer and the output layer). Each neuron in the input layer is connected to all neurons in the hidden layer and all neurons in the hidden layer are connected to the neuron in the output layer. In addition, the activation functions of hidden-layer neurons are a group of products of Euler polynomials $\{\varphi^E_{v_1}(x_1)\varphi^E_{v_2}(x_2)\varphi^E_{v_3}(x_3)\}$ [i.e., $\{\phi_k(x_1,x_2,x_3)\}$], the linking weights between hidden layer and output layer are $\{w_k\}$ and the output of the 3IEPN is y. Besides, for the purpose of simplifying the established neuronet and reducing the computational complexity, the linking weights between input layer and hidden layer are set to be 1, and all neuronal thresholds are set to be 0.

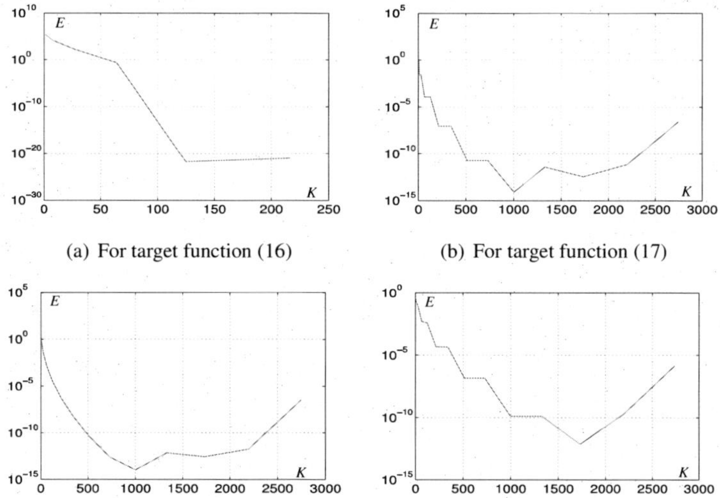

(a) For target function (16)

(b) For target function (17)

(c) For target function (18)

(d) For target function (19)

Figure 19. Relationship between the MSE (i.e., E) of training and the number of hidden-layer neurons (i.e., K). *Reproduced from Y. Zhang, W. Li et al., The 3-input Euler polynomial neuronet (3IEPN) with weights-and-structure-determination (WASD) algorithm, Figure 2, Proceedings of International Conference on Modelling, Identification & Control, pp. 1165-1170. ©[2012] IEEE. Reprinted, with permission.*

In order to obtain the optimal number of hidden-layer neurons and the optimal linking weights between hidden layer and output layer of the 3IEPN, a WASD algorithm based on the WDD method is presented as follows in details.

Let M be the total number of training samples, then the training-sample set can be expressed as $\{(X_p, \gamma_p)|X_p \in R^3, \gamma_p \in R, p = 1, 2, \cdots, Q\}$, where $X_p = [X_{p1}, X_{p2}, X_{p3}]^{\mathrm{T}}$ denotes the pth input vector, and γ_p denotes the corresponding desired output. Then, we can obtain the corresponding MSE function to evaluate the performance of the 3IEPN as below:

$$E = \sum_{p=1}^{Q} \left(\gamma_p - \sum_{k=1}^{K} w_k \phi_k^E(X_p) \right)^2 / Q, \tag{15}$$

where K denotes the total number of the hidden-layer neurons in Figure 18, and $\phi_k^E(\cdot)$ denotes the activation function of the kth hidden-layer neuron of the 3IEPN. It is worth mentioning that, for a given value of K, the MSE can be minimized by finding the optimal linking weights [i.e., the optimal $\{w_k\}$] based on the WDD method (Zhang et al., 2009a; Zhang et al., 2010a; Zhang et al., 2012a; Zhang et al., 2012d).

Furthermore, as mentioned above, the number of hidden-layer neurons can influence greatly the performance of the neuronet, and it is meaningful and important to obtain the optimal number of hidden-layer neurons. Moreover, the optimal number of hidden-layer neurons can be obtained by minimizing the values of the MSE function by considering different numbers of hidden-layer neurons. In order to illustrate the existence of the optimal number of hidden-layer neurons (i.e., K_{opt}) of the 3IEPN, a plenty of target functions have

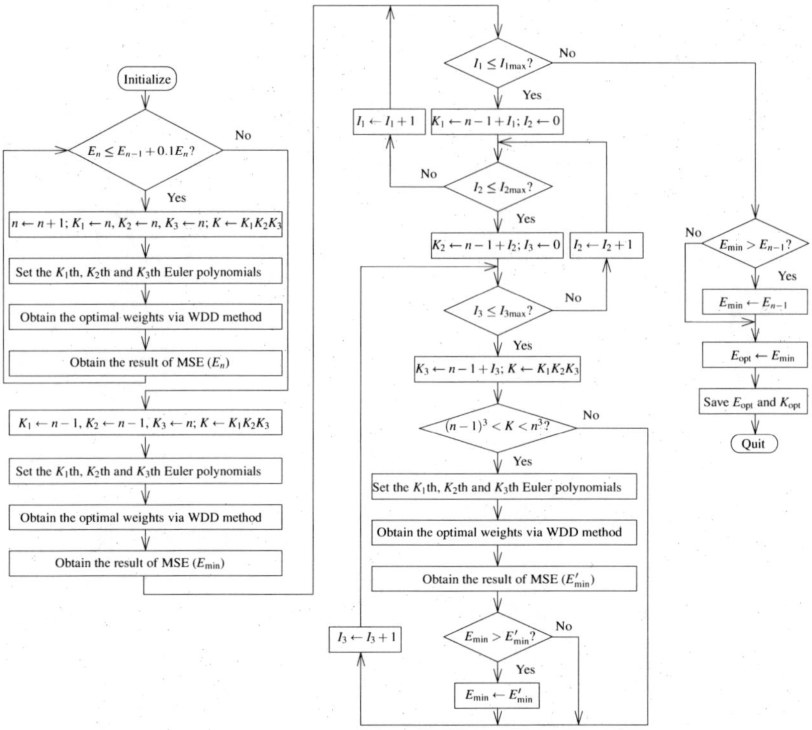

Figure 20. Flowchart of the WASD for the 3IEPN. *Reproduced from Y. Zhang, W. Li et al., The 3-input Euler polynomial neuronet (3IEPN) with weights-and-structure-determination (WASD) algorithm, Figure 5, Proceedings of International Conference on Modelling, Identification & Control, pp. 1165-1170. ©[2012] IEEE. Reprinted, with permission.*

been tested to investigate the relationship between the MSE and the number of hidden-layer neurons, such as the following four target functions:

$$f(x_1,x_2,x_3) = (x_1+5)^4 + x_1^3 + x_2^2 + x_3 + 30, \tag{16}$$

$$f(x_1,x_2,x_3) = \ln|(x_1+x_2+10)| + \sin(x_2x_3) + 20, \tag{17}$$

$$f(x_1,x_2,x_3) = \cos(x_2)e^{x_1} - 2\sin(x_3) + 20, \tag{18}$$

$$f(x_1,x_2,x_3) = \frac{x_1x_3 - x_2^2 + x_3^3}{10 - x_1^2} + \sin(x_1x_2 + x_1x_3) + 30. \tag{19}$$

Figure 19 present respectively the relationship between the MSE (i.e., E) and the number of hidden-layer neurons (i.e., K) about four target functions (16)-(19) of training (with the total number of training samples, $Q = 4096$). Note that all the curves shown in Figure 19 has shapes roughly like the letter "U" or "V", thus called "UV plots". Additionally, each of the UV plots has its own minimum point. Based on the similar characteristics of other testing target functions (which are not shown in this chapter), the WASD algorithm is thus designed to determine K_{opt} by obtaining the minimal value of E (i.e., E_{opt}) of the UV plot.

More specifically, the flowchart of the WASD algorithm is shown in Figure 20 and briefed as the following. Firstly, in order to acquire the minimal value of E, let n grow one by one

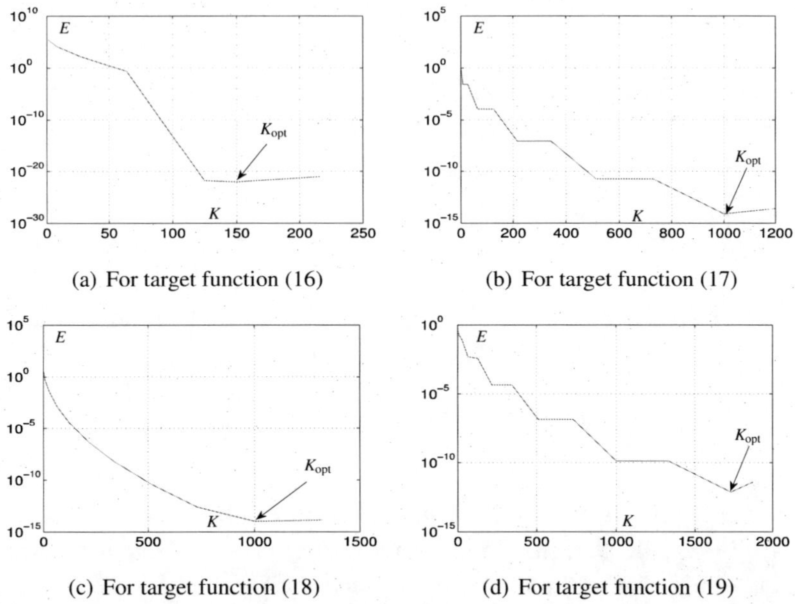

Figure 21. Searching process by exploiting the WASD algorithm to obtain K_{opt}. *Reproduced from Y. Zhang, W. Li et al., The 3-input Euler polynomial neuronet (3IEPN) with weights-and-structure-determination (WASD) algorithm, Figure 6, Proceedings of International Conference on Modelling, Identification & Control, pp. 1165-1170. ©[2012] IEEE. Reprinted, with permission.*

with $K = K_1K_2K_3 = n^3$. Secondly, K_{opt} is further obtained by calculating and minimizing the MSE of the 3IEPN with different structures. It means that the values of K_1, K_2 and K_3 are not completely the same, with inequality $(n-1)^3 < K = K_1K_2K_3 < n^3$ satisfied. In detail, three parameters (i.e., I_1, I_2 and I_3) are used to investigate different structures, where $I_1 = 0, 1, \cdots, I_{1\max}$, $I_2 = 0, 1, \cdots, I_{2\max}$ and $I_3 = 0, 1, \cdots, I_{3\max}$ ($I_{1\max}$, $I_{2\max}$ and $I_{3\max}$ are set to be 2). Finally, by following the flowchart of the WASD algorithm, the K_{opt} and E_{opt} can be obtained: the optimal structure and weights of the 3IEPN can be determined, and the best performance of the 3IEPN can be guaranteed.

3.1.3. Numerical Studies of 3IEPN

In this subsubsection, numerical results are presented to verify and substantiate the performance and efficacy of the proposed 3IEPN equipped with the WASD algorithm in terms of training, validation and prediction. Note that, due to results' similarity, only typical numerical results of training, validation and prediction about four target functions (16)-(19) are presented and analyzed here.

To verify the performance of the proposed 3IEPN equipped with the WASD algorithm, we sample uniformly over the region $[-1.5, 1.5]^3$ with gap-size 0.20 to generate a training sample set $\{(X_p, \gamma_p), p = 1, 2, \cdots, 4096\}$ (i.e., $Q = 4096$). In addition, the region $[-1.35, 1.35]^3$ with gap-size 0.18 is used for validation, while the untrained regions

Table 6. Training, validation and prediction results of the 3IEPN on four target functions. *Reproduced from Y. Zhang, W. Li et al., The 3-input Euler polynomial neuronet (3IEPN) with weights-and-structure-determination (WASD) algorithm, Table I, Proceedings of International Conference on Modelling, Identification & Control, pp. 1165-1170. ©[2012] IEEE. Reprinted, with permission.*

Target	K_{opt}	K_1	K_2	K_3	E_{tra}	E_{val}	E_{pre}
(16)	150	5	6	5	9.95×10^{-23}	6.68×10^{-23}	1.54×10^{-20}
(17)	1000	10	10	10	8.14×10^{-15}	8.33×10^{-16}	6.84×10^{-11}
(18)	1000	10	10	10	1.07×10^{-14}	2.72×10^{-15}	1.43×10^{-12}
(19)	1728	12	12	12	7.32×10^{-13}	5.23×10^{-13}	1.69×10^{-6}

Table 7. Training outputs of the 3IEPN on target function (18) with $K_{opt} = 1000$. *Reproduced from Y. Zhang, W. Li et al., The 3-input Euler polynomial neuronet (3IEPN) with weights-and-structure-determination (WASD) algorithm, Table II, Proceedings of International Conference on Modelling, Identification & Control, pp. 1165-1170. ©[2012] IEEE. Reprinted, with permission.*

x_1	x_2	x_3	γ	y
−1.50	−1.50	−1.50	22.010773576344675	22.010774912579347
0.90	0.10	−1.50	24.442305313740064	24.442305277013823
0.30	−1.50	−1.30	22.022601605528813	22.022601728443256
−0.50	0.10	−1.10	22.385915252905768	22.385915230950985
···	···	···	···	···
1.30	0.10	−0.90	25.217619287176895	25.217619243130677
0.70	−1.50	−0.70	21.430882605852602	21.430882600129681
−0.70	−1.50	−0.30	20.626167468102189	20.626167493865818
−1.50	0.10	−0.10	20.421682272040496	20.421682250937138
···	···	···	···	···
0.30	0.10	0.10	21.143448302781696	21.143448273846953
−0.30	−1.50	0.30	19.461362994552793	19.461362972860286
−1.10	0.10	0.50	19.372357037571792	19.372357005551564
1.10	0.10	1.30	21.062041336179039	21.062041267038989
···	···	···	···	···

$[-1.60,-1.50]^3$ and $[1.50,1.60]^3$ (i.e., $x_1,x_2,x_3 \in [-1.60,-1.50]\bigcup[1.50,1.60]$) with gap-size 0.02 is used for prediction.

For better understanding, the searching process of K_{opt} about four target functions (16)-(19) by exploiting the WASD algorithm for the 3IEPN is visually presented in Figure 21. With the determined optimal structures (i.e., $K_{opt} = K_1K_2K_3$) of the 3IEPN about four target functions (16)-(19), the corresponding minimal values of MSE (specifically, E_{tra}, E_{val} and

Table 8. Validation outputs of the 3IEPN on target function (18) with $K_{\text{opt}} = 1000$. *Reproduced from Y. Zhang, W. Li et al., The 3-input Euler polynomial neuronet (3IEPN) with weights-and-structure-determination (WASD) algorithm, Table III, Proceedings of International Conference on Modelling, Identification & Control, pp. 1165-1170. ©[2012] IEEE. Reprinted, with permission.*

x_1	x_2	x_3	γ	y
−1.35	−1.35	−1.35	22.008222066298512	22.008222998232910
0.81	0.09	−1.35	24.190256818543240	24.190256789003566
0.27	−1.35	−1.17	22.128392170037017	22.128392126443043
−0.45	0.09	−0.99	22.307099457454132	22.307099426755396
⋯	⋯	⋯	⋯	⋯
−0.99	−1.35	−0.81	21.529952128842019	21.529952195202871
1.17	0.09	−0.81	24.657526722827271	24.657526637362750
0.63	−1.35	−0.63	21.589498788500070	21.589498697796575
−0.09	0.09	−0.45	21.780163329978233	21.780163332580315
⋯	⋯	⋯	⋯	⋯
0.99	−1.35	−0.09	20.769155444375823	20.769155334876260
0.45	−1.35	1.35	18.392024140418531	18.392024060058706
0.99	0.09	1.17	20.838841132440070	20.838841091333723
1.35	−1.35	0.45	19.974870917963599	19.974870755027119
⋯	⋯	⋯	⋯	⋯

E_{pre}) are listed in Table 6. Note that E_{tra}, E_{val} and E_{pre} denote respectively the values of MSE function of the 3IEPN in terms of training, validation and prediction. The efficacy and accuracy of the proposed 3IEPN equipped with the WASD algorithm is substantiated by the tiny values of E_{tra}, E_{val} and E_{pre} with their orders of magnitude being $10^{-6} \sim 10^{-23}$.

In order to further substantiate the performance of the proposed 3IEPN equipped with the WASD algorithm, numerical results (with $K_{\text{opt}} = 1000$) about target function (18) are listed in Tables 7-9 (some other results are omitted due to space limitation). Compared with the desired outputs (i.e., γ) of target function (18), the 3IEPN output results (i.e., y) of training, validation and prediction demonstrate clearly the accuracy of the 3IEPN equipped with the WASD algorithm.

In summary, the above numerical results all demonstrate and substantiate the superior efficacy and performance of the proposed 3IEPN equipped with the WASD algorithm.

3.2. AN with Pruning-while-growing-type WASD

A new type of 3-input power-activation feedforward neuronet (3IPFN) is constructed and investigated in this subsection. For the 3IPFN, a novel weights and structure determination (WASD) algorithm is presented to solve data approximation and prediction problems, which can determine the optimal structure of the 3IPFN adaptively by growing and pruning hidden-layer neurons during the training process. Numerical results of illustrative examples

Table 9. Prediction outputs of the 3IEPN on target function (18) with $K_{\mathrm{opt}} = 1000$. *Reproduced from Y. Zhang, W. Li et al., The 3-input Euler polynomial neuronet (3IEPN) with weights-and-structure-determination (WASD) algorithm, Table IV, Proceedings of International Conference on Modelling, Identification & Control, pp. 1165-1170. ©[2012] IEEE. Reprinted, with permission.*

x_1	x_2	x_3	γ	y
−1.60	−1.60	−1.60	21.993251924203271	21.993248141165925
−1.54	−1.52	−1.60	22.010032296087459	22.010032370569537
1.50	1.54	−1.60	22.137144951712660	22.137145758078333
1.58	−1.60	−1.60	21.857384815601012	21.857386405975468
···	···	···	···	···
−1.58	−1.52	1.50	18.015468306312389	18.015468774585752
−1.52	1.54	1.50	18.011744484909830	18.011744492244365
1.54	−1.60	1.50	17.868806219147800	17.868809176208373
−1.52	1.54	1.60	18.007587252034927	18.007587110480316
···	···	···	···	···
1.50	1.54	−1.50	22.132987718837761	22.132988379701299
1.58	−1.60	−1.50	21.853227582726110	21.853229584666902
1.54	−1.60	1.60	17.864648986272897	17.864651788387718
1.60	−1.52	1.60	18.252340464124970	18.252340591381795
···	···	···	···	···

highlight the efficacy of the 3IPFN equipped with the so-called WASD algorithm. Note that the pruning technique has been investigated firstly in the WASD algorithm of the 2INCP1.

3.2.1. The Third Theoretical Basis and Analysis

In this subsubsection, the detailed descriptions of related essential theoretical basis and analysis are presented for constructing the 3IPFN. Firstly, the theorem of Taylor polynomial approximation (Zhang et al., 2012e) is given as follows.

Theorem 3. *Suppose that a target function f has the $(K+1)$th-order continuous derivative on interval $[a,b]$, and K is a nonnegative integer, then for $x \in [a,b]$,*

$$f(x) = P_K(x) + R_K(x),$$

where $R_K(x)$ is the error term and $P_K(x)$ is a polynomial used to approximate $f(x)$.

That is, with a fixed value $h \in [a,b]$, $f(x)$ can be approximated as

$$f(x) \approx P_K(x) = \sum_{i=0}^{K} \frac{f^{(i)}(h)}{i!}(x-h)^i,$$

where $i!$ denotes the factorial of i and $f^{(i)}(h)$ denotes the value of the ith-order derivative of $f(x)$ at the point h. Note that $P_K(x)$ is the so-called K-order Taylor series of function $f(x)$. In the special case with $h = 0$, $P_K(x)$ is called the Maclaurin series (Zhang et al., 2012e).

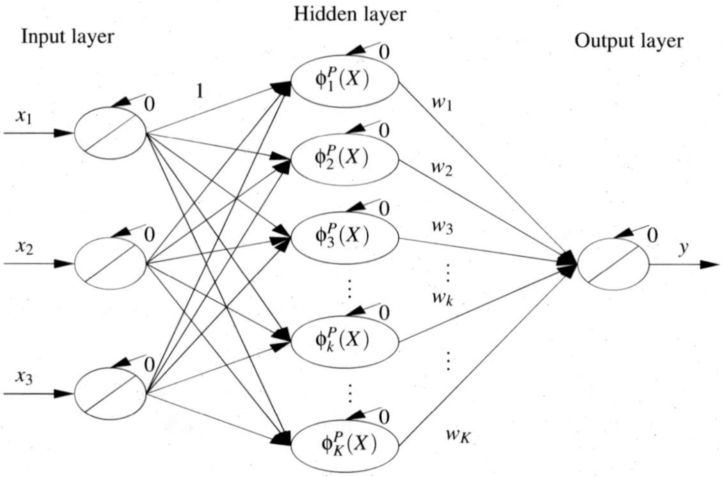

Figure 22. Model structure of the 3IPFN.

Proposition 3. The theorem of Taylor polynomial approximation can be generalized to functions with more than one variable (Zhang et al., 2012e). For a target function $f(x_1,\cdots,x_g)$ with g variables, which has the $(K+1)$th-order continuous partial derivatives in a neighborhood of the origin $(0,\cdots,0)$, the K-order Taylor series $P_K(x_1,\cdots,x_g)$ about the origin is

$$P_K(x_1,\cdots,x_g)=\sum_{i=0}^{K}\sum_{i_1+\cdots+i_g=i}\frac{x_1^{i_1}\cdots x_g^{i_g}}{i_1!\cdots i_g!}\left(\frac{\partial^{i_1+\cdots+i_g}f(0,\cdots,0)}{\partial x_1^{i_1}\cdots\partial x_g^{i_g}}\right),$$

where $i_1,\cdots,i_g$ are nonnegative integers.

In this section, a continuous real target function $f(x_1,x_2,x_3)$ with three variables is investigated. Note that the target function $f(x_1,x_2,x_3)$ has the $(K+1)$th-order continuous partial derivatives in a neighborhood of the origin $(0,0,0)$. According to the above result, the K-order Taylor series $P_K(x_1,x_2,x_3)$ of $f(x_1,x_2,x_3)$ about the origin $(0,0,0)$ is

$$P_K(x_1,x_2,x_3)=\sum_{i=0}^{K}\sum_{i_1+i_2+i_3=i}\frac{x_1^{i_1}x_2^{i_2}x_3^{i_3}}{i_1!i_2!i_3!}\left(\frac{\partial^{i_1+i_2+i_3}f(0,0,0)}{\partial x_1^{i_1}\partial x_2^{i_2}\partial x_3^{i_3}}\right),$$

which can be used to approximate $f(x_1,x_2,x_3)$, i.e.,

$$f(x_1,x_2,x_3)\approx P_K(x_1,x_2,x_3).$$

In addition, $P_K(x_1,x_2,x_3)$ can be rewritten in the form of K-order power series, thus

$$\begin{aligned}f(x_1,x_2,x_3)\approx\,&w_{0,0,0}x_1^0x_2^0x_3^0+w_{1,0,0}x_1^1x_2^0x_3^0+w_{0,1,0}x_1^0x_2^1x_3^0+w_{0,0,1}x_1^0x_2^0x_3^1+w_{2,0,0}x_1^2x_2^0x_3^0\\&+w_{1,1,0}x_1^1x_2^1x_3^0+w_{1,0,1}x_1^1x_2^0x_3^1+w_{0,2,0}x_1^0x_2^2x_3^0+w_{0,1,1}x_1^0x_2^1x_3^1+w_{0,0,2}x_1^0x_2^0x_3^2\\&+\cdots+w_{i_1,i_2,i_3}x_1^{i_1}x_2^{i_2}x_3^{i_3}+\cdots+w_{K,0,0}x_1^Kx_2^0x_3^0+\cdots+w_{0,0,K}x_1^0x_2^0x_3^K,\end{aligned}$$

where the weight for $x_1^{i_1}x_2^{i_2}x_3^{i_3}$ can be expressed as

$$w_{i_1,i_2,i_3} = \frac{1}{i_1!i_2!i_3!}\left(\frac{\partial^{i_1+i_2+i_3} f(0,0,0)}{\partial x_1^{i_1}\partial x_2^{i_2}\partial x_3^{i_3}}\right).$$

Thus, an unknown target function $f(x_1,x_2,x_3)$ can be best estimated via the optimal weights $\{w_{i_1,i_2,i_3}\}$ corresponding to basis functions $\{x_1^{i_1}x_2^{i_2}x_3^{i_3}\}$.

In addition, the basis functions $\{x_1^{i_1}x_2^{i_2}x_3^{i_3}\}$ are sequenced by a new method detailed as below, which is similar to the graded lexicographic order (Dunkl and Xu, 2001; Zhang et al., 2012e) to some extent.

Let $\phi_k^P(X) = \phi_k^P(x_1,x_2,x_3) = x_1^{i_1}x_2^{i_2}x_3^{i_3}$ and $\phi_{k'}^P(X) = \phi_{k'}^P(x_1,x_2,x_3) = x_1^{i'_1}x_2^{i'_2}x_3^{i'_3}$ be two different basis functions sequenced by the new method, where k and k' are two different nonnegative integers. Then, we say that $k > k'$ if either of the following two conditions is satisfied, i.e.,

condition 1: $(i_1+i_2+i_3) > (i'_1+i'_2+i'_3)$;

condition 2: $(i_1+i_2+i_3) = (i'_1+i'_2+i'_3)$ and the first nonzero element of the difference $(i_1-i'_1, i_2-i'_2, i_3-i'_3)$ is negative.

For better understanding, the first 10 basis functions $\{\phi_k^P(X)|k=1,2,\cdots,10\}$ sequenced by the new method are listed below:

$$\begin{cases} \phi_1^P(X) = x_1^0x_2^0x_3^0,\ \phi_2^P(X) = x_1^1x_2^0x_3^0,\ \phi_3^P(X) = x_1^0x_2^1x_3^0,\ \phi_4^P(X) = x_1^0x_2^0x_3^1, \\ \phi_5^P(X) = x_1^2x_2^0x_3^0,\ \phi_6^P(X) = x_1^1x_2^1x_3^0,\ \phi_7^P(X) = x_1^1x_2^0x_3^1,\ \phi_8^P(X) = x_1^0x_2^2x_3^0, \\ \phi_9^P(X) = x_1^0x_2^1x_3^1,\ \phi_{10}^P(X) = x_1^0x_2^0x_3^2. \end{cases}$$

In light of the above theoretical analysis, the unknown target function $f(x_1,x_2,x_3)$ can be approximated by an approximator $y = \sum_k^K w_k\phi_k^P(X)$, where K denotes the number of basis functions used to approximate $f(x_1,x_2,x_3)$ and $\{w_k|k=1,2,\cdots,K\}$ denote the corresponding weights for the basis functions $\{\phi_k^P(X)|k=1,2,\cdots,K\}$.

3.2.2. 3IPFN Model and WASD Algorithm

In this subsubsection, the model of the 3IPFN is constructed, and then the WASD algorithm is developed to achieve the superior performance of the 3IPFN.

According to the above theoretical basis and analysis, the 3IPFN is proposed with its structure shown in Figure 22. As seen from Figure 22, the input-layer and output-layer neurons are activated by simple linear identical functions. In addition, the hidden layer has K neurons, which are activated by a group of basis functions $\{\phi_k^P(X)|k=1,2,\cdots,K\}$. The weights linking the input-layer and hidden-layer neurons are set to be 1, and $\{w_k|k = 1,2,\cdots,K\}$ shown in Figure 22 denote the weights linking the hidden-layer and output-layer neurons which should be determined. All neuronal thresholds are set to be 0.

In order to achieve the superior performance of the proposed 3IPFN model, the WASD algorithm is presented. The WASD algorithm can obtain the optimal weights from hidden layer to output layer directly by utilizing the so-called WDD method. Moreover, the WASD algorithm can determine the optimal structure (i.e., the optimal number of hidden-layer neurons) of the 3IPFN adaptively by growing and pruning hidden-layer neurons during the training process.

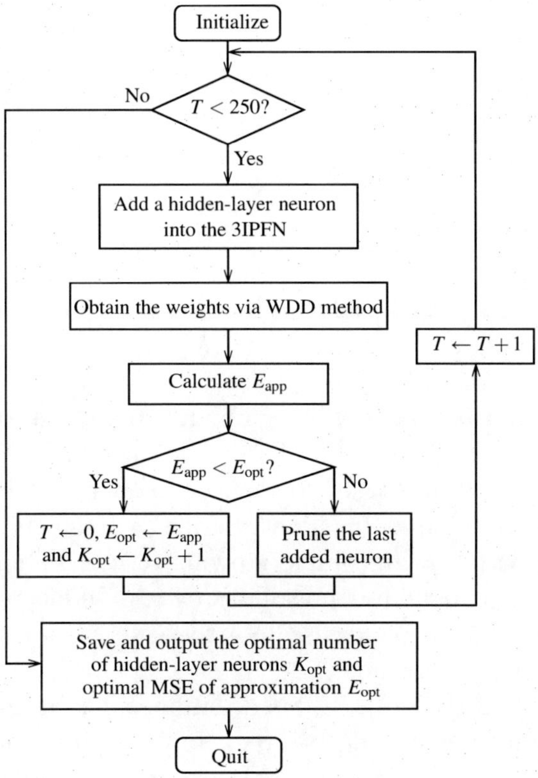

Figure 23. Flowchart of the WASD algorithm applied to the 3IPFN. *Reproduced from Y. Zhang, W. Lao et al., Weights and structure determination of pruning-while-growing type for 3-Input power-activation feed-forward neuronet, Figure 2, Proceedings of International Conference on Automation and Logistics, pp. 212-217. ©[2012] IEEE. Reprinted, with permission.*

For the purpose of illustration, a set of sample data is given as $\{(X_p, \gamma_p)|X_p \in R^3, \gamma_p \in R, p = 1, 2, \cdots, Q\}$, where $X_p = [X_{p1}, X_{p2}, X_{p3}]^{\mathrm{T}}$ denotes the pth input vector, and γ_p denotes the corresponding desired output. Note that Q is the total number of samples. Then, for the 3IPFN, the MSE function can be expressed based on (15). In addition, a novel WASD algorithm is designed to determine the optimal structure of the 3IPFN, which grows and prunes hidden-layer neurons according to the changes of the MSE of approximation (i.e., E_{app}) during the training process. For better understanding, the flowchart of the WASD algorithm is shown in Figure 23. It is worth pointing out that T shown in Figure 23 denotes a counter and is initially set to be zero. Besides, E_{opt} shown in Figure 23 is initially set to be large enough.

Actually before proposing the WASD algorithm, we have fully investigated the relationship between E_{app} (i.e., the approximation ability) and the number of hidden-layer neurons K of the 3IPFN via a variety of target functions. The following three target functions are chosen for the purpose of better illustration:

$$f(x_1,x_2,x_3) = 10x_2x_3\sin(x_1)e^{-2x_1^2-2x_2^2}+5, \tag{20}$$

$$f(x_1,x_2,x_3) = x_2e^{x_1\sin(\pi x_1x_3)}+x_1x_3^2+10, \tag{21}$$

$$f(x_1,x_2,x_3) = 0.05x_3e^{x_2}\sin(x_1)/(10-x_2^2)+\sin(x_1x_2+x_1x_3)+30. \tag{22}$$

The numerical results on target functions (20)–(22) are shown in Figure 24 which illustrate clearly the relationship between E_{app} and the number of hidden-layer neurons (i.e., K). As seen from Figure 24, all curves decline like the shape of ladder before the particular point $(K_{\min}, E_{\min})$, and, after the particular point $(K_{\min}, E_{\min})$, the curves show an increasing trend with oscillation, where $K_{\min}$ denotes the number of hidden-layer neurons with the minimal MSE of approximation $E_{\min}$. It is worth mentioning that such characteristics can be found in the training process of many other target functions. It means that there exists a structure of the 3IPFN with the best performance in approximation [i.e., the particular point $(K_{\min}, E_{\min})$] for most of target functions. Besides, we can obtain this particular point $(K_{\min}, E_{\min})$ easily via the following simple growing-type (GT) algorithm: 1) record the point with the minimal E_{app} found by far as the number of hidden-layer neurons increases one by one; 2) when E_{app} begins to increase, look forward 250 points (or say, 250 neurons) to make sure that $(K_{\min}, E_{\min})$ is found.

Unfortunately, the structure of the 3IPFN determined via the above-mentioned GT algorithm is fat, since the number of hidden-layer neurons is pretty large, which may lead to a very high computational complexity. To propose a more effective algorithm for the determination of the number of hidden-layer neurons, numerical results on a variety of target functions are further investigated. As seen from Figure 24, the increase of the hidden-layer neurons K may not result in the corresponding decrease in the E_{app} before the particular point $(K_{\min}, E_{\min})$, which implies that some hidden-layer neurons are redundant for data approximation. In view of the similar characteristics of many other target functions, the WASD algorithm is developed to prune the redundant hidden-layer neurons (while growing) during the training process. As mentioned above, the MSE is an important argument for measuring the performance of the 3IPFN. Thus, such a WASD algorithm of pruning-while-growing (PWG) type is designed to determine the optimal number of hidden-layer neurons by growing and pruning the hidden-layer neurons according to the changes of the MSE of approximation (i.e., E_{app}). Specifically speaking, if the newly-added hidden-layer neuron makes the 3IPFN achieve better performance in approximation (or say, makes the MSE of approximation decrease), reserve the new hidden-layer neuron; otherwise, prune this newly-added hidden-layer neuron. In this way, the WASD algorithm can determine the optimal structure (i.e., the optimal number of hidden-layer neurons K_{opt}) of the 3IPFN with its efficacy demonstrated in Figure 25.

3.2.3. Numerical Studies of 3IPFN

In this subsection, numerical experiments are performed to substantiate the efficacy of the 3IPFN equipped with the WASD algorithm in terms of data approximation, validation and prediction. In consideration of space limitation and similarity of results, numerical results are shown only for the aforementioned target functions (20)–(22).

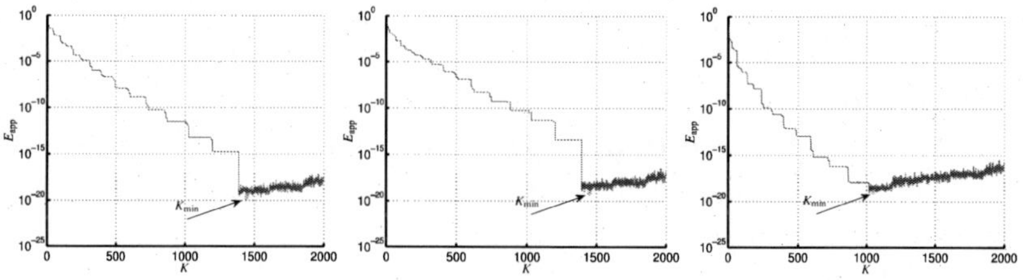

(a) For target function (20) (b) For target function (21) (c) For target function (22)

Figure 24. Relationship between the MSE of approximation (i.e., E_{app}) and the number of hidden-layer neurons (i.e., K). *Reproduced from Y. Zhang, W. Lao et al., Weights and structure determination of pruning-while-growing type for 3-Input power-activation feed-forward neuronet, Figure 3, Proceedings of International Conference on Automation and Logistics, pp. 212-217. ©[2012] IEEE. Reprinted, with permission.*

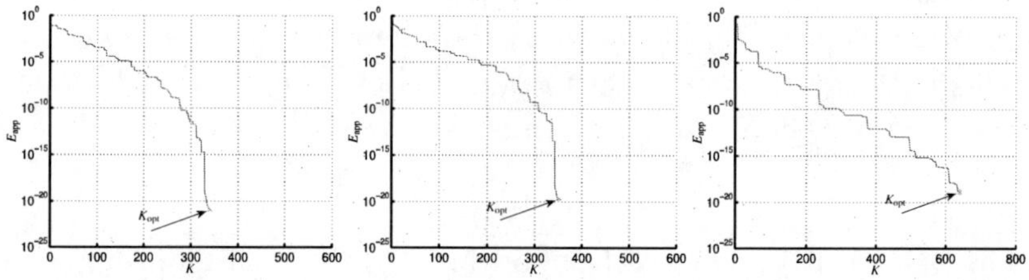

(a) For target function (20) (b) For target function (21) (c) For target function (22)

Figure 25. Finding the optimal number of hidden-layer neurons for the 3IPFN via the WASD algorithm. *Reproduced from Y. Zhang, W. Lao et al., Weights and structure determination of pruning-while-growing type for 3-Input power-activation feed-forward neuronet, Figure 4, Proceedings of International Conference on Automation and Logistics, pp. 212-217. ©[2012] IEEE. Reprinted, with permission.*

1) Approximation

To verify the superior approximation ability of the 3IPFN equipped with the WASD algorithm, numerical results about data approximation are presented and analyzed. For each target function, we sample uniformly over the region $[-1,0.8]^3$ with gap size 0.2 to generate a set of sample data $\{(X_p,\gamma_p)|X_p \in R^3, \gamma_p \in R, p=1,2,\cdots,1000\}$ (i.e., the number of samples $Q=1000$) for training. For convenience, the 3IPFN equipped with the PWG-type WASD algorithm and the 3IPFN equipped with the pure GT algorithm are expressed as the 3IPFN-W and the 3IPFN-G, respectively. Via numerical experiments on target function (20)–(22), the comparison of approximation performance between the 3IPFN-W and the 3IPFN-G is provided in Table 10. As shown in the table, the optimal number of hidden-layer neurons determined via the WASD algorithm for each target function (i.e., K_{opt}) is less than 650, while the number of hidden-layer neurons determined via the GT algorithm for each target function (i.e., $K_{\min}$) is more than 1000. Moreover, the MSE of approximation corresponding to K_{opt} (i.e., E_{opt}) is smaller than that corresponding to $K_{\min}$ (i.e., $E_{\min}$) for each target function. In other words, the 3IPFN-W achieves better performance on ap-

Table 10. Approximation results of the 3IPFN with the WASD or GT algorithm. *Reproduced from Y. Zhang, W. Lao et al., Weights and structure determination of pruning-while-growing type for 3-Input power-activation feed-forward neuronet, Table I, Proceedings of International Conference on Automation and Logistics, pp. 212-217. ©[2012] IEEE. Reprinted, with permission.*

Target function	PWG-type WASD algorithm		Pure GT algorithm	
	K_{opt}	E_{opt}	$K_{\min}$	$E_{\min}$
(20)	340	1.2717×10^{-21}	1447	1.9324×10^{-20}
(21)	351	1.8979×10^{-20}	1447	7.3417×10^{-20}
(22)	640	1.3804×10^{-19}	1048	1.2433×10^{-18}

Table 11. Validation results of the 3IPFN with the WASD or GT algorithm. *Reproduced from Y. Zhang, W. Lao et al., Weights and structure determination of pruning-while-growing type for 3-Input power-activation feed-forward neuronet, Table II, Proceedings of International Conference on Automation and Logistics, pp. 212-217. ©[2012] IEEE. Reprinted, with permission.*

Target function	PWG-type WASD algorithm		Pure GT algorithm	
	t_{val} (s)	E_{val}	t_{val} (s)	E_{val}
(20)	2.24	1.6471×10^{-6}	32.03	2.4867×10^{-5}
(21)	2.29	2.0294×10^{-7}	32.54	9.0050×10^{-7}
(22)	7.30	1.0279×10^{-12}	17.21	1.2193×10^{-10}

proximation with simpler structure (or say, fewer hidden-layer neurons) compared with the 3IPFN-G. Together with the above analysis, the numerical results have verified the efficacy of the 3IPFN-W on data approximation.

2) Validation and Prediction

To further demonstrate the superiority of the 3IPFN-W on validation (or say, generalization), a set of sample data with 1331 samples is generated by sampling uniformly over region $[-0.85, 0.75]^3$ with gap size 0.15. The numerical results on validation are shown in Table 11. From the table, we can see that the 3IPFN-W achieves superior performance in terms of validation with the MSE of validation (i.e., E_{val}) being very small (i.e., of order $10^{-6} \sim 10^{-12}$), which also demonstrates the good generalization ability of the 3IPFN-W. Besides, Table 11 shows that the 3IPFN-W has smaller MSE of validation and much shorter validation time (i.e., t_{val}) than the 3IPFN-G for each target function, which further verifies the better efficacy and the less computational complexity of the 3IPFN-W on validation.

Following the above numerical experiments about the generalization ability, we further investigate the prediction ability of the 3IPFN-W. To perform numerical experiments about prediction, some samples are selected beyond region $[-1, 0.8]^3$. Due to space limitation,

Table 12. Prediction results of the 3IPFN with the WASD for function (22). *Reproduced from Y. Zhang, W. Lao et al., Weights and structure determination of pruning-while-growing type for 3-Input power-activation feed-forward neuronet, Table III, Proceedings of International Conference on Automation and Logistics, pp. 212-217. ©[2012] IEEE. Reprinted, with permission.*

x_1	x_2	x_3	y_{pre}	γ_{pre}
0.85	0.85	0.85	31.00010686784	31.00015019266
0.85	0.90	0.90	31.00805437999	31.00821618498
0.90	0.90	0.95	31.00508232559	31.00552443671
0.90	0.95	0.95	31.00040031355	31.00090214440
0.95	0.95	1.00	30.97094431295	30.97214275323
−1.05	0.85	0.85	29.01362877198	29.01355709640
−1.05	−1.05	0.85	30.20708959075	30.20700998407
−1.10	0.90	−0.85	29.95519878420	29.95516490634
0.90	−1.05	−1.10	29.06335040214	29.06389783763
1.00	−1.10	0.90	29.80251202694	29.80276463520
−1.05	−1.05	−1.05	30.80756705825	30.80733487673
−1.10	−1.05	−1.10	30.70376596239	30.70278082275
−1.15	−1.15	−1.10	30.52928221080	30.52800377075
−1.20	−1.15	−1.15	30.37803454799	30.37435458838
−1.20	−1.20	−1.20	30.26924215768	30.26058704670

Table 13. Numerical results of further simplification on the 3IPFN-W. *Reproduced from Y. Zhang, W. Lao et al., Weights and structure determination of pruning-while-growing type for 3-Input power-activation feed-forward neuronet, Table IV, Proceedings of International Conference on Automation and Logistics, pp. 212-217. ©[2012] IEEE. Reprinted, with permission.*

	Target function		
	(20)	(21)	(22)
K_{sim}	79	101	635
E_{app}	2.1513×10^{-23}	6.4547×10^{-21}	1.8787×10^{-19}
E_{val}	1.6471×10^{-6}	2.0294×10^{-7}	1.0279×10^{-12}
t_{val} (s)	0.17	0.26	7.18

prediction results of the 3IPFN-W are shown only for target function (22), which are given in Table 12. From the table, we can easily find that the output of the 3IPFN-W (i.e., y_{pre}) is approximately equal to the desired output (i.e., γ_{pre}) for each sample. Note that, for the selected samples shown in Table 12, the MSE of prediction E_{pre} is less than 10^{-4}, which further substantiates the efficacy of the proposed 3IPFN-W on prediction.

In summary, we can conclude from Tables 11 and 12 that the proposed 3IPFN-W achieves good performances in terms of validation and prediction.

3) Further Simplification

As mentioned above, the WASD algorithm is designed to prune redundant hidden-layer neurons during the training process in order to determine the structure of the 3IPFN. Besides, the above numerical results have substantiated that the 3IPFN-W has better performances and a simpler structure than the 3IPFN-G which does not include a pruning process. As an extension and exploration, we further prune the redundant hidden-layer neurons of the 3IPFN-W in expectation for even better performances and a even simpler structure, i.e., the further simplification of the 3IPFN-W. Specifically speaking, first of all, for a determined structure of the 3IPFN-W (i.e., K_{opt}), the influence factor δ_k of the kth hidden-layer neuron to the neuronet output is defined as

$$\delta_k = \frac{|\text{sum}(w_k \boldsymbol{C}_k)|}{\sum_{n=1}^{K_{\text{opt}}} |\text{sum}(w_n \boldsymbol{C}_n)|},$$

where $\boldsymbol{C}_k = [\phi_k^P(X_1), \phi_k^P(X_2), \cdots, \phi_k^P(X_Q)]^{\text{T}}$ denotes the output vector of the kth hidden-layer neuron, and $\text{sum}(w_k \boldsymbol{C}_k)$ denotes the summation of the elements of the weighted output vector $w_k \boldsymbol{C}_k$; so do $\boldsymbol{C}_n$ and $\text{sum}(w_n \boldsymbol{C}_n)$; and $|\cdot|$ denotes the absolute value of a scalar argument. Then, the 3IPFN with a further simpler structure can be obtained by pruning hidden-layer neurons, of which the influence factors are small enough (e.g., less than 10^{-12}). Moreover, the corresponding numerical results are given in Table 13. Comparing Table 10 and Table 13, we see that more than 70% of the hidden-layer neurons have been deleted for target functions (20) and (21), and 5 hidden-layer neurons have been deleted for target function (22). In addition, the 3IPFN has better approximation performance for target functions (20) and (21), which is illustrated via Table 10 and Table 13. Moreover, with the structure being further simpler, the 3IPFN has shorter validation time and barely-changed validation performance for target functions (20)-(22), which is illustrated via Table 11 and Table 13.

4. WASD of AN with More Inputs

Based on the above neuronets with two or three inputs, we can further investigate multiple-input neuronets. In this section, two 4-input neuronets are proposed and then two neuronets with any number of inputs are presented.

4.1. Neuronet of 4-input Bernoulli Polynomials

This subsection constructs a novel 4-input Bernoulli polynomials neuronet (4IBPN) on the basis of function approximation theory. Then, a corresponding WASD algorithm can also be obtained to further determine the optimal weights and structure for the 4IBPN.

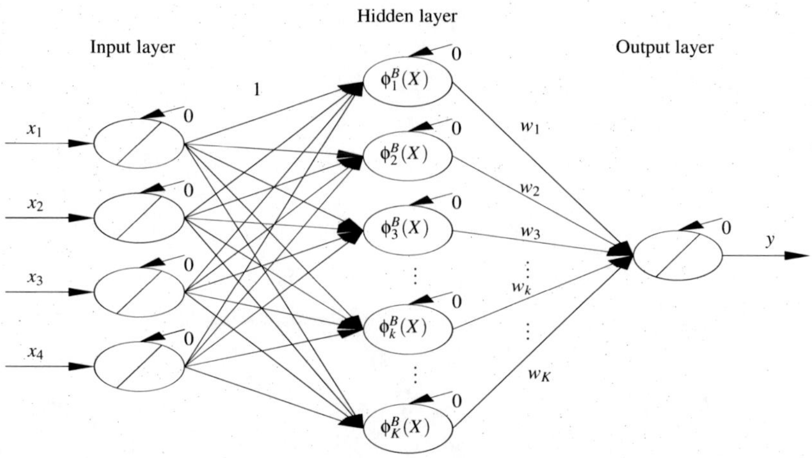

Figure 26. Model structure of the 4IBPN.

4.1.1. Theoretical Basis and Model

According to the previous work (Zhang and Ruan, 2009), Bernoulli polynomial can be defined as follow with $\varphi_0^B(x) = 1$:

$$\varphi_j^B(x) = x^j - \frac{1}{j+1}\sum_{k=0}^{j-1}\left(\frac{(j+1)!}{k!(j+1-k)!}\varphi_k^B(x)\right),$$

where $\varphi_j^B(x)$ denotes a Bernoulli polynomial of degree j, with $j = 1, 2, 3, \cdots, n, \cdots$.

Considering the excellent approximation ability of the Weierstrass polynomial (Liang and Li, 2005; Peng and Li, 2009; Liu et al., 2009), we can use a known polynomial (e.g., Bernoulli polynomial) to construct the Weierstrass polynomial. Following this recurrence expression of Bernoulli polynomial, we can obtain a proposition for the 4IBPN.

Proposition 4. For the given values of q and v_q, a polynomial constructed by Bernoulli polynomials can be used to approximate the target function $f(x_1, x_2, \cdots, x_k)$, once the v_qth degree of the qth group of Bernoulli polynomial [i.e., $\varphi_{v_q}^B(x_q)$] replaces $p_{n_q,v_q}(x_q)$ in (14), where $x_q \in [a, b]$ ($a, b \in R$ and $a \leq b$).

Without loss of generality, we focus on the function with four variables in this subsection. That is, according to above the proposition, we have

$$\begin{aligned} f(x_1, x_2, x_3, x_4) &\approx \sum_{v_1=0}^{K_1-1}\sum_{v_2=0}^{K_2-1}\sum_{v_3=0}^{K_3-1}\sum_{v_4=0}^{K_4-1} w_{v_1,v_2,v_3,v_4}\varphi_{v_1}^B(x_1)\varphi_{v_2}^B(x_2)\varphi_{v_3}^B(x_3)\varphi_{v_4}^B(x_4) \\ &= \sum_{k=1}^{K} w_k \phi_k^B(x_1, x_2, x_3, x_4), \end{aligned}$$

where K_1, K_2, K_3 and K_4 denote the number of Bernoulli polynomials used to respectively replace $\{p_{n_1,v_1}(x_1),\ v_1 = 0, 1, \cdots, n_1\}$, $\{p_{n_2,v_2}(x_2),\ v_2 = 0, 1, \cdots, n_2\}$, $\{p_{n_3,v_3}(x_3),\ v_3 = 0, 1, \cdots, n_3\}$ and $\{p_{n_4,v_4}(x_4),\ v_4 = 0, 1, \cdots, n_4\}$, w_{v_1,v_2,v_3,v_4} corresponds to

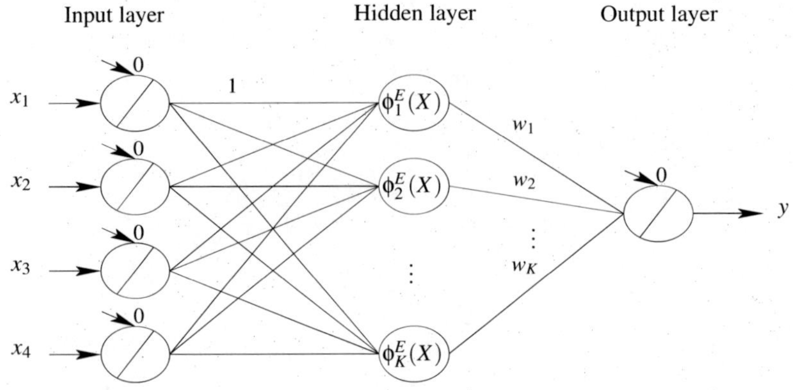

Figure 27. Model structure of the 4IEPN.

$f(v_1/n_1, v_2/n_2, v_3/n_3, v_4/n_4)$ in (14) and $K = K_1K_2K_3K_4$ is the total number of products of Bernoulli polynomials. In addition, $w_k = w_{v_1,v_2,v_3,v_4}$ denotes the weight for $\phi_k^B(x_1,x_2,x_3,x_4)$, and $\phi_k^B(X) = \phi_k^B(x_1,x_2,x_3,x_4) = \varphi_{v_1}^B(x_1)\varphi_{v_2}^B(x_2)\varphi_{v_3}^B(x_3)\varphi_{v_4}^B(x_4)$.

As a result, the target function $f(x_1,x_2,x_3,x_4)$ can be uniformly approximated by the polynomial constructed above, once we obtain the optimal weights $\{w_k\}$ for basis functions $\{\phi_k^B(X)\}$ as well as their most fitting number. Figure 26 shows the model structure of the 4IBPN. The neurons in input and output layers are activated by simple linear function $h(x) = x$, whereas the activation function in each hidden-layer neuron is the product of four Bernoulli polynomials. To simplify the circuit implementation and the computational complexity, all neuronal thresholds in the 4IBPN are set to be 0 and the weights between input and hidden layers are set to be 1.

4.1.2. WASD Algorithm for 4IBPN

Based on the authors' previous work, the WDD method can determine the weights between hidden and output layers directly within one step, which eschews the lengthy and inefficient iterative computation. By exploiting WDD method, the corresponding WASD algorithm can be developed for the 4IBPN, which is similar to the investigation of neuronets with two or three inputs. It is worth pointing out that the 4IBPN equipped with the WASD algorithm also has superior performance in data approximation and generalization.

4.2. Neuronet of 4-input Euler Polynomials

In this subsection, a novel 4-input Euler-polynomial neuronet (4IEPN) is established and discussed. Based on the 3IEPN, we can further present the 4IEPN, which is shown as Figure 27. To simplify the neuronet model and its computation, all the linking weights between input layer and hidden layer are set to be 1 and all the thresholds of neurons are fixed to 0. The linking weights between hidden layer and output layer are $\{w_k\}_{k=1}^K$, which can be determined directly by the WDD method. Besides, the hidden-layer neurons are activated by the basis functions $\{\phi_k^E(X) = \phi_k^E(x_1,x_2,x_3,x_4) = \varphi_{k_1}^E(x_1)\varphi_{k_2}^E(x_2)\varphi_{k_3}^E(x_3)\varphi_{k_4}^E(x_4)\}_{k=1}^K$ and the neurons of the input and output layers are activated by the linear function $h(x) = x$.

Based on the 4IEPN, a WASD algorithm can be designed to obtain the optimal number of hidden-layer neurons. As a matter of fact, numerical results can also further verify the superior performance of the proposed 4IEPN equipped with the WASD algorithm in terms of training, testing and prediction.

4.3. Neuronet of Multiple-input Chebyshev Polynomials

A new type of multiple-input neuronet activated by Chebyshev polynomials (Class 1 or 2) (MINCP) is presented in this subsection. By combining the presented WDD method with pruning technique, a WASD algorithm can also be obtained.

Based on the above definition of the Chebyshev polynomials and the Weierstrass approximation theorem, we can expand two-input or three-input neuronets (the 2INCP1, the 2INCP2 and the 3IEPN) to multiple-input neuronets, which is shown in Figure 28. As seen from Figure 28, the MINCP model adopts a conventional three-layer structure, including the input layer, the hidden layer and the output layer. For the MINCP with D inputs and one output, it has D neurons in the input layer and one neuron in the output layer, which are activated by a simple linear function. The hidden-layer neurons of the MINCP are activated by a group of basis functions $\{\phi_k^C(X) \mid k = 1, 2, \cdots, K\}$, where K denotes the number of hidden-layer neurons, $\phi^C(\cdot)$ denotes the product of Chebyshev polynomials with the graded lexicographic order and $X := [x_1\ x_2\ \cdots\ x_D]^{\mathrm{T}} \in [0,1]^{D\times 1}$. Besides, the linking weights between the input-layer and hidden-layer neurons are all fixed to be 1. In addition, all neuronal thresholds are all fixed to be 0. Based on the WDD method and the pruning technique, a WASD algorithm can be obtained for the MINCP. Note that the multiple-input neuronets (i.e., MINCP) equipped with WASD algorithm can also be applied on the classification.

4.4. Neuronet of Multiple-input Hermit Polynomials

In this subsection, we propose a new type of multiple-input Hermit orthogonal polynomials neuronet (MIHOPN) model by giving the definition of Hermit orthogonal polynomials (HOP). The HOP can be defined as below (Zhang et al., 2012f).

Definition 5. The jth Hermit orthogonal polynomial $\varphi_j^H(x)$ with $0 \leq x < +\infty$ satisfies the following recursive formula:

$$\varphi_j^H(x) = 2x\varphi_{j-1}^H(x) - 2(j-2)\varphi_{j-2}^H(x),$$

where $\varphi_1^H(x) = 1,\ \varphi_2^H(x) = 2x,\ \varphi_3^H(x) = 4x^2 - 2,\ \varphi_4^H(x) = 8x^3 - 12x,$

$$\varphi_5^H(x) = 16x^4 - 48x^2 + 12,\ \varphi_6^H(x) = 32x^5 - 160x^3 + 120x,\ \cdots$$

Based on the above definition, we can further put forward a new type of MIHOPNN model consisting of three layers, which is shown in Figure 29. The weights from input layer to hidden layer are randomly generated between 0 and 1 (Zhang et al., 2010a; Zhang et al., 2012f). Besides, the thresholds of all neurons are set to be 0. The input-layer and output-layer neurons are activated by the linear function $h(x) = x$, while the hidden-layer neurons are activated by a group of HOP.

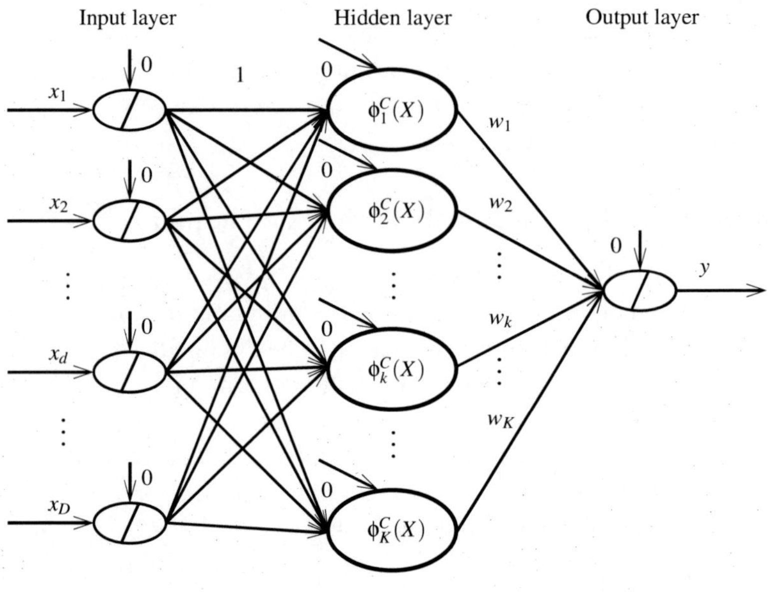

Figure 28. Model structure of the MINCP.

As mentioned, the performance of BP-type neural networks varies with the number of hidden-layer neurons. Thus, a WASD algorithm is designed to determine the optimal number of hidden-layer neurons as well as the hidden-to-output weights (Zhang et al., 2012f). Numerical results shown in (Zhang et al., 2012f), have verified the conclusion that the MI-HOPN equipped with the WASD algorithm has good training and generalization abilities.

Conclusion

As self-organizing systems, artificial neuronets (AN) have been investigated in terms of modeling the optimal structure with best linking weights for three-layer AN. Differing from algorithmic improvements about the BP iterative-training procedure, the method of using linearly-independent or orthogonal polynomials as activation functions and improving the conciseness of the neuronet structure to make AN achieve better performance has been adequately investigated, analyzed and verified in this chapter. Note that the neuronets proposed and investigated in the chapter are two-, three- or multiple-input AN, which are consistent with the fact that multiple-input systems are the most frequently encountered in practical applications. Based on the weights-direct-determination (WDD) method, the authors have proposed a corresponding weights and structure determination (WASD) algorithm for each presented neuronet. In addition, for better performance (e.g., more efficiency and conciseness in self-organizing systems), the authors have proposed and investigated pruning techniques in the neuronet structure determination. Numerical results have further substantiated the efficacy of the proposed neuronets equipped with the corresponding WASD algorithms, which shows the better performance in terms of training (or say, approximation or learning), generalization (or say, testing or validation) and prediction. In summary, it has

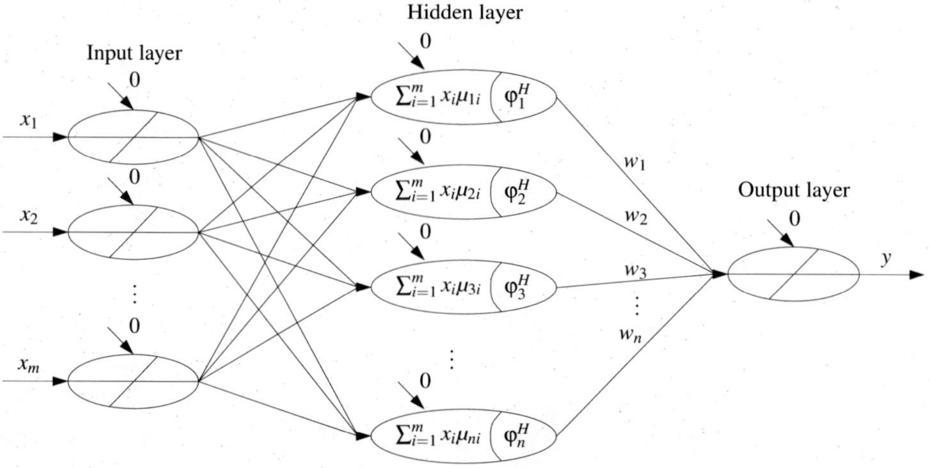

Figure 29. The MIHOPN model with input-to-hidden weights being positively randomly generated. *Reproduced from Y. Zhang, J. Chen et al., Weights and structure determination (WASD) of multiple-input Hermit orthogonal polynomials neural network, Figure 1, Proceedings of 24th Chinese Control and Decision Conference, pp. 1106-1111. ©[2012] IEEE. Reprinted, with permission.*

been adequately investigated and verified that designing and adopting the WASD method for the presented various new types of neuronets to overcome the conventional weaknesses is efficacy in the self-organizing systems of neuronets.

Acknowledgments

This work is supported by the National Natural Science Foundation of China (under grants 61075121 and 60935001), the Specialized Research Fund for the Doctoral Program of Institutions of Higher Education of China (with project number 20100171110045), and also the National Innovation Training Program for University Students (under grant 201210558042). The authors would like to thank the editor (WenJun Zhang) of the book, "Self-organizing: Theories and Methods", for his invitation to write this chapter.

References

Ding S.F., Su C.Y., Yu J.Z. 2011. An optimizing BP neural network algorithm based on genetic algorithm. *Artificial Intelligence Review*, 36: 153-162.

Dunkl C.F., Xu Y. 2001. Orthogonal Polynomials of Several Variables, Cambridge University Press, Cambridge.

Fu Y., Chai T.Y. 2007. Nonlinear multivariable adaptive control using multiple models and neural networks. *Automatica*, 43: 1101-1110.

Jenkins W.M. 2006. Neural network weight training by mutation, *Computers and Structures*, 84: 2107-2112.

Liang X.Z., Li Q. 2005. Multivariate Approximation. 1st Chinese Edition, National Defence Industry Press, Beijing, China.

Liu L.Y., Li C.H., Wu Z.H. 2009. Weierstrass approximation theorem in the promotion of multi-function. *Journal of Nanyang Institute of Technology*, 1(6): 85-87.

Mathews J.H., Fink K.D. 2007. Numerical Methods Using MATLAB. 4th Edition, Publishing House of Electronics Industry, Beijing, China.

Narendra K.S., Mukhopadhyay S. 1993. Adaptive control of nonlinear multivariable systems using neural networks. In: Proceedings of the 32nd Conference on Decision and Control, pp. 3066-3097.

Peng P.R., Li Z.Q. 2009. Extension and application on Weierstrass approximation theorem. *Journal of Henan Institute of Education (Natural Science)*, 18(1): 1-2.

Sadeghi B.H.M. 2000. A BP-neural network predictor model for plastic injection molding process. *Journal of Materials Processing Technology*, 103: 411-416.

Su J.M., Ruan S.Y. 2008. MATLAB Practical Guide, Publishing House of Electronics Industry, Beijing, China.

Wang L., Sui T.Z. 2007. Application of data mining technology based on neural network in the engineering. In: Proceedings of International Conference on Wireless Communications, Networking and Mobile Computing, pp. 5544-5547.

Wilson D.R., Martinez T.R. 2001. The need for small learning rates on large problems. In: Proceedings of International Joint Conference on Neural Networks, pp. 115-119.

Xu L.Z., Yang J.X. 1987. A survey of recent development of multivariate approximation. *Advances in Mathematics*, 16(3): 241-249.

Yu X.H., Chen G.A. 1995. On the local minima free condition of BP learning. *IEEE Transactions on Neural Networks*, 6(5): 1300-1303.

Zhang Y.N., Chen J.H., Guo D.S., Yin Y.H., Lao W.C. 2012b. Growing-type weights and structure determination of 2-input Legendre orthogonal polynomial neuronet. In: Proceedings of IEEE International Symposium on Industrial Electronics, pp. 852-857.

Zhang Y.N., Chen J.W., Fu S.B., Xiao L., Yu X.T. 2012f. Weights and structure determination (WASD) of multiple-input Hermit orthogonal polynomials neural network. In: Proceedings of the 24th Chinese Control and Decision Conference, pp. 1106-1111.

Zhang Y.N., Lao W.C., Yin Y.H., Xiao L., Chen J.H. 2012e. Weights and structure determination of pruning-while-growing type for 3-Input power-activation feed-forward neuronet. In: Proceedings of IEEE International Conference on Automation and Logistics, pp. 212-217.

Zhang Y.N., Li J., Zhang Z.J., Ruan G.Q., Jiang X.H. 2010b. A weights-direct-determination method and structure-automatic-determination algorithm for SIMO trigonometrically-activated fourier neural networks. *Information and Control*, 39(4): 1-8.

Zhang Y.N., Li L.F., Yang Y.W., Ruan G.Q. 2009a. Euler neural network with its weight-direct-determination and structure-automatic-determination algorithms. In: Proceedings of the Ninth International Conference on Hybrid Intelligent Systems, pp. 319-324.

Zhang Y.N., Li W.B., Yu X.T., Xiao L., Chen J.H. 2012d. The 3-input Euler polynomial neuronet (3IEPN) with weights-and-structure-determination (WASD) algorithm. In: Proceedings of International Conference on Modelling, Identification and Control, pp. 1165-1170.

Zhang Y.N., Li W., Cai B.H., Xiao X.C. 2009b. Number determination of hidden-layer neurons in weights-directly-determined Legendre neural network. *Journal of Chinese Computer Systems*, 30(7): 1298-1301.

Zhang Y.N., Li W., Yi C.F., Chen K. 2008a. A weights-directly-determined simple neural network for nonlinear system identification. In: Proceedings of the IEEE International Conference on Fuzzy Systems, pp. 455-460.

Zhang Y.N., Ruan G.Q. 2009. Bernoulli neural network with weights directly determined and with the number of hidden-layer neurons automatically determined. *Lecture Notes in Computer Science*, 5551: 36-45.

Zhang Y.N., Xiao X.C., Chen Y.W., Zou A.J. 2010a. Number determination of hidden-layer nodes for Hermite feed-forward neural network. *Journal of Zhejiang University (Engineering Science)*, 44(2): 271-275.

Zhang Y.N., Yin Y.H., Yu X.T., Guo D.S., Xiao L. 2012c. Pruning-included weights and structure determination of 2-input neuronet using Chebyshev polynomials of Class 1. In: Proceedings of the 10th World Congress on Intelligent Control and Automation, pp. 700-705.

Zhang Y.N., Yu X.T., Guo D.S., Li J., Fan Z.P. 2012a. Weights and structure determination of feed-forward two-input neural network activated by Chebyshev polynomials of Class 2. In: Proceedings of the 24th Chinese Control and Decision Conference, pp. 1100-1105.

Zhang Y.N., Zhong T.K., Li W., Xiao X.C., Yi C.F. 2008b. Growing algorithm of Laguerre orthogonal basis neural network with weights directly determined. *Lecture Notes in Computer Science*, 5227: 60-67.

In: Self-organization: Theories and Methods
Editor: WenJun Zhang
ISBN: 978-1-62618-865-5

Chapter 6

LOW-DIMENSIONAL STRUCTURES EMBEDDED IN HUMAN LOCOMOTION: DATA ANALYSIS AND MODELING

Shinya Aoi [1,2*]
[1] Dept. of Aeronautics and Astronautics, Graduate School of Engineering, Kyoto University, Yoshida-honmachi, Sakyo-ku, Kyoto, Japan
[2] JST, CREST, Chiyoda-ku, Tokyo, Japan

Abstract

Humans produce adaptive locomotion through dynamic interactions among the nervous system, the musculoskeletal system, and the environment in a self-organized manner. A human musculoskeletal system has more degrees of freedom (DOFs) than necessary for locomotion and humans solve the redundancy problem in some way to establish locomotion. It has been suggested that individual DOFs are not manipulated independently, but some DOFs are functionally connected by object tasks to reduce the number of DOFs. These relationships among DOFs appear as low-dimensional structures in the DOFs. This chapter shows such low-dimensional structures in joint movements and muscle activities (kinematic and muscle synergies) during human locomotion based on the analysis of measured data. In addition, it shows a constructive approach using a computer simulation with a neuromusculoskeletal model based on anatomical and physiological findings to examine the functional roles of the low-dimensional structures that generate adaptive locomotor behaviors through dynamic interactions among the nervous system, the musculoskeletal system, and the environment.

Keywords: Human locomotion; Synergy; Central pattern generator; Phase resetting; Neuromusculoskeletal model

1. Redundant Systems in Human Locomotion

Humans produce adaptive behaviors in diverse environments by cooperatively and skillfully manipulating their complicated musculoskeletal systems. Locomotion, which is an essen-

*Correspondence: shinya_aoi@kuaero.kyoto-u.ac.jp

tial whole body movement among the activities of daily living, emerges from dynamic interactions among the nervous system, the musculoskeletal system, and the environment in a self-organized manner [43]. In human locomotion, motor commands from the nervous system produce muscle activity, and muscle tensions around joints generate joint movements. The human skeleton consists of more than 200 bones, and the human body has more than 400 skeletal muscles. Various parts of the nervous system, such as the cerebral cortex, the basal ganglia, the cerebellum, the brainstem, and the spinal cord, contribute to generating motor commands for locomotion by integrating such sensory information as visual, vestibular, and somatosensory information. Humans use many degrees of freedom (DOFs) and much information to achieve locomotion. How do humans manipulate such a large amount of DOFs and information during locomotion?

Locomotion involves moving the center of mass (COM) of the body using the legs. Humans have more DOFs in their joints than necessary for producing such a movement and have more DOFs in their muscles than in their joints. It is obvious that such redundancy plays an important role in achieving adaptive behaviors. However, is such a large amount of DOFs and information necessary for locomotion? To put it extremely simply, a body with two links like a compass can produce locomotion, as observed in passive dynamic walking [32]. Humans have an extremely redundant system for locomotion and must solve the redundancy problem in some way.

2. Low-Dimensional Structures in Human Locomotion

To overcome the redundancy problem in humans, it has been suggested that individual DOFs are not manipulated independently, but some DOFs are functionally connected by object tasks. In other words, humans use conditions of constraint to reduce the number of DOFs to solve motor redundancy. In locomotion, this is associated with simultaneous movements in several joints and co-variations of muscle activities.

These relationships among DOFs are represented as low-dimensional structures and appear by analyzing measured data during human locomotion. Low-dimensional structures in joint movements and muscle activities are called kinematic and muscle synergies, respectively. In addition to locomotion, these low-dimensional structures are observed in various tasks from simple to complex (e.g., a single arm reaching task and whole body standing quietly and sit-to-stand tasks) and also observed in many animals [11, 13, 30, 44, 45]. Therefore, these synergies are viewed as one solution to handle the redundancy problem in biological systems.

2.1. Kinematic Synergy in Joint Movements

When elevation angles at the thigh, shank, and foot during human locomotion are plotted, they describe regular loops that are almost on a plane (Fig. 1) [6, 29, 38]. This suggests that these three DOFs have one linear relationship during locomotion that constrains these motions on a plane. This relationship represents the intersegmental (intralimb) coordination of limb movements during locomotion. In addition, such planar motion, which is observed in various gaits, such as walking, running, hopping, and stepping over an obstacle, is associated with the control of the limb axis length and orientation (i.e., the position of the limb

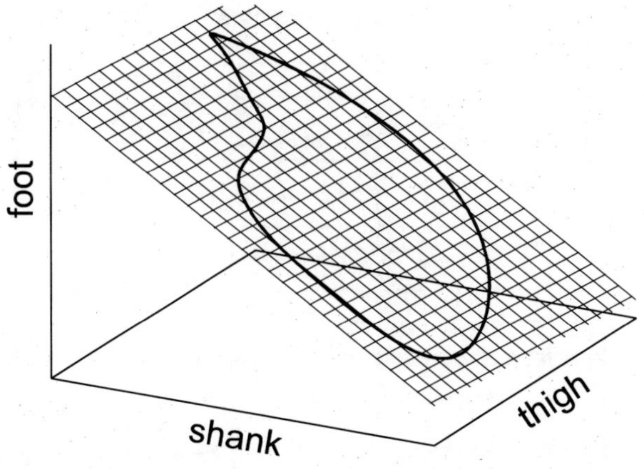

Figure 1. Regular loop by elevation angles of thigh, shank, and foot during human locomotion from kinematic data [20]

endpoint relative to the root and its direction) [26]. The plane orientation changes depending on the locomotion condition, such as the locomotion speed, suggesting a change in the intersegmental coordination of limb movements. The planar motion of the limb segmental movements is also observed in the locomotion of other animals, such as cats [38] and monkeys [27, 35].

Funato et al. [20] used seven elevation angles (thighs, shanks, feet, and trunk) in the sagittal plane to investigate low-dimensional structures in the whole body movements of human locomotion. They prepared a matrix by putting the time-series data of each angle as a column (the time was normalized by step cycles) and extracted the principal spatiotemporal patterns by performing singular value decomposition (SVD) on the matrix (Fig. 2). The right singular vectors obtained from the SDV correspond to the principal groups of the simultaneously active joints (i.e., intersegmental coordination), which represent the kinematic synergy. The left singular vectors exhibit the activation patterns (i.e., temporal coordination) of the joint groups. When SVD is individually performed on the data of the single- or double-support phase, the cumulative proportion exceeds 99% after two elements. In contrast, for the data of one step including both the single- and double-support phases, the cumulative proportion exceeds 99% after three elements. These results suggest that whole body movements by seven angles during the single- or double-support phase are constrained on a plane whose axes correspond to two kinematic synergies and that whole body movements during locomotion are explained by switching these two planes (Fig. 3).

In this analysis, the spatiotemporal patterns extracted by SVD show specific functional roles in the locomotor behavior and suggest human locomotion strategies. The first kinematic synergy (Synergy 1) swings the leg forward (orientation of the limb axis), and the second kinematic synergy (Synergy 2) changes the distance between the hip and the foot (length of the limb axis). Even when the environmental situation (such as the locomotion speed and slope angle) varies, the low-dimensional plane by the whole body movement appears and the plane orientation changes, similar to the three angle movements described above. The kinematic synergies vary depending on the environmental situation. However,

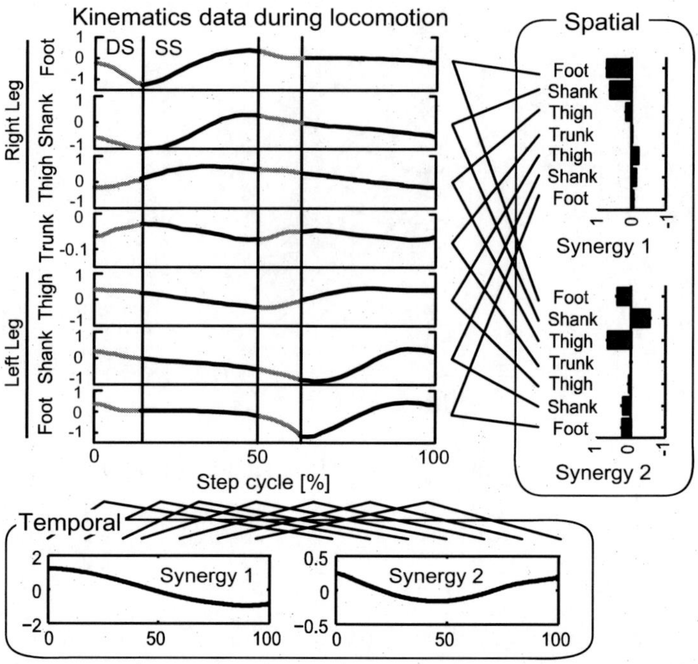

Figure 2. Extraction of principal spatiotemporal patterns by singular value decomposition on time-series data of seven elevation angles during human locomotion (modified from [20]). SS and DS indicate single- and double-support phases.

unlike the kinematic synergies that exhibit spatial patterns, the temporal patterns have no significant change, suggesting that humans modulate spatial patterns to adapt to environmental variations while maintaining the temporal patterns in steady walking.

2.2. Muscle Synergy in Muscle Activities

In addition to the joint movements, low-dimensional structures during human locomotion have been investigated in muscle activities using such statistical techniques as principal components analysis (PCA) and factor analysis (FA) on the electromyographic (EMG) data measured during locomotion.

Ivanenko et al. [24] examined the EMG data of 25 muscles recorded during human walking and reported that, although the spatiotemporal patterns of the EMG data are complicated, they can be explained by a combination of only five basic patterns. Figure 4 shows the five basic patterns extracted from the EMG data of 25 muscles during human locomotion measured by Winter [46], whose combinations can explain over 90% of the EMG data. In addition to normal walking, the EMG data in voluntary tasks during locomotion (such as stepping over an obstacle and kicking a ball) were investigated. When voluntary tasks were performed during locomotion, the five basic patterns were invariantly present, and the voluntary tasks were accompanied by an additional sixth pattern [25]. When the EMG data were compared between walking and running gaits, one basic pattern shifted during running [9].

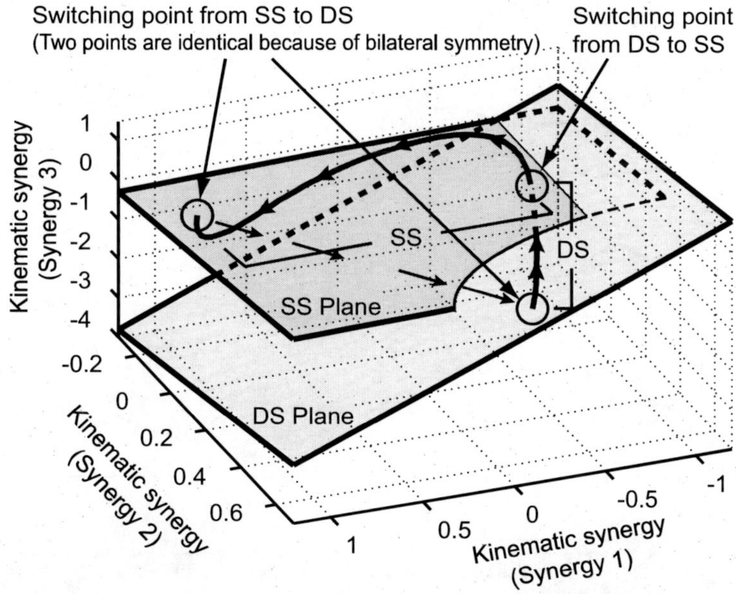

Figure 3. Whole body movement by seven angles during human locomotion constrained on two planes (modified from [20])

These low-dimensional structures in muscle activities have also been investigated in the locomotion of various animals, including rats, cats, and monkeys, and similar basic patterns were observed in humans and animals [14]. Frogs share some basic patterns for producing muscle activities among various movements (e.g., jumping, swimming, and walking) and produce them by adding other patterns [12].

3. Control of Low-Dimensional Structures in the Nervous System

From the analysis of the measured data, such as joint angles and muscle EMG data, various low-dimensional structures have been observed. The activities of the joints and muscles are sensed and controlled in the nervous system. How are such low-dimensional structures expressed and controlled in the nervous system during locomotion?

3.1. Limb Axis Length and Orientation

Spinocerebellar neurons receive sensory information from proprioceptors and cutaneous receptors through the ventral spinocerebellar tract. The investigation of spinocerebellar neuron activity by moving the hindlimb of a cat using a robot showed that neuron activity encodes the global information of limb kinematics, such as the limb axis length and orientation [7, 37, 38]. This suggests the contribution of low-dimensional structures in joint movements to the generation and the modulation of motor commands in the nervous system.

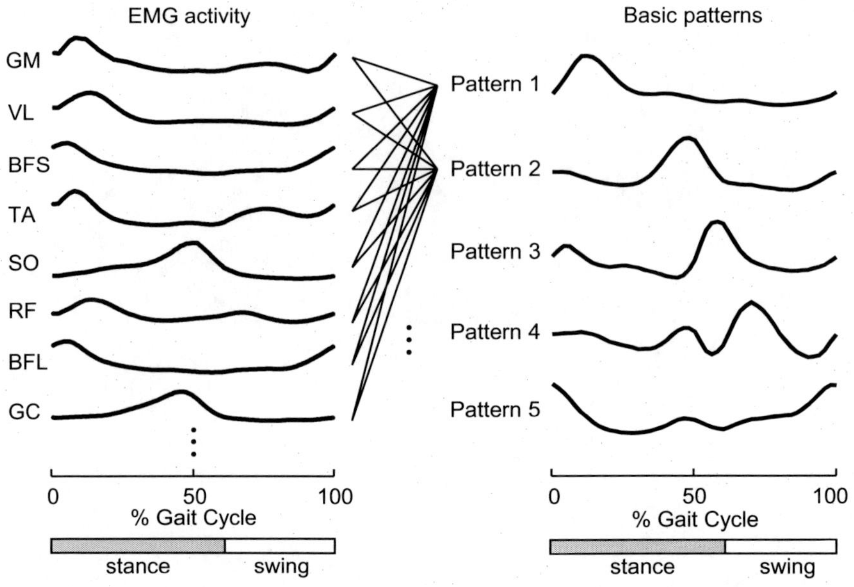

Figure 4. Five basic patterns extracted from EMG data of 25 muscles during human locomotion

3.2. Central Pattern Generator

Physiological evidence suggests that the central pattern generator (CPG) in the spinal cord strongly contributes to rhythmic limb movement, such as locomotion [22, 36, 42]. Although CPG organization remains largely uncertain, physiological findings suggest that CPG consists of hierarchical networks composed of rhythm generator (RG) and pattern formation (PF) networks [8, 31, 39, 40]. The RG network generates the basic rhythm and alters it in response to sensory information. The PF network shapes the rhythm into spatiotemporal patterns of motor commands. CPG separately controls the locomotor rhythm and motor commands in the RG and PF networks.

The measured EMG data represent the muscle activities activated by motor commands from the nervous system. Therefore, the low-dimensional structures extracted from the analysis of EMG data suggest that the nervous system uses modular control to produce motor commands and that CPG plays an important role in this control. Ivanenko et al. [25] revealed that the five basic patterns extracted from the EMG data during human locomotion are characterized by a narrow peak of activation (Gaussian-like) at a particular phase of the gait cycle and proposed a CPG model that generates five basic patterns as a sequence of activation pulses delivered to motoneurons through interneurons.

3.3. Sensory Regulation of Motor Commands

CPG can produce oscillatory signals even without rhythmic input and proprioceptive feedback. However, sensory feedback is required to produce adaptive and effective locomotion. Spinal cats produce locomotor behaviors on treadmills and their gait changes among walking, trotting, and galloping gaits based on the belt speed [18, 36], suggesting that tactile sensory information between their feet and the belt influences the locomotor rhythm and phase generated by CPG [16]. In addition, the locomotor rhythm and phase are modulated by producing phase shift and rhythm resetting based on sensory afferents and perturbations (phase resetting) [10, 15, 31, 39, 41].

Based on the CPG model by Ivanenko et al. [25], basic motor commands are generated by combining five activation pulses that are produced at a particular phase of the gait cycle. It has been suggested that CPG manages the timing to produce these pulses based on kinematic events, such as foot contact and foot off.

4. Constructive Approach Using Neuromusculoskeletal Model

The analysis of the measured data during human locomotion has provided hypotheses about the contribution of the nervous system to the generation of adaptive locomotor behaviors. However, because locomotion is a well organized motion generated by dynamic interactions among the nervous system, the musculoskeletal system, and the environment in a self-organized manner, it is difficult to fully clarify locomotion mechanisms only from the data analysis in physiological studies. It is also crucial to elucidate the dynamic characteristics in the neuro-mechanical interactions during locomotion.

To overcome such a limitation of physiological studies, a constructive approach using a computer simulation has recently attracted attention [4, 17, 43, 48, 49]. From anatomical studies, human skeletal characteristics (e.g., skeletal structure and dimensions and mass distribution) and muscular characteristics (e.g., attachment points and physiological cross-sectional areas of muscles) have been clarified. In addition, physiological findings address the understanding of the roles of the nervous system in locomotion. These allow us to construct reasonably realistic models of musculoskeletal and nervous systems. By integrating musculoskeletal and nervous system models and by comparing simulation results with the measured data (such as joint angles, EMG data, and ground reaction forces), this approach aims to improve the understanding of the mechanism of information processing in the nervous system to establish adaptive locomotor behaviors through dynamic interactions among the nervous system, the musculoskeletal system, and the environment. This approach also helps verify physiological hypotheses from a dynamic viewpoint.

The following sections show research that uses a neuromusculoskeletal model of humans to investigate the functional roles of the low-dimensional structure of muscle synergy and the sensory regulation of motor commands based on the phase resetting mechanism in generating adaptive locomotor behaviors [4].

4.1. Musculoskeletal Model

Figure 5**A** shows a skeletal model that consists of seven rigid links that represent the HAT (head, arms, and trunk), thighs, shanks, and feet. Each joint (hip, knee, and ankle) is modeled as a pin joint and the ground reaction force is modeled by a linear spring and damper system. Figure 5**B** shows a muscle model that uses nine principal muscles for each leg. Each muscle consists of a contractile element and passive elastic and damping elements parallel to the contractile element. The muscle activation for the contractile element is given

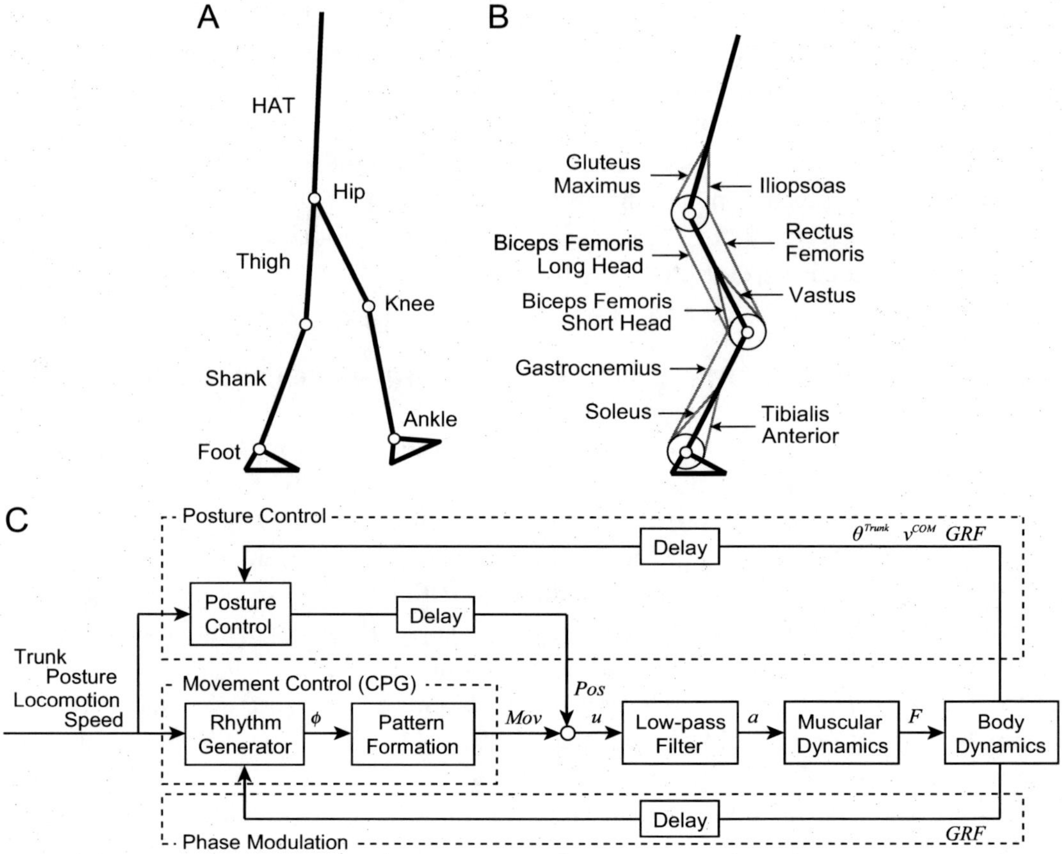

Figure 5. Musculoskeletal model (**A**: skeletal model and **B**: muscle model) and nervous system model (**C**)

by a second-order low-pass filter to the command signal from the nervous system model, and the muscle tension depends on the force-length and force-velocity relationships.

4.2. Nervous System Model

Figure 5**C** shows a nervous system model based on physiological findings. To focus on specific contributions in the nervous system, it mainly consists of the following three components:

1. Movement control (spinal cord level), which produces five activation pulses based on muscle synergy in a feedforward fashion to create periodic limb movements for forward motion;

2. Phase modulation (spinal cord level), which modulates the timing to produce feedforward signals in the movement control based on the phase resetting mechanism using foot-contact information; and

3. Posture control (brainstem and cerebellar levels), which creates command signals

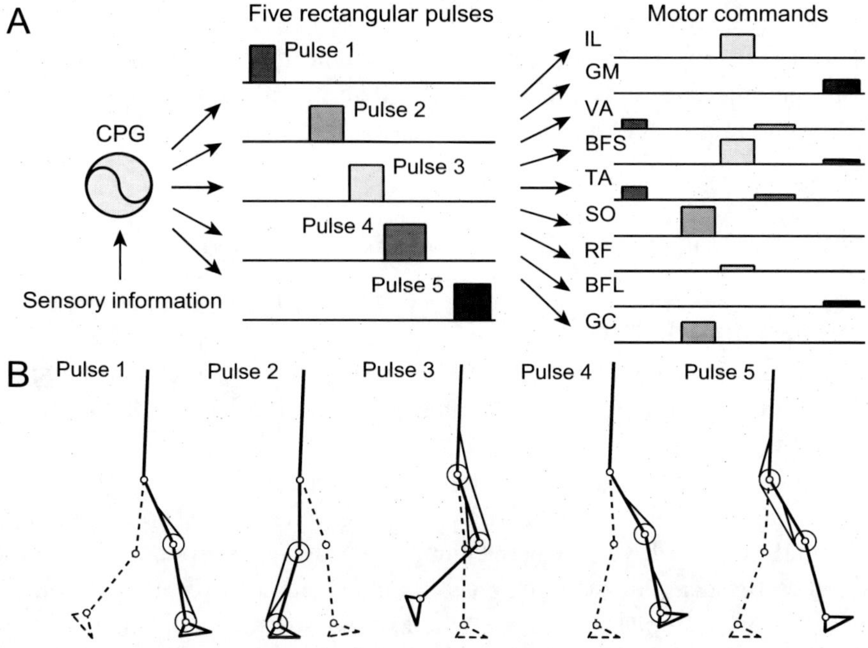

Figure 6. CPG produces five rectangular pulses and manages production timing based on sensory information (**A**: motor commands for muscles composed of combinations of five rectangular pulses and **B**: activated muscles by five rectangular pulses)

in a feedback fashion based on somatosensory information to control and regulate postural behavior.

4.2.1. Movement Control with Five Activation Pulses Based on Muscle Synergy

The movement control is modeled based on the two-layered hierarchical network composed of the RG and PF networks in CPG. The RG model used two simple phase oscillators to produce the basic locomotor rhythm and phase for the corresponding leg. The PF model used five rectangular pulses (Fig. 6) based on the low-dimensional structure of muscle synergy; the rectangular pulse becomes a Gaussian-like pulse through the second-order low-pass filter [28]. The timing and duration of the rectangular pulses are determined based on the oscillator phase in the RG model. The five activation pulses are delivered to motoneurons with adequate weight, and the motor commands are given by the combination of the five rectangular pulses.

4.2.2. Phase Modulation of Five Activation Pulses Based on Phase Resetting

In the RG model, the two oscillator phases interact with each other to maintain antiphase to produce alternating leg movements. For sensory regulation, the oscillator phase is modulated based on the phase resetting mechanism. More specifically, the oscillator phase is reset to the nominal value by the foot-contact information. The nominal value is determined so that the amount of the phase reset disappears when steady walking is achieved on flat ground without any disturbance. The oscillator phase is shifted by the phase difference at the foot-contact and foot-off events between steady and disturbed walking. Because the timing to produce the five rectangular pulses is determined by the oscillator phase, this phase resetting modulates the production timing of the pulses based on the foot-contact information.

4.2.3. Posture Control Based on Somatosensory Information

Locomotion requires that leg movements move the COM of the body and stabilize the posture to prevent falling over. The generation of leg movements and the stabilization of the posture interact and the mechanism that consistently satisfies both remains largely unclear. Elucidating the underlying mechanism to integrate the control of the leg movements and posture is also crucial.

In human locomotion, it is crucial to move the COM forward at a desired speed and maintain a vertical trunk pitch. By focusing on these two factors for posture control, a simple feedback control is employed using the antagonistic muscles around the hips and ankles to regulate them. Because this posture control is managed at the brainstem and cerebellar levels, somatosensory information and command signals are delayed, which forces the gain parameters in the feedback control to be small.

4.3. Generation of Adaptive Walking by Sensorimotor Integration

By integrating the five rectangular pulses in the movement control, the sensory regulation by phase resetting, and the posture control, steady walking is established in computer simulations. The simulation results were compared with the measured data during human locomotion and exhibit similar patterns to humans (Fig. 7).

In particular, the functional roles of the sensory regulation based on the phase resetting mechanism were investigated by focusing on the following two types of adaptations.

1. Adaptation against perturbing forces: After the model achieved steady walking, a perturbing force was added to the COM of the HAT in the horizontal direction (forward or backward). Various perturbation magnitudes and timings were used to examine if the walking behavior recovers after being disturbed and how quickly.

2. Adaptation against sudden environmental changes: After the model achieved steady walking, an environmental situation was suddenly changed to examine if the model continued to walk without falling down by adapting to the environmental change.

Figure 8 shows the adaptation results against the perturbing forces by comparing four cases: without phase resetting, with phase resetting only at foot off, with phase resetting only at foot contact, and with phase resetting at both foot off and foot contact. Using phase resetting at both foot off and foot contact yields the greatest degree of robustness among these four cases and demonstrates that phase resetting contributes to generating adaptive walking against perturbing forces.

Figure 9 shows the adaptation results against sudden environmental changes (loading on the trunk and change in the ground from flat to upslope) by comparing with and without phase resetting at both foot off and foot contact. After the environmental change, the model

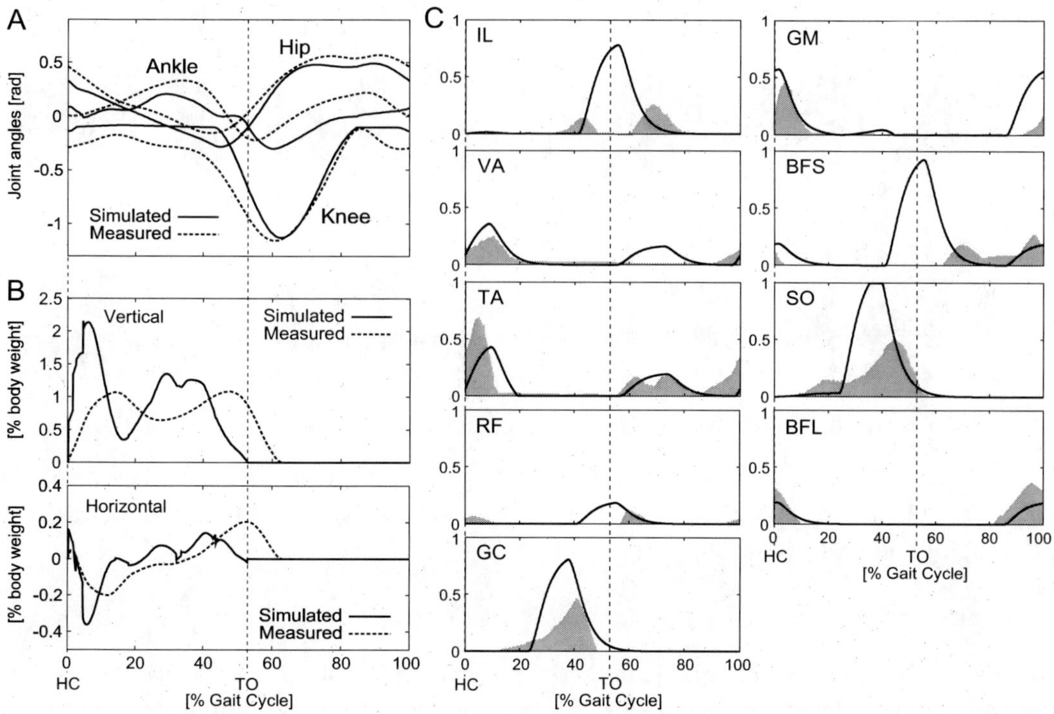

Figure 7. Simulation result of steady walking and comparison with measured data during human locomotion (**A**: joint angles, **B**: ground reaction forces, and **C**: muscle activation patterns). Measured data for joint angles and ground reaction forces are from Winter [47], and muscle activation patterns are compared with measured EMG data in Inman [23] (filled gray areas). HC and TO indicate heel contact and toe off in simulation results.

with phase resetting kept walking, but the model without phase resetting easily fell down. In the model with phase resetting, the joint kinematics, the gait cycle, and the locomotion speed adapted to the environmental changes through the interactions among the nervous system model, the musculoskeletal model, and the environment. These results show the usefulness of phase resetting to generate robust walking to sudden environmental variations.

4.4. Functional Roles of Phase Resetting in Generation of Adaptive Walking

Leg movements during locomotion consist of swing and stance phases. The swing leg is raised and swung forward, while the stance leg supports the body and generates a propulsive force from the ground. Since the swing and stance legs play completely different roles in locomotion, adequate timing based on foot-contact information is crucial for producing motor commands. Phase resetting contributes to the modulation of this timing.

The spatiotemporal patterns of motor commands determine the locomotor behavior, and phase resetting manages the temporal modulation based on foot-contact and foot-off events. Because it models part of the roles of the nervous system, this model has limitations in locomotor adaptations relative to actual humans. However, since the musculoskeletal and nervous system models are simple, they increase understanding of the dynamic char-

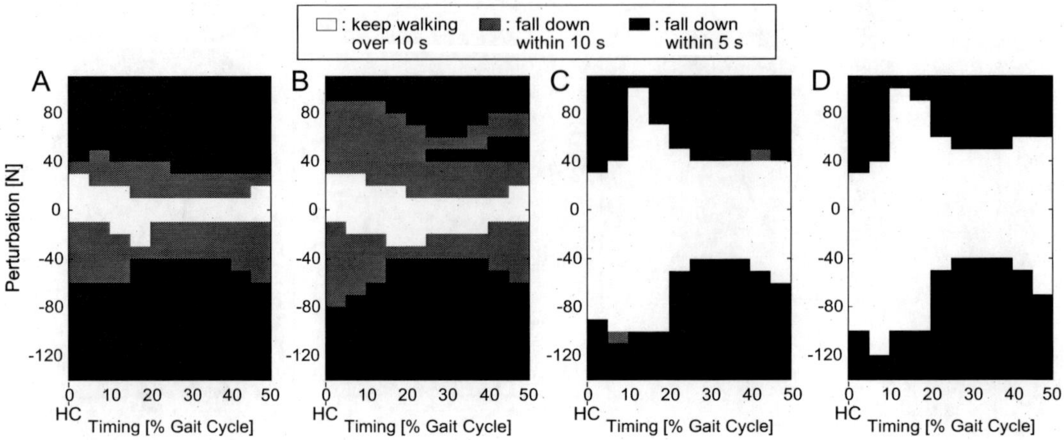

Figure 8. Tolerance of perturbing forces (**A**: without phase resetting, **B**: with phase resetting only at foot off, **C**: with phase resetting only at foot contact, and **D**: with phase resetting at both foot off and foot contact)

acteristics in the neuro-mechanical interactions during locomotion. Furthermore, this sensory regulation model has been successfully used not only for neuromusculoskeletal models of biological systems [4, 48, 49] but also for the locomotion control systems of biped robots [1, 2, 3, 5, 33, 34].

It remains unclear if humans use such sensory regulation in locomotion. However, analysis of the measured data during recovery from perturbed walking [19, 21, 49] might provide clues to clarify the sensory regulation mechanism in the future.

5. Conclusion

This chapter discussed how humans manipulate a large amount of DOFs and information to produce adaptive locomotion from the analysis of joint angles and muscle EMG data measured during human locomotion and from computer simulations using a neuromusculoskeletal model. So far, physiological and biomechanical studies have been independently developed to elucidate the locomotion mechanisms in humans. In addition, a simulation study using a neuromusculoskeletal model would be a useful tool by integrating the analysis of measured data to clarify the adaptability mechanism in locomotion through dynamic interactions among the nervous system, the musculoskeletal system, and the environment.

Acknowledgments

This chapter is based on collaborative studies with Kazuo Tsuchiya, Tetsuro Funato, Naomichi Ogihara, and Yasuhiro Sugimoto. This chapter is supported in part by a Grant-in-Aid for Scientific Research (B) No. 23360111 from the Ministry of Education, Culture, Sports, Science, and Technology of Japan.

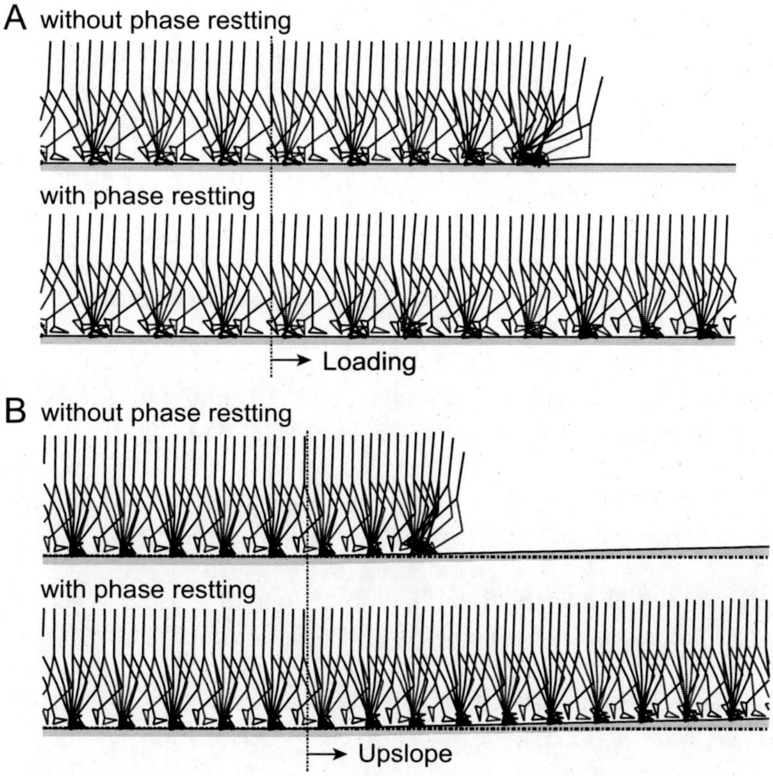

Figure 9. Simulated walking behavior with sudden environmental changes comparing with and without phase resetting (**A**: loading on the trunk and **B**: change in the ground from flat to upslope)

References

[1] S. Aoi and K. Tsuchiya, Locomotion control of a biped robot using nonlinear oscillators, *Auton. Robots*, 19(3):219–232, 2005.

[2] S. Aoi and K. Tsuchiya, Stability analysis of a simple walking model driven by an oscillator with a phase reset using sensory feedback, *IEEE Trans. Robotics*, 22(2):391–397, 2006.

[3] S. Aoi and K. Tsuchiya, Adaptive behavior in turning of an oscillator-driven biped robot, *Auton. Robots*, 23(1):37–57, 2007.

[4] S. Aoi, N. Ogihara, T. Funato, Y. Sugimoto, and K. Tsuchiya, Evaluating functional roles of phase resetting in generation of adaptive human bipedal walking with a physiologically based model of the spinal pattern generator, *Biol. Cybern.*, 102(5):373–387, 2010.

[5] S. Aoi, Y. Egi, R. Sugimoto, T. Yamashita, S. Fujiki, and K. Tsuchiya, Functional roles of phase resetting in the gait transition of a biped robot from quadrupedal to bipedal locomotion, *IEEE Trans. Robotics*, 28(6):1244–1259, 2012.

[6] L. Bianchi, D. Angelini, G.P. Orani, and F. Lacquaniti, Kinematic coordination in human gait: Relation to mechanical energy cost, *J. Neurophysiol.*, 79:2155–2170, 1998.

[7] G. Bosco and R.E. Poppele, Proprioception from a spinocerebellar perspective, *Physiol. Rev.*, 81:539–568, 2001.

[8] R.E. Burke, A.M. Degtyarenko, and E.S. Simon, Patterns of locomotor drive to motoneurons and last-order interneurons: Clues to the structure of the CPG, *J. Neurophysiol.*, 86:447–462, 2001.

[9] G. Cappellini, Y.P. Ivanenko, R.E. Poppele, and F. Lacquaniti, Motor patterns in human walking and running, *J. Neurophysiol.*, 95:3426–3437, 2006.

[10] B.A. Conway, H. Hultborn, and O. Kiehn, Proprioceptive input resets central locomotor rhythm in the spinal cat, *Exp. Brain Res.*, 68:643–656, 1987.

[11] A. d'Avella, P. Saltiel, and E. Bizzi, Combinations of muscle synergies in the construction of a natural motor behavior, *Nat. Neurosci.*, 6:300–308, 2003.

[12] A. d'Avella and E. Bizzi, Shared and specific muscle synergies in natural motor behaviors, *Proc. Natl. Acad. Sci. USA*, 102(8):3076–3081, 2005.

[13] T. Drew, J. Kalaska, and N. Krouchev, Muscle synergies during locomotion in the cat: a model for motor cortex control, *J. Physiol.*, 586(5):1239–1245, 2008.

[14] N. Dominici, Y.P. Ivanenko, G. Cappellini, A. d'Avella, V. Mondì, M. Cicchese, A. Fabiano, T. Silei, A. Di Paolo, C. Giannini, R.E. Poppele, and F. Lacquaniti, Locomotor primitives in newborn babies and their development, *Science*, 334:997–999, 2011.

[15] J. Duysens, Fluctuations in sensitivity to rhythm resetting effects during the cat's step cycle, *Brain Res.*, 133(1):190–195, 1977.

[16] J. Duysens, F. Clarac, and H. Cruse, Load-regulating mechanisms in gait and posture: Comparative aspects, *Physiol. Rev.*, 80:83–133, 2000.

[17] Ö. Ekeberg and K. Pearson, Computer simulation of stepping in the hind legs of the cat: An examination of mechanisms regulating the stance-to-swing transition, *J. Neurophysiol.*, 94:4256–4268, 2005.

[18] H. Forssberg and S. Grillner, The locomotion of the acute spinal cat injected with clonidine i.v., *Brain Res.*, 50:184–186, 1973.

[19] A.G. Feldman, T. Krasovsky, M.C. Baniña, A. Lamontagne, and M.F. Levin, Changes in the referent body location and configuration may underlie human gait, as confirmed by findings of multi-muscle activity minimizations and phase resetting, *Exp. Brain Res.*, 210(1):91–115, 2011.

[20] T. Funato, S. Aoi, H. Oshima, and K. Tsuchiya, Variant and invariant patterns embedded in human locomotion through whole body kinematic coordination, *Exp. Brain Res.*, 205(4):497–511, 2010.

[21] T. Funato, T. Hosokawa, S. Aoi, N. Tomita, and K. Tsuchiya, *Isochron of human walking derived from the perturbation of floor*, Proc. 5th Int. Symp. Adaptive Motion of Animals and Machines, pp. 35–36, 2011.

[22] S. Grillner, *Locomotion in vertebrates: central mechanisms and reflex interaction*, Physiol. Rev., 55(2):247–304, 1975.

[23] V.T. Inman, *The pattern of muscular activity in the lower extremity during walking (Technical Report Series II, No. 25)*, Prosthetic Devices Research Project, Institute of Engineering Research, University of California, Berkeley, Calif, 1953.

[24] Y.P. Ivanenko, R.E. Poppele, and F. Lacquaniti, Five basic muscle activation patterns account for muscle activity during human locomotion, *J. Physiol.*, 556:267–282, 2004.

[25] Y.P. Ivanenko, R.E. Poppele, and F. Lacquaniti, Motor control programs and walking, *Neuroscientist*, 12(4):339–348, 2006.

[26] Y.P. Ivanenko, G. Cappellini, N. Dominici, R.E. Poppele, and F. Lacquaniti, Modular control of limb movements during human locomotion, *J. Neurosci.*, 27(41):11149–11161, 2007.

[27] Y.P. Ivanenko, A. d'Avella, R.E. Poppele, and F. Lacquaniti, On the origin of planar covariation of elevation angles during human locomotion, *J. Neurophysiol.*, 99:1890–1898, 2008.

[28] S. Jo and S.G. Massaquoi, A model of cerebrocerebello-spinomuscular interaction in the sagittal control of human walking, *Biol. Cybern.*, 96:279–307, 2007.

[29] F. Lacquaniti, R. Grasso, and M. Zago, Motor patterns in walking, *News Physiol. Sci.*, 14:168–174, 1999.

[30] M.L. Latash, *Synergy*, Oxford University Press, 2008.

[31] M. Lafreniere-Roula and D.A. McCrea, Deletions of rhythmic motoneuron activity during fictive locomotion and scratch provide clues to the organization of the mammalian central pattern generator, *J. Neurophysiol.*, 94:1120–1132, 2005.

[32] T. McGeer, Passive dynamic walking, *Int. J. Robotics Res.*, 9(2):62–82, 1990.

[33] J. Nakanishi, J. Morimoto, G. Endo, G. Cheng, S. Schaal, and M. Kawato, Learning from demonstration and adaptation of biped locomotion, *Robot. Auton. Syst.*, 47(2-3):79–91, 2004.

[34] M. Nakanishi, T. Nomura, and S. Sato, Stumbling with optimal phase reset during gait can prevent a humanoid from falling, *Biol. Cybern.*, 95:503–515, 2006.

[35] N. Ogihara, T. Kikuchi, Y. Ishiguro, H. Makishima, and M. Nakatsukasa, Planar co-variation of limb elevation angles during bipedal walking in the Japanese macaque, *J. R. Soc. Interface*, 9(74):2181–2190, 2012.

[36] G.N. Orlovsky, T. Deliagina, and S. Grillner, *Neuronal control of locomotion: from mollusc to man*, Oxford University Press, 1999.

[37] R.E. Poppele, G. Bosco, and A.M. Rankin, Independent representations of limb axis length and orientation in spinocerebellar response components, *J. Neurophysiol.*, 87:409–422, 2002.

[38] R.E. Poppele and G. Bosco, Sophisticated spinal contributions to motor control, *Trends. Neurosci.*, 26:269–276, 2003.

[39] I.A. Rybak, N.A. Shevtsova, M. Lafreniere-Roula, and D.A. McCrea, Modelling spinal circuitry involved in locomotor pattern generation: insights from deletions during fictive locomotion, *J. Physiol.*, 577(2):617–639, 2006.

[40] I.A. Rybak, K. Stecina, N.A. Shevtsova, and D.A. McCrea, Modelling spinal circuitry involved in locomotor pattern generation: insights from the effects of afferent stimulation, *J. Physiol.*, 577(2):641–658, 2006.

[41] E.D. Schomburg, N. Petersen, I. Barajon, and H. Hultborn, Flexor reflex afferents reset the step cycle during fictive locomotion in the cat, *Exp. Brain Res.*, 122(3):339–350, 1998.

[42] M.L. Shik and G.N. Orlovsky, Neurophysiology of locomotor automatism, *Physiol. Rev.*, 56(3):465–501, 1976.

[43] G. Taga, Y. Yamaguehi, and H. Shimizu, Self-organized control of bipedal locomotion by neural oscillators in unpredictable environment, *Biol. Cybern.*, 65:147–159, 1991.

[44] L.H. Ting and J.M. Macpherson, A limited set of muscle synergies for force control during a postural task, *J. Neurophysiol.*, 93:609–613, 2005.

[45] E. Todorov and M.I. Jordan, Optimal feedback control as a theory of motor coordination, *Nat. Neurosci.*, 5:1226–1235, 2002.

[46] D.A. Winter, *The biomechanics and motor control of human gait: normal, elderly and pathological,* Waterloo Biomechanics Press, 1991.

[47] D.A. Winter, *Biomechanics and motor control of human movement (3rd edition)*, New York: John Wiley & Sons, 2004.

[48] S. Yakovenko, V. Gritsenko, and A. Prochazka, Contribution of stretch reflexes to locomotor control: A modeling study, *Biol. Cybern.*, 90:146–155, 2004.

[49] T. Yamasaki, T. Nomura, and S. Sato, Possible functional roles of phase resetting during walking, *Biol. Cybern.*, 88:468–496, 2003.

In: Self-organization: Theories and Methods
Editor: WenJun Zhang
ISBN: 978-1-62618-865-5

Chapter 7

SELF-ORGANIZATION IN MOTION OF A SET OF LIVING INDIVIDUALS

Milovan Živanović[1,*] ***and Ivan Stojković***[2,]
[1]Digital Control Systems Oziris, Sopot, Serbia
[2]Mihajlo Pupin Institute, Belgrade, Serbia

ABSTRACT

This paper discusses self-organized movement of a set of some individuals. It is adopted that individuals perceive reality as its projection on coordinate system that is firmly attached to the individual. Movement of individuals is interpreted as the movement of points in this coordinate system according to the algorithms by which each individual is equipped. The projection of reality and other individuals simultaneously changing the parameters of algorithms and simultaneously impose constraints on the movement of these points. Set of individuals moves so that each individual performs arbitrary motion within the allowed constraints in one of two ways: a) individuals follow one or more leaders, b) individuals preserve system configuration during movement, i.e. not change certain typical magnitude of set. Motion of individuals or set of individuals from one position to another always takes place according to the functions of natural impulses. Shortly we'll give comparison between function of the natural impulse and other custom-fit functions.

Keywords: Self-organization, Self-organization of living individuals, algorithms, natural impulse AMS Subject Classification: 70E55

[*] E-mail address: dsu.oziris@open.telekom.rs.

1. Introduction

The following definitions of self-organization are commonly accepted in the literature.

Self-organization refers to a process where a system changes its internal organization to adapt to changes in its goals and the environment without explicit external control. Self-organization often results in emergent behavior that can be either desirable or undesirable. In cybernetics, it is commonly adopted that self-organization is a process where the organization (constraint, redundancy) of a system spontaneously increases, i.e. without this increase being controlled by the environment or an encompassing or otherwise external system. Similar definition states that self-organization is the mechanism or the process enabling a system to change its organization without explicit external command during its execution time [1]. Strong self-organizing systems are those systems where there is no explicit central control either internal or external. Weak self-organizing systems are those systems where, from an internal point of view, there might be a re-organization in process under an internal (central) control or planning.

Self-organization is a process where some form of global order or coordination arises out of the local interactions between the components of an initially disordered system (global definition). This process is spontaneous: it is not directed or controlled by any agent or subsystem inside or outside of the system; however, the laws followed by the process and its initial conditions may have been chosen or caused by an agent. It is often triggered by random fluctuations that are amplified by positive feedback. Resulting organization is wholly decentralized or distributed over all the components of the system. Previous definitions are such that self-organization Common examples are: crystallization, the emergence of convection patterns in a liquid heated from below, chemical oscillators, the invisible hand of the market, swarming in groups of animals or robot pattern formation, and the way neural networks learn to recognize complex patterns occurs in a variety of physical, chemical, biological, social and cognitive systems [2-5].

Without diminishing significance of recent results from the research of self-organizing, in this work we suggest approach that in a specific way interprets and implements the principles and laws applied in mechanics, control theory, and other fields of physics.

The basic idea is simple. Motion of anything that has mass obeys the laws of mechanics (dynamics). That anything represents the controlled mechanical object from the perspective of control theory. Motion of the controlled object is determined and specified by its dynamic characteristics, and by characteristics of its force [6, 7]. Force (compulsion) can be uncontrolled, internally or externally controlled or disturbance, independently produced by mechanical control object's external environment. Uncontrolled force (compulsion) is not an explicit function of the mechanical object's states or states of the environment. Under the effect of force or uncontrolled disturbance, the motion of pure (uncontrolled) mechanical object is accomplished. If the force is function of object's states and the states of environment, then mechanical object is equipped with control. Force function is then called a control law. Record of control law will be called mechanical object algorithm. Any sensible motion of the object can be performed only if it is equipped with an appropriate algorithm, which is explicit or implicit function of its states and the states of the environment. This indicates that the definition of self-organization does not take above specified sufficiently precise and explicit. Existence of algorithm excludes the spontaneity of movement and/or the development of a process. External environment impact is implicitly embedded in the algorithms. For example, growth of plants in spring is not spontaneous, but is directly driven by the environment in a way that environment defines the temperature, humidity, acidity, sunlight and other parameters essential for the beginning of growth. By achieving certain values of the previous parameters, algorithms have controlled outer information for start of algorithms implemented in plants, i.e. for the growth process.

This paper first, in a specific way, presents the dynamics of mechanical object through a brief overview of the theory given on the basic principles and laws of mechanics (see website http://en.wikipedia.org/wiki/Hamilton's_principle). Interpretation of these principles and laws is given in a form suitable for implementation in this work. It is adopted that all motion occurs in accordance with natural impulse function. It is shown that the function of natural impulses and trigonometric sinusoidal functions are similar, but their derivatives are essentially different. Then using the logic of principles and laws of mechanics, interpretation of autonomous motion of one individual in set is given mechanical control system. We started from the assumption that each individual is equipped with its eigenfunction (in control theory that function is control law, record of that function is an algorithm) that defines the manner of motion in a similar way as in mechanics and/or control theory. The authors consider that algorithms are formed exclusively based on projections of the external environment and the projection of their own movements on individual mobile system that is firmly attached to the individual (body system). It is adopted that normal motion continuously occurs pulsed periodically. It is allowed for eigenfunction and its constraints to be changed by learning and by environmental impacts. Analysis of the normal motion of a set of individuals that follows leader or a common goal described by some characteristic values of set is then conducted.

2. The Dynamics of the Mechanical Object - Theoretical Background

Mechanics claims that each and every mechanical system is characterized by its eigenfunction. That function is called Lagrangian function, kinetic potential or "live force". Hamilton's principle (principle of least action) asserts that the true (real) motion $q(t)$ of a system described by n generalized coordinates $q=(q_1,q_2,...,q_n)$ between two specified states $q_1=q(t_1)$ and $q_2=q(t_2)$ at two specified times t_1 and t_2 is a stationary point (a point where the variation is zero), of the action functional

$$S(q)=\int_{t_1}^{t_2} L(\dot{q}(t),q(t),t)dt \tag{1}$$

where $L(\dot{q}(t),q(t),t)$ is the Lagrangian function for the system. In other words, any first-order perturbation of the true motion results in (at most) second-order changes <u>in</u>. The action is a functional, i.e., something that takes as its input a function and returns a single number, a scalar ($S(q)\in R^1$). In terms of functional analysis, Hamilton's principle states that the true motion of a physical system is a solution of the functional equation $\partial S / \partial q(t) = 0$. Requiring that the true trajectory $q(t)$ be a stationary point of the action functional is equivalent to a set of differential equations for $q(t)$ (the Euler-Lagrange equations). Hamilton's principle requires that this first-order change is zero for all possible perturbations, i.e., the true path is a stationary point of the action functional (a minimum, maximum or saddle point). This requirement can be satisfied if and only if

$$\frac{d}{dt}\frac{\partial L}{\partial \dot{q}} - \frac{\partial L}{\partial q} = \frac{d}{dt} grand_{\dot{q}} E_k - \frac{\partial L}{\partial q} = \mathbf{0}, \qquad L = E_\kappa(\dot{q}) - E_p(q) \tag{2}$$

where E_k and E_p are the kinetic and potential energy. These equations are the equations of motion (described dynamics of mechanical system) and are known as the Euler-Lagrange equations.

Hamilton's principle does not require the total energy of the system to be constant during the motion. Therefore, in the general case, the Newton, Lagrange and Hamilton equations are used, respectively, in form

$$\frac{d}{dt}(Mv) = Q, \qquad \frac{d}{dt}\frac{\partial E_k}{\partial \dot{q}} = \frac{d}{dt} grand_{\dot{q}} E_k = \frac{\partial E_k}{\partial q} + \Theta = \bar{\Theta}, \qquad \begin{array}{l} \dot{q} = \dfrac{\partial H}{\partial p}; \; p = \dfrac{\partial E_k}{\partial \dot{q}} \\ \dot{p} = -\dfrac{\partial H}{\partial \dot{q}} + \Theta \end{array} \tag{3}$$

where: $\mathbf{v} = v(\dot{q})$ vector of (translational and rotational) velocity in inertial space, the vector Q (or Θ, $\bar{\Theta}$) which symbolically represents the resulting vector of forces/moments that balances derivative of gradient vector $d/dt(grad_{\dot{q}} E_k)$ kinetic energy E_k when $\partial E_k / \partial q = 0$, $H = H(p, q)$ Hamilton function, p generalized momentum, $M = diag(M_1, .., M_i, ..., M_n)$, $M_i = diag(diag(m, m, m), J)$ generalized inertia matrix of a rigid body, m mass, J inertia matrix and $\bar{Q}$ vector that balances derivative of kinetic energy gradient. Hamilton's function defines the total energy and is defined as $H = p\dot{q} - L = E_p + E_k$, where $L = E_k - E_p$ kinetic potential (Lagrange function). Derivative of Hamilton's function by time is obtained or consumed power of the system $\dot{H} = \dot{E}_k + \dot{E}_p = \Theta\dot{q}$. The equations implicitly or explicitly indicate that the total kinetic energy gradient derivative by time is equal to the sum of some of the forces or moments that may, but need not be a function of generalized coordinates and/or their derivatives. In [8] it is shown that (3) are the same equations given in different coordinate systems. Equations (3) can be considered as filter equations which input is force Q, and output is vector of generalized coordinates q. They can be realized as appropriate electrical filter too, rather than just a mechanical system. Thus, the property that each mechanical system is characterized by eigenfunction L is equivalent to the property that the mechanical system acts as an appropriate filter where the input force is Q.

If the mechanical system of constant mass moves in the gravitational field, the fragment of equation (3) related to the translational motion can be written as [9]

$$\frac{dV}{dt} = G + F_{external}$$
$$m(a-g) = ma_{measurable} = F_{external} \quad (4)$$

where $a = dV/dt$ net kinematic acceleration vector, $a_{measurable} = a - g$ a vector of total measurable acceleration (acceleration which is read by sensors), $G = mg$ weight vector, g the gravitational acceleration vector and $F_{external}$ vector of resulting external load. This representation separates the load where only characteristics of the measurable mechanic system $ma_{measurable}$ figure, from external load $F_{external}$ conditioned by state outside of the mechanical system. In a short time interval the motion of living being can be considered as the motion of mechanical systems with constant mass and variable inertial properties.

The simplest and basic description of the kinetic energy function is a quadratic form of generalized coordinates derivative, i.e. velocity $2E_k = \dot{q}^{\mathrm{T}}\mathrm{I}\dot{q}$, where $M = I$ is inertial tensor. Sum of complete squares (when I is a diagonal matrix) is obtained when for the generalized coordinates are chosen system's own coordinates (the coordinate axes along the principal axes of the hyper ellipsoid of inertia). In geometric terms, kinetic energy is norm of the radius vector of some point on the hyper ellipsoid $\dot{q}^{\mathrm{T}}\mathrm{I}\dot{q} = \left(\sqrt{2\mathrm{E}_{\kappa}}\right)^2$ in the space of generalized coordinates' derivatives. Any application of the previous procedure does not change the numerical value of the kinetic energy, but only the coordinates by which it is described are more or less complex. All other descriptions of kinetic energy can be obtained from the basic description of the application after a coordinate system transformation.

The previous method formulated in mechanics is applied in many branches of physics and science with specific formulation of Lagrangian function for each branch [10]. The method provides sufficient accurate results for usual technical applications.

Interestingly, the Lagrangian function can be considered in different ways as: homogeneous, relativistic, optical or metric function. If time is adopted as an independent $n+1$ generalized coordinate, then the action becomes

$$S = \int_{\nu} L_o(\dot{\overline{q}}, \overline{q})d\tau = \int_0^{\tau} \overline{L}(\frac{\overline{q}}{\dot{t}}, \overline{q}, t)\dot{t}d\tau \quad (5)$$

where: $\overline{q} = col(q,t) \in R^{n+1}$ new generalized coordinates, $L_o(\dot{\overline{q}}, \overline{q}) = \overline{L}(\dot{\overline{q}}/\dot{t}, \overline{q}, t)\dot{t}$ new Lagrangian function, t eigentime, $\nu = \nu(\overline{q}_1, \overline{q}_2)$ curve in the space of generalized coordinate $\overline{q}$ with the new parameter τ , $0 \le \tau \le \tau_2$ and $\dot{\overline{q}} = d\overline{q} / dt$, $\dot{t} = dt / d\tau$ derivatives of new generalized coordinates. Action element $dS = L_o(\dot{\overline{q}}, \overline{q})d\tau = L_o(d\overline{q}, \overline{q}) = d\mu$ can be viewed as a "length" of element $d\overline{q}$ line ν. In this way, the action S is the length of lines ν measured with metrics μ given with L_o. Least action is the minimum line length ν connecting the two points with a given $\overline{q}_1 = col(q_1, t_1)$ and $\overline{q}_2 = col(q_2, t_2)$. Such an interpretation provides a relatively simple way for implementation of dynamic programming (see website http:// en.wikipedia.org/wiki/Dynamic_programming).

Logic that each system is characterized by a function we will use in this paper, where the description of the motion equations will be in the form (4).

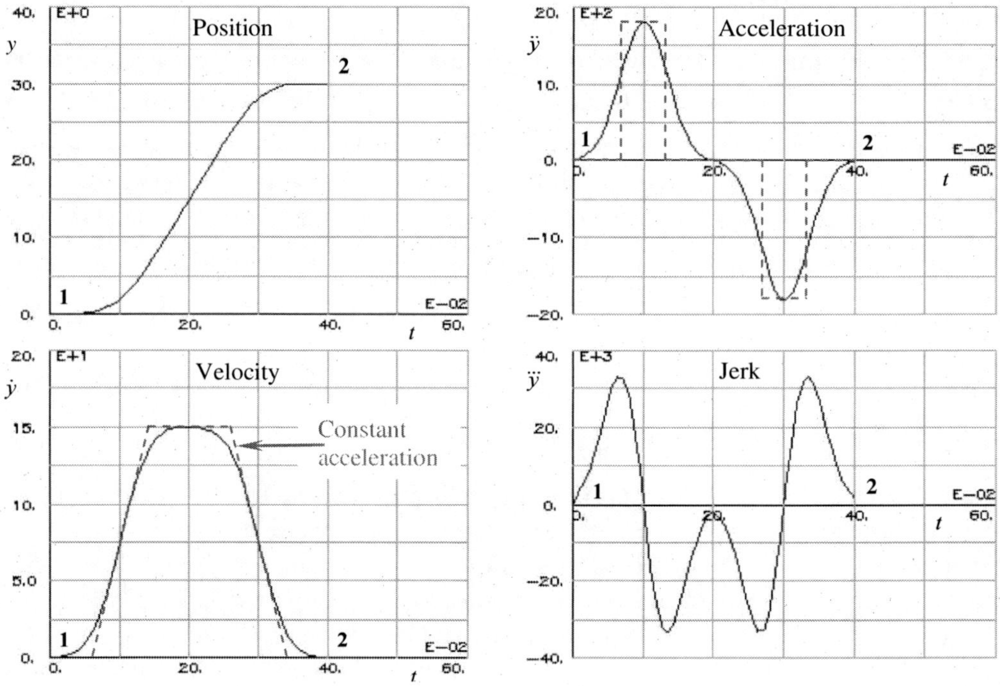

Figure 1. Function of natural impulses.

Every shifting from one state of relative rest to the other state of relative rest is performed in the same way in nature. All derivatives of displacement are zero at rest. Any displacement in the nature starts from the rest on this way that all derivatives of displacement start from zero simultaneously. Any shift in the nature of the ends in a way that all derivatives of displacement become zero simultaneously. A function that has these features can be called a function of natural impulses. In mathematical terms, the displacement is the integral of lowest derivative function of natural impulses (Figure 1).

The function of the natural impulse is named after the graph of velocity function. Velocity function graph (symbols $\dot{y}$ and INTe at Figures 1 and 2) has a pulse shape with rounded edges. With increase in amplitude, this function takes the form of pulses as a smooth continuous function. Movement takes place and can be observed if there is a change in velocity. Integral of movement velocity (symbols y and IInt) is a position that can be directly observed. Differentials are acceleration (symbols $\ddot{y}$ and ex2) and jerk (symbols $\dddot{y}$ and ex2S). There are two characteristics that distinguish the functions of natural impulses (from others). The first feature of functions of natural impulses is that at the beginning and at the end of movement functions themselves and their derivatives are equal to zero. Second feature is the very function of the natural impulse (velocity) and its integral (position) smooth and continuous functions. Integral of natural impulse is a function with the asymptotic approaching the initial and the final values.

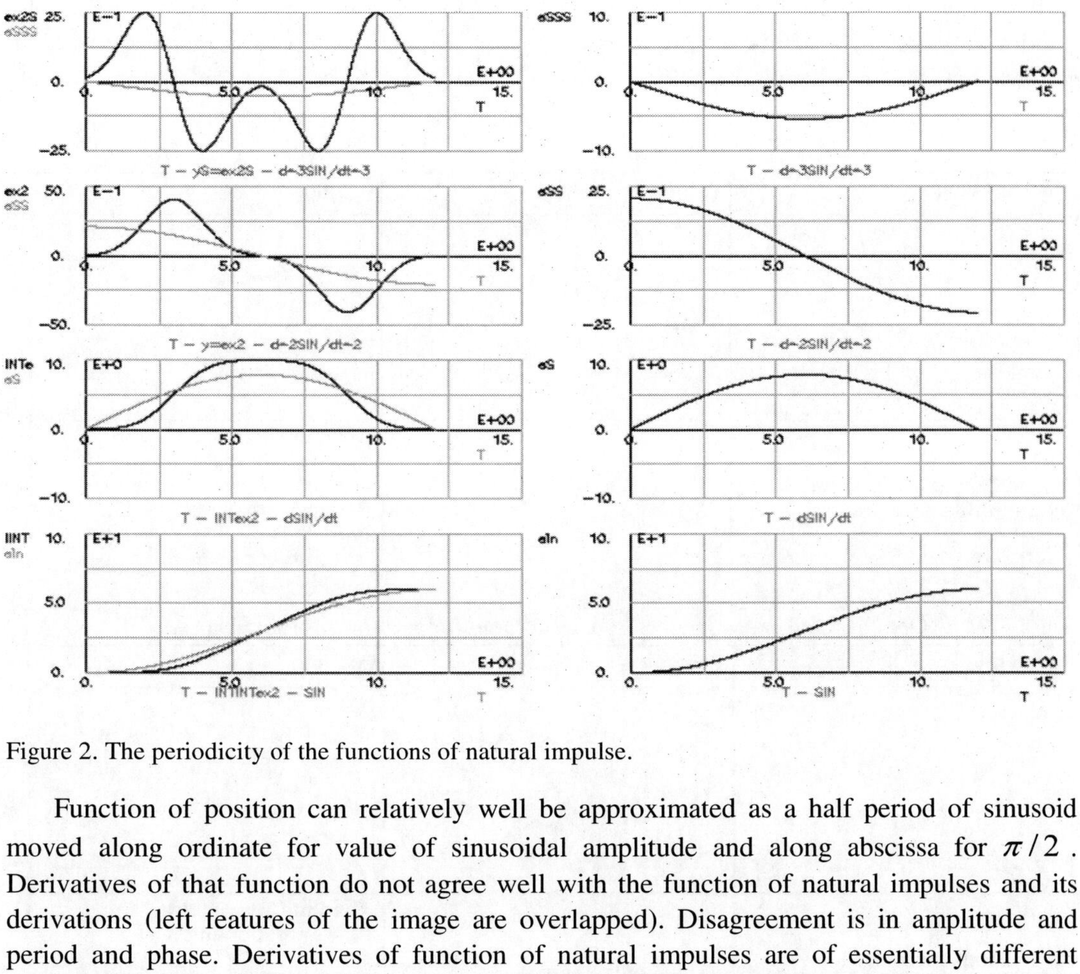

Figure 2. The periodicity of the functions of natural impulse.

Function of position can relatively well be approximated as a half period of sinusoid moved along ordinate for value of sinusoidal amplitude and along abscissa for $\pi/2$. Derivatives of that function do not agree well with the function of natural impulses and its derivations (left features of the image are overlapped). Disagreement is in amplitude and period and phase. Derivatives of function of natural impulses are of essentially different characteristics in amplitude and in relation to the sinusoid's derivatives. This is primarily related to the functions of acceleration (symbol ex2) and jerk (symbol ex2S). Approximated function of acceleration in initial and end times shows significant values. In reality, that property of acceleration functions in mechanical systems would manifest as a force that exists in initial and end moment of motion, although in reality such a force does not exist.

Derivatives of sine function do not change the characteristics of functions periodicity (right side Figure 2 shows only half-period sinusoidal and its derivatives). Periodicity of natural impulses function derivatives can be seen as the number of repetitions of deformed and possibly rotated position functions. In this sense, the period halves with each derivation. For case depicted in Figure 2 position function period would be 24, velocity period 12, acceleration period 6, and jerk 3. For this reason, approximation of the natural impulse function and its derivatives with trigonometric functions can agree neither per period nor per phase.

Significance of characteristics of movement according to the natural impulse function is well known among manufacturers of servo actuators. For example, the eminent manufacturer of high-quality electric servo actuators MAXON, uses profiles of position, velocity and acceleration given on Figure 3 and 4 [11]. Acceleration profiles in both modes of working possess discontinuation of the first type (acceleration derivatives are discontinued). This type of discontinuation is less pronounced in velocity profiles, with the aim of getting a smooth function of position, which is very similar to integral of functions natural impulses profile. In spite of that, the profiles of the selected position, velocity and acceleration are only rough approximation of natural impulse functions, especially the function of acceleration.

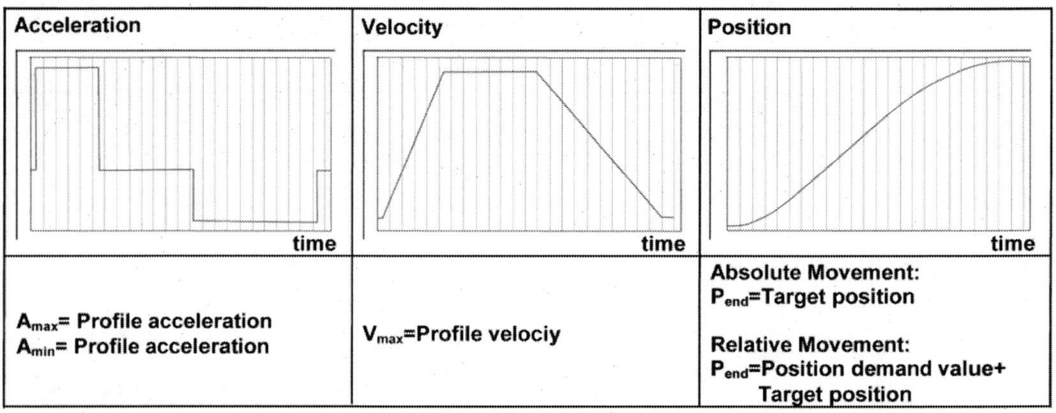

Figure 3. MAXON: Profile Position Trajectory Linear ramp (trapezoidal profile) [11].

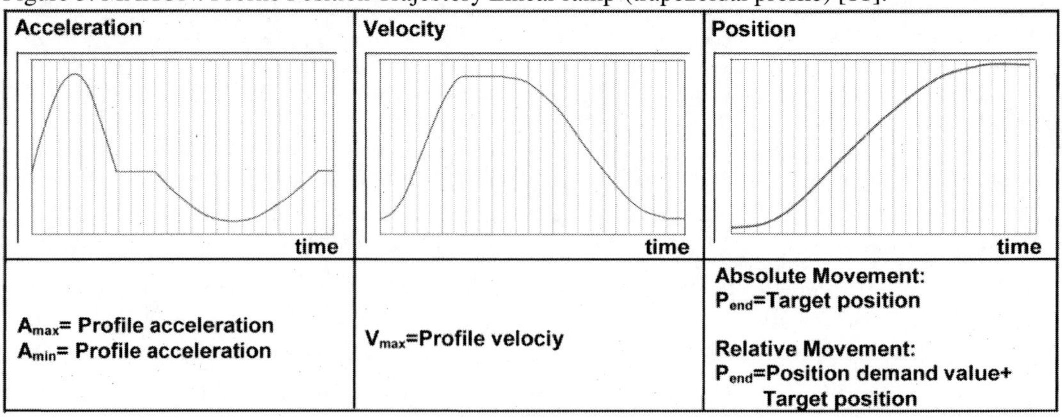

Figure 4. MAXON: Profile Position Trajectory Sin2 ramp (sinusoidal profile) [11].

Movement that is in accordance with the functions of natural impulse can easily be explained on the example of moving objects on the Earth from one position to another. According to current theory, each point on the Earth is in constant motion. That means that two close positions (points) 1 and 2 on Earth have approximately the same absolute velocity and its derivatives. Let an object occupies position 1 on Earth. If we want to move the object from position 1 to position 2 it is necessary to change the absolute velocity of the object, and then bring it back to initial value. If we consider the relative motion between positions 1 and 2, then it is necessary to change relative velocity from zero to some value, and then again back to zero. Additionally, all derivatives should simultaneously start from zero in position 1 and at the same time to return to zero at position 2 (Figure 1). This can be easily observed and understood by observation and/or measurement of extremity movements of living beings or their total movement.

Movement in accordance with the functions of natural impulse is not in conflict with the previously stated theory from mechanics. It determines the way of movement from one position to another. Navigation according to the functions of natural impulse is provided on one of the following ways: by selecting the appropriate compulsion Q ($\bar{Q}$) from equation (3) or providing profiles of generalized coordinates or profiles of some of its derivatives or providing power changes profile $dE_k / dt = \dot{q}^{\mathrm{T}}\mathrm{I}\ddot{q}$.

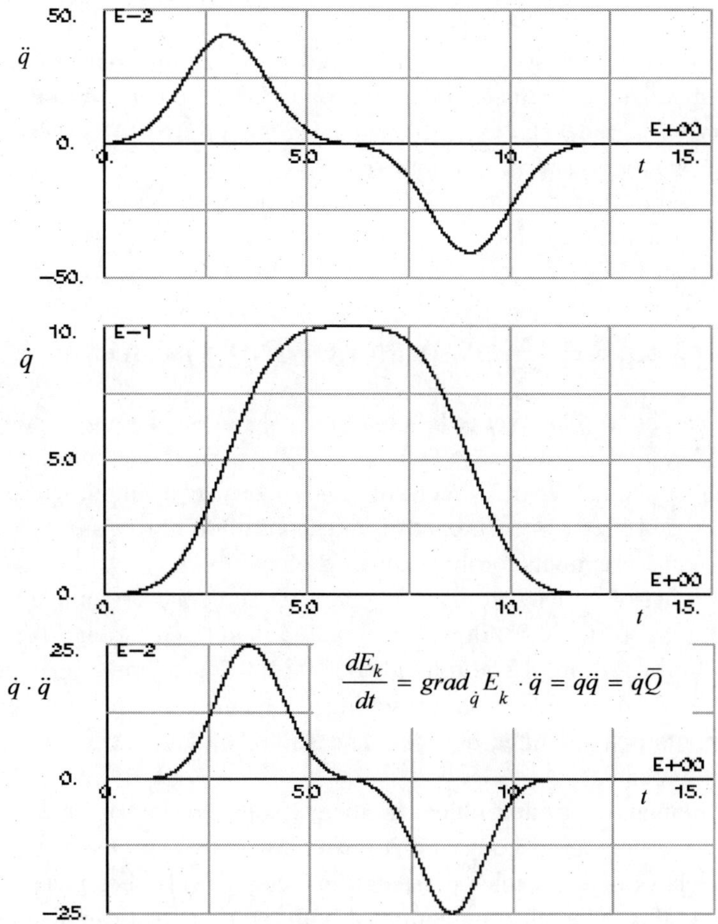

Figure 5. Power changes during the events of natural impulses.

We will explain the case where inertia matrix I is constant. In this case, the $grad_{\dot{q}}\mathrm{E}_{\kappa} = I\dot{q}$, $d/dt(grad_{\dot{q}}\mathrm{E}_{\kappa}) = \mathrm{I}\ddot{q} = \overline{\Theta}$. Total extract kinetic energy by the time [12], [13], [14], [15]

$$\frac{dE_k}{dt} = grad_{\dot{q}}\mathrm{E}_{\kappa} \cdot \ddot{q} = \dot{q}^{\mathrm{T}}\mathrm{I}\ddot{q} = \dot{q}^{\mathrm{T}}\overline{\Theta} \tag{6}$$

defines a change of power. This change can be described by only using the generalized derivatives coordinates $dE_k / dt = \dot{q}^{\mathrm{T}}\mathrm{I}\ddot{q}$ or by using extracts generalized coordinates and generalized external load $dE_k / dt = \dot{q}^{\mathrm{T}}\overline{\Theta}$. In geometrical sense, total kinetic energy derivative can be evaluated by means of the angle between the gradient of the kinetic energy $grad_{\dot{q}}\mathrm{E}_{\kappa}$ and the generalized coordinates velocity derivatives $\ddot{q}$ [12-15]. If the angle is sharp, change in power is growing and kinetic energy is increased. If the angle is obtuse kinetic energy decreases. If the angle is straight (90°) kinetic energy remains unchanged during the movement.

Change of the external load can be monitored by observing only the first and second derivatives of generalized coordinates or their products. In the case of unitary inertia matrix and unitary changes in the velocity, the change of the product of generalized coordinate velocity and its derivative is shown in Figure 5.

This paper assumes that the movement of each individual takes place only in accordance with the function of natural impulses.

3. The Movement of Individuals

Each individual, as well as a set of live individuals, can be imagined and represented as a complex control system with many feedbacks. Some strictly controlled processes are achieved using this system. Generally, control system consists of: inputs (usually equal to the desired output of controlled system), control unit, controlled object and possibly feedbacks. Outputs of control unit are inputs for the controlled object.

Inputs to the control system are usually given by someone or something. Input can be pre-programmed. Goal of the control system function is defined through the input; here it is the goal of the control system of one individual. The starting moment of input action can be determined in accordance with external setting, but can also be programmed. Start of programmable inputs is determined by ratios and values of some states of the object and of the external environment.

With living beings, controlled object is always some elastic or deformable body with many degrees of freedom. The body and its parts can move in space, more or less. That motion obeys the laws of mechanics presented in Section 2. Bodies were designed to take care of themselves; therefore, they are equipped with the parts that can also be considered as controlled objects. These parts are intended as devices for providing communication (e.g.

voice, typing with fingers...), devices that allow survival (food processing equipment) and the devices for reproduction.

If there is no feedback in system then the system works programmed in advance. Control systems usually possess a feedback. Feedback can be external, internal and cross. Information for functioning of the feedback loop is the information on current states of the object and environment. Living beings have a lot of feedbacks of all kinds, making it difficult to analyze their behavior as a control system. It is convenient to detect sequence of feedback loop closing. After that, the object within the first inner loop is a new control structure, which provides a recursive process to analyze the dynamic behavior of the system.

In this paper, we will limit ourselves only to the consideration of the actual mechanical movement of living individuals, illustrated by the movement of a humanoid. This movement consists of macro movement of individual as a whole (quasi-rigid body), and sequence of movements of the parts of individuals, which provides movement of individual as a whole. The difference between these two movements is in feedback used to control movement. To control the macro motion, sensors that locate target in space (usually the eyes, sense of smell, or sense of hearing) are used. To control the movement of body parts, sensors that locate the mutual positions of body parts in space and sensors which detect the load on parts of individual are used.

We will first consider the macro movements (Figure 6).

Each living individual is equipped with the sensors that are firmly attached to it. Actual environment and the impact of the environment on the individual are projected on the measurement axis of a sensor. The image is mapped in the state space of individual. Individual state space is a space that is firmly attached to it. It means that the system is able to move and it moves along with the individual.

Someone or something (e.g. programmed conditions) defines the goal of movement.

If we are spectators that define the task and/or stand far away from the individual as an executor of task, during its motion we shall see Figure 6. Individual moves along a surface or flying in 3D space above a surface with some obstacles around or above which other individuals move. Goal towards which the individual is moving is at some distance from it. Goal may be within sight, but not necessarily. This means that the line connecting the individual and the goal (line of sight) may or may not have to go through obstacles (here through an obstacle O_3, so the goal is not in sight). Between individual and goal there exist obstacles: close immediately to individuals O_3 and j_1 (e.g. in reach of extremities), distant (out of reach in the extremities but in sensors sight) O_1 and j_2 and far (they may be out of sight of sensors) O_2, j_3, O_4, O_5, j_6, O_7 and O_8. The obstacles are moving (e.g. other individuals j_1, j_2, j_3, j_6) and fixed (O_1 to O_8) in relation to the surface on which the movement is performed. An individual is moving along a fictitious leading line (pivotal path) avoiding collisions with obstacles. It boils down to the fact that the leading line doesn't pass through the obstacles and that it passes far enough away from it (at a distance of not less than $d_p \geq r_0$).

An individual is not a spectator. It concludes of all elements of the movement on the basis of information in the moving system $Oxyz$ which is firmly attached to it (body coordinate system). Coordinates x, y, z of moving coordinate system are the parts of the coordinates of the individual's state vector.

The goal is a point in individual's state space. For example, a man is standing at some distance from a tree. Somebody or he himself orders him to go to the tree. The man accepts an order through sensors of hearing or sight sensors (if order is in sign language). The obtained information about the order is processed.

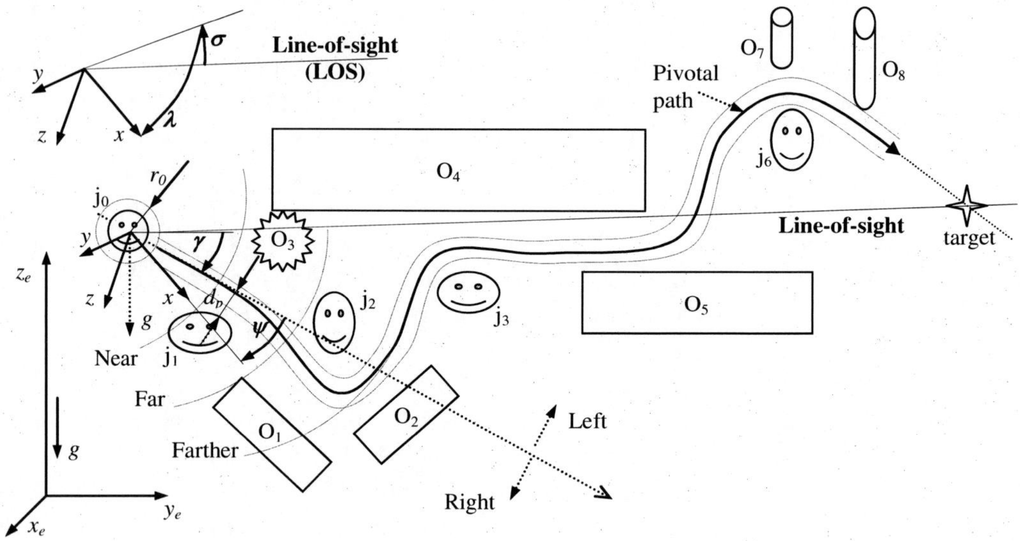

Figure 6. External layout of task motion of individual.

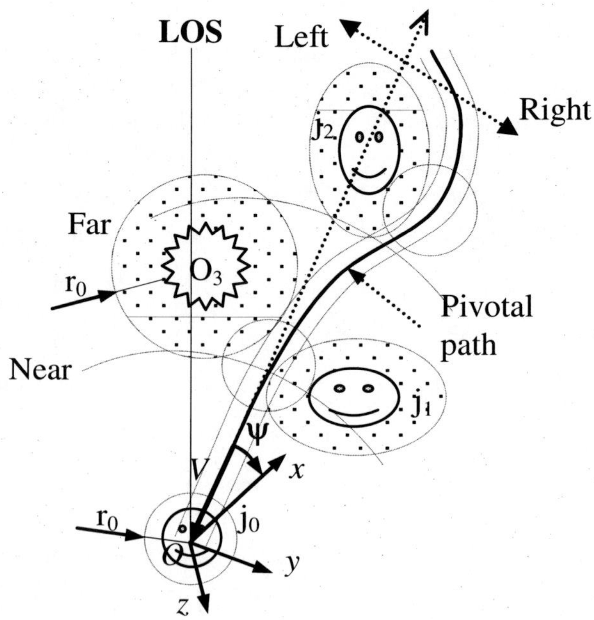

Figure 7. Layout of task motion of individual in body coordinate system, view from above.

After that, a man looks for images of tree in his state space obtained for example with his eyes. Then he compares that image with the image of tree from his own database (memory). When he finds such image, he then memorizes its position in state space. That position in state space is a target point. The task reduces to bringing the target point to a point in state space in which it is currently located. In the body coordinate system $Oxyz$ task boils down to bringing goal in the origin, while taking into account the fixed and moving obstacles.

Next, an individual plans the desired path to reach the tree according to task layout in the body coordinate system. Task layout in body coordinate system is not the same as an external task layout. An individual sees its environment according to Figure 7. Individual sees that goal is on the left in relation to the longitudinal axis x at an angle λ and up, at an angle σ. Pivotal path lies on surface at the angle γ to the horizontal. In direction of the tangent to the path is instantaneous velocity vector V of individual's movement. Longitudinal symmetry plane xz lies at an angle ψ compared to the velocity of movement. Information about the movable and immovable obstacles is used for correct path planning. The path is planned so that during the passage of the body near the obstacles there is no contact (the distance to the obstacle d_p is not less than r_0, $d_p \geq r_0$, Figure 6, 7. Depending on the characteristic of surface where path of movement lies and on the characteristics of fixed or moving obstacles, motion path planning is performed.

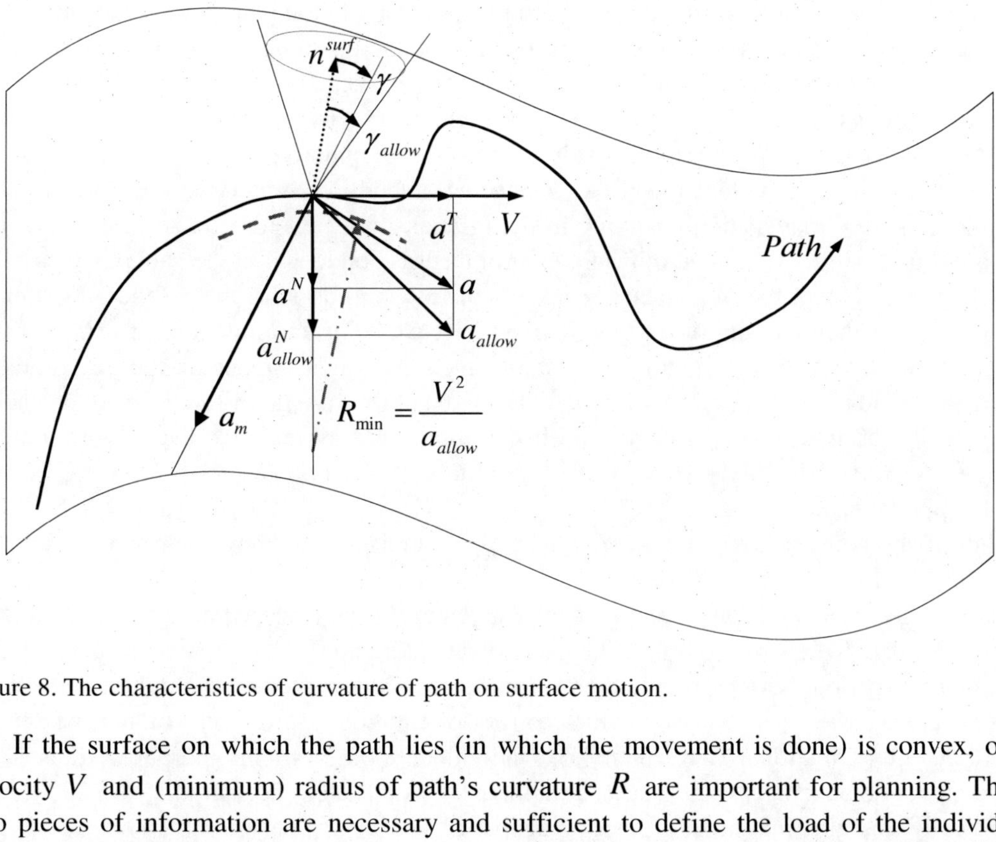

Figure 8. The characteristics of curvature of path on surface motion.

If the surface on which the path lies (in which the movement is done) is convex, only velocity V and (minimum) radius of path's curvature R are important for planning. These two pieces of information are necessary and sufficient to define the load of the individual (module of the central acceleration is $a_N = V^2 / R$). Total kinematic acceleration is

$\vec{a} = \vec{a}_T + \vec{a}_N$, where the module tangential acceleration $a_T = \dot{V}$, V is the velocity vector module. Total load individual proportional to the module of vector-measurable acceleration $\vec{a}_m = \vec{a}_{measurable} = \vec{a} - \vec{g}$, where $\vec{g}$ is the vector of gravitational acceleration (see equation (4)). Each individual has a restriction on the allowed load, which is expressed as allowed acceleration $\vec{a}_{allow}$. When planning the path, the load that an individual suffers must be taken into account. Actual acceleration module of measurable $|\vec{a}_{measurable}|$ and/or its components in different directions should not be greater than the module of allowable acceleration $\vec{a}_{allow}$ or its components. If there is no contact between and if movement velocities are not extreme, then the load limit is the only limit in choosing of the movement path. If there is a contact, then the contact type imposes constraints $|a^s_{allow}|$ in the tangential plane of the surface at which movement is performed. With these limitations $|a^s_{allow}|$, the projection of the total measurable acceleration $\vec{a}^s$ is compared at the point of contact in the tangential plane of the surface on which it performs movement. This is manifested as geometric constraint on motion path curvature R_{min} or γ_{allow} of tilt γ of vector $\vec{a}_m$ compared to normal n^{surf} on the surface at the point of contact (Figure 8).

If the surface is not convex (has bumps, pockets or is stepped), then the planning of path is performed in accordance with pockets or stairs dimension in the plane of movement. In the case that the pockets and steps are small in dimensions the planning can be done in the same way as in the case of convex surface. Otherwise, the pockets and steps must be treated as obstacles and planning must circumnavigate them. While bypassing obstacles maximum overload and the radius of the path and velocity must be taken into account.

When moving, individual images of obstacles in body coordinate system are moving regardless of were they static or moving in the inertial frame. Therefore, the progress of the individual in the immovable and movable environment is conducted in the same manner. The only difference is that the movement of images of moving obstacles in the body coordinate system is faster than moving of images of fixed obstacles. For an individual, it is practically irrelevant whether you are moving in a stationary outer environment or the environment among other individuals. For an individual it is essential to estimate distances from obstacles and velocity of images of obstacles in body coordinate system. Collision with nearby obstacles is avoided directly. Location and distribution of distant obstacles is used to correctly plan the pivotal path.

Information on the movement of obstacles is input to the movement body parts of individual, to avoid its collision with an obstacle.

Let us consider micro movement. If the movement takes place without the individual's own drive, the pivotal path is also the path of an individual. Otherwise, the path of the individual is periodic path around the pivotal path. Generally periodic micro motion is performed in a way that it periodically changes orientation of the body of the individual (performed by rotational movement of individual body) which produce translatory acceleration components. Zeroing these components of translatory acceleration is achieved by periodically changing individual's position in space.

We define special dynamically balanced state (reset state) and special dynamically balanced motion (reset motion) of individual as a state and motion for which natural trihedrons, of two lines along which chosen points (which are also origins of trihedrons) move, are parallel [9]. Special case of this motion is the motion when vectors $\vec{a}_{mB}$ and $\vec{a}_{mD}$, at two chosen points, B and D, are mutually collinear and lie on the line segment connecting these points, $\overline{BD}$, which is equivalent to choosing vanishing point on that direction (Figure 9). Considering the vector of rotational acceleration of one point with respect to other, kinematic relationship for the motion of direction defined by these points along reset directions is found. This relationship comes from the condition that the vector of rotational acceleration does not have orthogonal component to direction defined by the chosen points $\vec{a}_{B/D}^{T} = 0$. Condition is applicable in both gravitational field and out of it. It is also shown that the same condition is applicable if the direction defined by points' moves parallel to its previous position during transition from one reset path to other. This transition can be achieved by slight periodic motion around some pivotal path. This transition can be achieved by some sort of kinematic couple formed by defined direction and direction connecting one of two points and some point on the ground over which the individual moves.

Using previously mentioned method allows for easy execution of any movement of individual. Individual is approximated by any body, and in the special case of a living being. The concrete applications are not the subject of this paper, but nevertheless we can say the following. The methodology set out in this paper allows for easy planning of arbitrary individual motion path; simply separate the control loop of individual and performance of the movement in accordance with current demand of motion and current state of individual, but not repetition of memorized movements and appropriate path. This means that the nominal value is not necessarily stored since it is easy and simple can simultaneously determine in accordance with the current state of individual. We are implicitly set to the normal movement and control of normal movement of individual occurs in the same way that the movement of bicyclist or aircraft occurs. This means that it firstly establishes a 'balance', i.e. individual (direction BD) leads in direction $\vec{a}_m$ so that individual does not rotate in relation to that direction. Further movement normally takes place in a manner that the direction $\vec{a}_m$ is either permanently placed in the direction (coordinated turn aircraft, cyclists in the corner...) or periodically leads to the direction $\vec{a}_m$ (walking, swimming and flying living beings). Tasks about movement control are not the subject of this paper. Based on previous experience of the author (see web site http://dsuoziris.coolpage.biz), the tasks of movement control are: simultaneous calculation of nominal, track of the path of individual as a 3D line, the stabilization of individual on a path (which amounts to stabilize the direction BD in relation to direction $\vec{a}_m$) and control of the internal kinematics of individual (in the case of robots and living beings) that will ensure that the contact point and the direction BD located where they are needed. This setup allows the use of simple algebraic control law. The proposed laws require only algebraic computation. Therefore, the application of our method does not set strict technical requirements for hardware in terms of its velocity or size of storage space. Practically every PIC (programmable interrupt controller, price is a few dollars) can handle all the computation of control law using the methods based on relation $\vec{a}_{B/D}^{T} = 0$.

We consider now motion which is realized as series of reset states connected by transitional process. Transitional processes should be realized by programmed natural impulse function for which both amplitude and duration is programmed. Numerical integrations may be necessary only during transitional processes (which are relatively short) with initial conditions defined by starting reset state.

We illustrate phases which generate a motion of individual by means of relative periodic lateral motion of the direction $\overline{DB}$ with respect to pivotal path. Pivotal path is the average reset path of the point D (Figure 9). Motion of individual along pivotal path is considered as its carrying motion. Around pivotal path, point of individual performs additional relative periodic motion. In this example we consider starting motion, motion, and stopping of the individual. It is assumed that pivotal path is straight line. This example is inspired by recorded motion of a woman given by file AudioMotion Female Walk.bvh (http://acabar.cafwap.net).

Defined dynamically balanced, reset, state of individual assumes that total measurable acceleration at points B and D are collinear and also collinear to the direction $\overline{BD}$. If individual possesses symmetry plane it is useful to choose points D and B on this plane. In that case if individual is in motion such that velocity vector tangents symmetry plane, then in reset state oscillatory plane tangents symmetry plane. It is, then, enough to observe acceleration in one point D, and with internal kinematic ensure that oscillatory plane of the path of this point is tangent to symmetry plane of the individual, and that reset direction is tangent to some line on that plane. Following this logic in example we analyze only kinematics of point D. Here we do not consider internal kinematic of individual, but it is assumed that this kinematics ensures parallel motion of direction $\overline{DB}$ during transition between reset states.

As point B during reset motion we choose point $B = B_{pren}$ of the penetration of axis Dz through surface above which motion fulfills. For the description of periodic motion we choose point $B = B_{rel}$ on the axis Dz (parallel to reset direction on pivotal path), so segment $\overline{DB}$ represents intensity of increment vector $\Delta\vec{\rho}$.

Dynamical equilibrium state of individual is realized by removing of components of total measurable acceleration in points B and D perpendicular on direction $\overline{BD}$, i.e. ensuring collinearity of total measurable acceleration and vector $\vec{\rho} = \overrightarrow{DB}$.

Information $\vec{a}^{T}_{B/D} = 0$ (for example $\vec{\varepsilon}_D = \mathbf{0}$ and $\vec{\omega}_D = \mathbf{0}$) ensures that all measurable accelerations are collinear if overlap direction $\vec{\rho}$. Condition $\vec{a}^{T}_{B/D} = 0$ does not guarantee that measurable accelerations lie on direction $\overline{DB}$. Information that acceleration in point D is collinear with direction $\overline{DB} = Dz$ means that there are no components of acceleration perpendicular on this direction. There is infinitely large number of ways to achieve this. The simplest way is to choose arbitrary point on the individual and then to rotate and deform individual until reset state is achieved.

Let at initial moment individual is in reset state (pose 00, Figure 9), thus direction $\overline{DB}$ overlaps with the direction of total measurable acceleration $\vec{a}_{mD} = -\vec{g}$ at point D. At first individual lowers itself (pose 0*S*) for $d\rho$ to future pivotal path (Figures 9b and 10). With this point D is brought in reset state at pivotal path, this state is initial state for longitudinal individual motion. During lowering dynamic equilibrium is lost because it is altered $d\rho$, $d\dot{\rho}$ and $d\ddot{\rho}$, so that $\vec{a}^{T}_{B/D} \equiv 0$.

Individual can move differently in different directions, and these directions do not need to be mutually perpendicular. Here we consider motion of individual in longitudinal direction along axis x directed toward pivotal path.

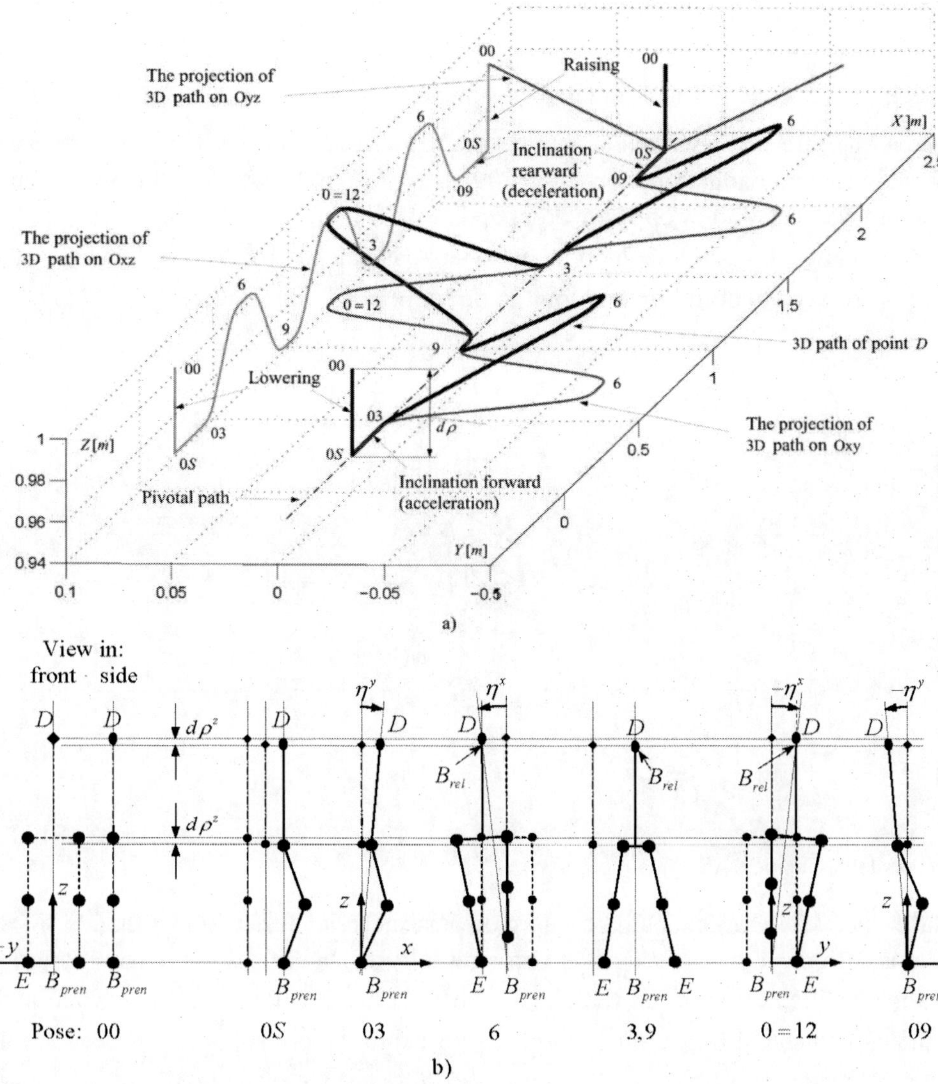

Figure 9. 3D path of the point *D* and its projections.

In this example we assume, firstly, that carrying motion of individual is generated, and then its relative motion. The carrying motion is generated when individual leaves initial reset state in the direction of pivotal path (poses 00, 0*S* and 03, Figure 9).

To generate self-propelled motion of individual, an impulse of measurable total acceleration da^x_{Drel} is generated at point D in the direction of pivotal path Point D will make an angular displacement for angle η^y in the direction of future carrying motion, so that total acceleration in point D in direction of Dx axis will be $da^x_D = 0.5 \cdot g \cdot \sin 2\eta^y + da^x_{Drel}$ (pose 03, Figure 9). Point D gets velocity in the direction of x-axis. After that by rotation of direction $\overline{DB}$ about point D point B is moved from its initial position to the current reset state. With this motion inclination of the individual is lost ($\eta^y = \mathbf{0}$). Consequence is that total measurable acceleration da^x_D along direction of x-axis drops to zero, because relative acceleration da^x_{Drel}, velocity and angular movement of point D with respect to point D in initial reset state are lost. Point D continues to move along x-axis with reached constant velocity. To stop individual motion (pose 09, 0*S* i 00) process is analogous to start of transition, but with rotation of opposite order and sense (Figure 10).

After stopping non-periodic motion, individual is raised for $d\rho$ with which starting reset state is achieved but on different point in different position in space.

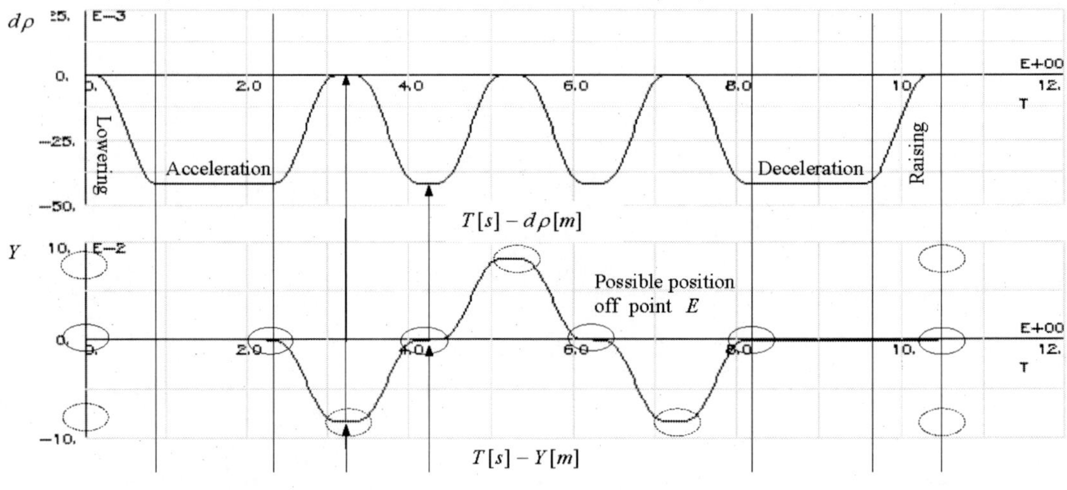

Figure 10. Suitable positions of contact points.

Relative periodic motion could be interpreted as successive transitional motion by which individual is periodically disposed of and brought to reset state. During relative motion point D performs additional motion in the plane arbitrary placed with respect to pivotal path. In this example the plane of transitional motion is normal to the pivotal path direction. Point D can move arbitrary in this plane. Path of point D in transverse plane represents projection of its three-dimensional motion to this plane. For illustration it is taken that the projection of the path is consisted of two intersecting segments with intersection points at the pivotal path (Figure 9). Due to visibility in this example only change of kinematical quantities are given for one and half steps (three half steps). Pivotal path can be generated in two ways. First, it connects saddle points of transition motion, and second it connects extreme points of the path. In this example it connects minimums.

In order to preserve reset motion equations $d\vec{a}_{B/D}^{T}=0$ should be fulfilled and accelerations in point D and B should be collinear with direction $\overline{DB}$. During transitions between reset states with translation of the direction $\overline{DB}$ must be $d\vec{a}_{B/D}^{T}=0$. Transition is done in such way that accelerations in points D and B gets components non-collinear with direction $\overline{DB}$. The $\overline{BD}$ direction is impulsively (programmed) discharged from its dynamic equilibrium, and then is relaxed in order to gain new equilibrium state. If initial state of individual is on pivotal path (pose 6 or 0=12) discharge is performed from pivotal path (for example, from poses 03 or 3 to the left into pose 6). After that individual is impulsively returned back to the pivotal path, where dynamically balanced again. The same procedure is applied for discharge of the $\overline{BD}$ direction in opposite direction form the pivotal path (into pose 0=12) and return to it. During parallel translation of the direction $\overline{DB}$ it is allowed its length variation. In the example it is taken that $d\rho=\overline{DB}_{rel}$ changes its length from 0 on the pivotal path up to the height at initial state (i.e. up to pose 00). The amplitude along y-axis is twice as large.

Impulse functions are chosen such that accelerations in direction Y and Z axis simultaneously drop to zero. Since direction DB is parallel translated simultaneously it is fulfilled $d\vec{a}_{B/D}^{T}=0$ and collinearity of total accelerations with direction $\overline{DB}$, thus, in that moment reset state is fulfilled. Reset states correspond to mean and extreme deflections of $\overline{BD}$ from pivotal path. During normal walking contact E of individual with the ground could be penetration point of the reset direction through the ground surface (Figure 10).

Hence, normal movement of individual is performed analogous to normal movement of man. Normal movement of man takes place in the way that man first sets the plane of symmetry of his body (i.e. a line BD) in direction $\vec{a}_m$. If a man standing on the ground, then is $\vec{a}_m=-\vec{g}$, and the direction $\vec{a}_m$ and direction $\vec{g}$ are same. If a man stands on a platform that performs arbitrary motion (i.e. standing on the plane or on a platform that makes turn) the plane of symmetry of his body is put in the direction of vector $\vec{a}_m=\vec{a}-\vec{g}$. After this, movement (walking) is carried such that body (i.e. direction BD) ejects and again comes back in the direction $\vec{a}_m=\vec{a}-\vec{g}$. The position of direction BD is not the same as in space, and by that the movement in space is realized. Penetrations of these directions through the base may (but need not) be the foot contacts with the base during walking. Direction BD can be defined in different ways. We selected two points for definition of the direction BD. The rest of the humanoid mechanism must move in a way that the laws of motion of direction BD defined in this paper are not broken. This means that the robot dynamics and positions of contact points must simultaneously be planned and controlled in order to preserve the defined laws.

Individual dynamics is considered in a specific way. The conditions for movement direction vector $m\vec{a}_m = \vec{F}_s$, $\vec{F}_s = \vec{F}_{external}$ external load is considered in this paper. Obtained conditions are subsequently used to properly generate and control movement, for example, for individual. Component perpendicular to the direction of vector $m\vec{a}_m$ or component perpendicular to the direction of vector $\vec{F}_s$ should be controlled. On aircraft only direction vector $\vec{F}_s$ are controlled, and vector $m\vec{a}_m$ takes one value that dictates the change of the vector $\vec{F}_s$. Changing the vector $\vec{F}_s$ is dictated by reaction of supports and/or through external action, while the change of vector $m\vec{a}_m$ is dictated by a change of inertial properties of the assembly of individual. Change of the previous vector is achieved by the command to change orientation individual, which results in a change of vector $\vec{F}_s$ and $m\vec{a}_m$.

Methods (set of control laws, i.e. algorithm) that changes the previous vectors are the eigenfunctions of each individual that moves. These methods are stored in each individual. These methods generate movements, so that they do not remember nor repeat the trajectories of individual body parts, but the algorithms provide execution of the trajectory. For example, if a $d\vec{a}^T_{B/D} \neq 0$ then in order to maintain balance it is necessary to produce acceleration of opposite direction. This is done by redeployment of the individual parts of the body weight in the opposite direction of the $d\vec{a}^T_{B/D}$. As already noted, the information on the movement of obstacles is input to the movement of individual body parts in order to avoid its collision with an obstacle. The task of avoiding obstacles determines the direction of desired velocity and acceleration vectors of pivotal path. Earlier we saw that the movement acceleration $d\vec{a}^T_{B/D} \neq 0$ should be intentionally periodically produced to produce a pivotal path and periodic movement of individuals around it. The motions are produced by relocating the masses of individual body parts in the desired direction and acceleration $d\vec{a}^T_{B/D}$. These movements are either programmed or taught during the life of the individual. For example if the obstacle moves towards the origin, then crash is realized upon arrival of obstacle in the environment r_0 (Figures 6 and 7). In order to avoid a collision, picture of obstacle should pass next to the circle of radius r_0. Learning of the individual consists of good estimation of obstacle distance in function of its velocity, and in learning trajectories of their body parts (i.e. learning movements of body parts) that should be promptly executed (according to natural impulse function) in order to pass beside the obstacle at a distance of at least r_0. Spectator would observe that individual is moving away or stepping aside from the obstacle.

Tests carried out in order to improve training of pilots have shown that the characteristics of their behavior depend on their level of training [16]. For example, reaction delay of a well trained pilot is not greater than 0.15s, for information detected with eyes. Reaction delay of an average man is not less than 0.35s, for the same excitation. Training (learning) ensures change in the pilot (i.e. individual) control system parameters. This allows the change in parameters of control systems and individual depending on the current developments in the external environment. In that way, obstacle that moves fast requires faster reaction and slower reaction for the obstacles that moves slow. That is, only the external environment affects the change of individual system parameters, if the individual is capable of it.

4. THE SET OF INDIVIDUALS MOTION

The essence of normal independent movement of individuals is to avoid collisions with obstacles and other individuals. Under free independent motion it is considered a motion where individual sets a motion goal for oneself. This is accomplished in a way that no obstacles come in the vicinity of an individual, so there is no contact or sectional cross between individual and obstacle forms. Maintenance of requirement is that individual distance from obstacles and other individuals d_p is not less than r_0, $d_p \geq r_0$ (Figures 6, 7). Generalization of conditions for normal movement of individual is to avoid hurting itself during movement. It comes down to the requirement where the individual strains do not exceed the limits. Viewed from the side, requirement is that the individual shape should be in acceptable limits. The requirement for individual deformation usually is present when the other individual or obstacle is too close ($d_p < r_0$) and there is a touch or sectional undeformed shape of the individual and undeformed shape of the obstacle. In order for an individual to remain compact and connected it must necessarily adjust its shape to shape of obstacle. These conditions determine the meaning and manner of movement of set of individuals motion.

During motion of set of individuals, individuals that are not leaders (slave entities) get their goals from outer environment. Motion goal of a set is always given by one or more leaders. We will analyze several cases of movement of the set of individuals.

Simplest movement of the set of individuals is moving in formation (i.e. hold system configuration). Moving in formation is strictly regulated movement, i.e. without chaotic movement of some individuals within the set. This movement involves maintaining approximately constant mutual distances of individuals. This does not mean that all distances between each other are the same, but that they are roughly constant between neighboring individuals in the movement. Typical examples are the movement of flocks of birds or schools of fish, troops of soldiers or aircraft formations. During formational movement, each individual unconditionally follows neighboring units. Path of articulated individuals is determined by the leader who is on outer boundary of set. This position provides him a good overview of its environment, and thus also the ability to correctly and easily plan its own path. The condition of leader's movement is again maintenance of not less than its distance obstacles $d_p \geq r_0$, but now r_0 does not apply in an individual, but to a set and usually denotes diameter of a set. During this kind of motion, individuals don't plan their pivotal path; instead it is a consequence of distance maintained during following the leader.

Seen from aside, self-organization during formational movement, i.e. preservation of system configuration, would be seen in a following way. At initial moment, individuals are randomly deployed in space (location). After getting the task to move in formation, individuals take place at the nodes of a fictitious spatial elastic grid. After that we would see formational movement of individuals as movement of a fictitious spatial elastic grid represented through movement of its nodes in which individuals are placed.

More complex are movements of the set of individuals, i.e. individuals in a set of individuals who are moving toward the common goal, tracking of common goals or leaders. The goal of motion is communicated to individual in all these cases. According to its current position in the external environments and current position within the set, every individual comprehends and marks the goal, according to its own performance, as a position within the coordinate system firmly attached to it.

The problem of complex movements is that each individual does not realize the same goal in the same manner. Mathematically it can be described in the following way. The common goal is mapped to motion target for each individual in the coordinate system firmly attached to it. If we would do the inverse mapping of comprehended goals of each individual in the coordinate system attached to them, then those goals wouldn't map into the same position in inertial space, but in more different goals that are generally not given. This means that even if it the same common goal is specified, it does not mean that individuals will move towards the same common goal.

Further movement of individual occurs as independent movement toward comprehended common goal. Position of a goal is usually defined with regard to the obstacles. Such obstacles are the landmarks. They are used for the synthesis of the desired pivotal path of an individual and quality tracking of desired pivotal path during the motion. Control consists of determining the position of the individual in relation to the landmarks, i.e. determine where the landmarks are, in front, behind, left, right, up or down in relation to the individual, during the motion.

For an individual it means approximate determination of quadrant of the coordinate system firmly attached to it, where a specific landmark is. When moving together as set of individuals, other individuals appear as moving obstacles in the coordinate system firmly attached to individual.

These obstacles hinder or prevent visibility of target during the course of a movement towards him. Therefore, the additional problem is the loss of the goal during the course of a movement towards a common goal. When observed from the side, moving of set of individuals can be seen as chaotic movements of individuals. It does not mean that the movement is really chaotic, but that movement of an individual is independently carried out in accordance with their own image of environment.

When moving towards a common goal path shafts of most individuals end in the near vicinity of that goal. Seen from the side, the common goal and misguided common goals are sinks for pivotal path of most individuals. In the coordinate system firmly attached to an individual, comprehended common goal is brought in the immediate vicinity of origin of the coordinate system. Chaotic paths are paths of individuals who have lost their motion target and failed to define it again during the movement. Examples of moving towards a common goal are, e.g. movement of livestock to watering or to the pen.

For the tracking of a common goal or common leaders it is characteristically that the goal is mobile and that individuals control their position in relation to common goal or common leader and in accordance with current mobile environment. Seen from aside the individual moves in order to avoid collisions with other individuals and to keep distance from leader not to change significantly. The trajectories of individuals can be understood as currents that carry individuals in the watercourse. The result is that the set of individuals moves in a way to maintain fictional geometric figure around the followed target/leader. Typical examples are hunting the same prey by the pack of hunters, pursuing of queen bee in bees swarming, etc.

Seen from aside, self-organization during movement towards a common goal, following of a common goal or common leader would be seen in a following way. At initial moment, individuals are randomly deployed in space. After getting the task to move towards the common goal, individuals begin to move as a rigid particles carried by fluid in its turbulent sink flow.

CONCLUSION

The paper considered the self-organization that refers to living individuals. It was pointed out that all living individuals should be considered as control systems, whose control object is a mechanical system. First we was discussing the dynamics of mechanical object and moving in accordance with the functions of natural impulses. As a function of the natural impulses we considered smooth functions whose all derivatives start and end at the same time in zero. The description of the dynamics is adopted such that it separates external load caused by characteristics of environment of a mechanical object and load which depends only on characteristics of the mechanical system as a function of measurable acceleration ($m(\vec{a}-\vec{g})=m\vec{a}_{measurable}$). Further, by using control theory logic, a point of view of individual autonomous motion was given. We assumed that each individual is equipped with its eigenfunction (i.e. control law whose record we call the algorithm) that defines how it will move on the basis of "projection of its environment" on the coordinate system firmly attached to it. Normal motion is where the individual with given goal (self-defined or communicated by leader) plans a pivotal path in accordance with the image of environment in coordinate system firmly attached to it. Sense of normal independent motion is to avoid obstacles during the motion toward a goal. By self-propelled normal motion it was considered a motion where an individual, in impulses, periodically leaves its dynamically equilibrium state (reset state) around the pivotal path and backs in it. It was pointed out that the motion of a set of individuals is different from isolated motions of individual only by characteristics of the motion goal. Ordered motion of a set of individuals, each of which is considered as a control system, is a motion where desired inputs of a system are coordinated, in a certain way. Such motion can be seen as self-organized motion of a set of individuals, observed from aside.

REFERENCES

[1] G.Di Marzo Serugendo, M.-P. Gleizes, and A. Karageorgos, "Self-organization in multi-agent systems," *The Knowledge Engineering Review*, vol. 20, no. 2, p. 165–189, 2005. doi:10.1017/S0269888905000494.

[2] W. Elmenreich, R. D'Souza, C. Bettstetter, and H. De Meer, "A survey of models and design methods for self-organizing networked systems," *Self-Organizing Systems, Lecture Notes in Computer Science*, vol. 5918, pp. 37–49, 2009.

[3] H. Guo, Y. Meng, and Y. Jin, "Swarm robot pattern formation using a morphogenetic multi- cellular based self-organizing algorithm," in *IEEE International Conference on Robotics and Automation, Shanghai International Conference Center*, no. 978-1-61284-385-8 in 11, (Shanghai, China), May 9-13 2011.

[4] D. Kim, "Self-organization for multi-agent groups," *International Journal of Control, Automation, and Systems*, vol. 2, pp. 333–342, September 2004.

[5] W.-M. Shen, P. Will, A. Galstyan, and C.-M. Chuong, "Hormone-inspired self-organization and distributed control of robotic swarms," *Autonomous Robots*, no. 17, p. 93–105, 2004.

[6] M. Živanović and M. Vukobratović, "Control of multi-arm cooperating robots with elastic interconnection at the contacts," *ROBOTICA*, vol. 18, no. 2, pp. 183–193, 2000.

[7] M. Živanović and M. Vukobratović, "A generalized approach to tracking the nominals of mechanical systems," *Archive of Applied Mechanics*, vol. 75, pp. 125 – 137, January 2006.

[8] V. Velichenko, *Matrichno-geometricheskie metod v mekhanike s prilozheniyami k zadacham robototekhniki*. Moskva: Nauka, 1988.

[9] M. Živanović, Z. Petrović, and M. Živanović, "Possible kinematic law of living being motion," *Robotics and Autonomous Systems*, no. doi:10.1016/j.robot.2011.06.009, 2011.

[10] L. Polak, *Variacionnie principi mehaniki*. Moskva: Gasudarstvennoe izdateljstvo fiziko-matematiceskoj literaturi, 1959.

[11] M. motor control, *EPOS Positioning Controller, Documentation, Firmware Specification*. Maxon motor ag., Brünigstrasse 220 P.O. Box 263 CH-6072 Sachseln, maxon document #798675-01 ed., May 2007.

[12] M. Živanović, M. Bajović, P. Stojaković, and G. Memon, "New general approach to the airplane rotation analysis on all angle of attack" *Naučnotehnički Pregled*, vol. XLVIII, pp. 70–79, novembar 1998 (in serbian).

[13] M. Živanović, M. Bajović, P. Stojaković, and B. Rašuo, "New flight dynamics model of aircraft rotation" in *24th Congres of Theoretical and Applied Mechanics*, (Faculty of Mechanical Engineering, Belgrade), p. CD, October 2003.

[14] M. Živanović, M. Bajović, P. Stojaković, and B. Rašuo, "New general approach to the airplane rotation analysis" in *ICAS 2004, 24th Congress of International Council of the Aeronautical Sciences*, (Yokohama,Japan), pp. 1–20, CD:ICAS 2004–4.9.1, Aug.-Sept. 2004.

[15] M. Bajović, M. Živanović, B. Rašuo, and P. Stojaković, "New general approach to airplane rotation analysis", *Transactions of the Japan Society for Aeronautical and Space Sciences*, vol. 53(2010), no. 180, pp. 130–137, 2010.

[16] V. Bodner, *Operator i Letateljnij Aparat*. Moskva: Masimostroenie, 1976.

In: Self-organization: Theories and Methods
Editor: WenJun Zhang
ISBN: 978-1-62618-865-5

Chapter 8

A Cellular Automata Method for Species Migration Process in a Heterogeneous Environment

***WenJun Zhang*[*1,2], *YanHong Qi*[1] *and ZhiGuo Zhang*[1]**
[1]Sun Yat-sen University, Guangzhou, China
[2]International Academy of Ecology and Environmental Sciences, Hong Kong

Abstract

Habitat diversity is an important mechanism for species migration and distribution. A stochastic model for species migration in a heterogeneous environment is described in this chapter. This model is used to describe various types of dynamics in biological processes, such as periodic oscillation, monotonic increase and decline, and fluctuation. The species extinction likelihood in the migrated area largely determines the dynamics and pattern of species migration. There is a significant difference in species migration between heterogeneous and homogeneous habitats. The increasing fitness of species will lead to a faster migration and higher percentage of migrated areas. Existence of boundary areas may retard species migration.

Keywords: species migration; cellular automata; habitat heterogeneity

1. Introduction

Habitat heterogeneity refers to the diversity of soil, climate, vegetation, landscape and land use, etc. Habitat heterogeneity forms an important mechanism for species migration and distribution. Habitat heterogeneity helps to form stable food webs, ecosystems and different types of barriers to species migration (Zhang and Schoenly, 1999). To study species migration in a heterogeneous habitat is important for understanding species distribution.

[*] E-mail: zhwj@mail.sysu.edu.cn, wjzhang@iaees.org

Cellular automata are a kind of agent-based modeling (Griebeler, 2011; Liu and Zhang, 2011; Ballestores and Qiu, 2012). They have been widely used, particularly in land uses, etc. (White and Engelen, 1993; Wu, 1996; Clarke et al., 1997; Batty et al., 1999; Ward et al., 2000; Li and Yeh, 2002, 2004; Qi and Zhang, 2002; Chen and Mynett, 2003; Jiao and Boerboom, 2003; Cheng and Masser, 2004; Ménard and Marceau, 2005; Liu et al., 2007; Moreno et al., 2009; Zhang, 2012). In the present chapter, we describe the cellular automata according to some basic rules of species migration. Various factors influencing species migration were studied using the cellular automata and some general conclusions were described.

2. Cellular Automata

In this chapter, we use the cellular automata of Zhang (2012). Suppose the study region is divided into $c \times f$ grids with equal area, a_{ij}, i=1,2,…,c; j=1,2,…,f. Species migration will occur in this region. Given the fitness of each grid to species migration, q_k ($q_k \in \boldsymbol{R}$, k=1,2,…,m), the corresponding probability of being migrated p_k ($p_k \in [0,1]$, k=1,2,…,m), and the initial distribution of species (that is, whether or not the species occurs in a grid before the simulation (t=0)). The initial distribution is expressed as:

$v_{ij}=1$, if grid a_{ij} has been migrated;

$v_{ij}=0$, if grid a_{ij} is not yet migrated.

$i=1,2,\ldots,c; j=1,2,\ldots,f.$

And
$v_{0i}= v_{f+1\ i}=0,\ i=1,2,\ldots,c+1;$

$v_{i0}= v_{i\ c+1}=0,\ i=1,2,\ldots,f+1.$

Meanwhile, the fitness distribution is also known:

$u_{ij}=q_k$, if the fitness of grid a_{ij} is q_k (k=1,2,…,m).

$i=1,2,\ldots,c; j=1,2,\ldots,f.$

Thus the probability distribution of migration is

$p_{ij}=p_k$, if $u_{ij}= q_k$ (k=1,2,…,m).

$i=1,2,\ldots,c; j=1,2,\ldots,f.$

Suppose each migration process occurs at unit time, and the species migrates into those grids with which the present grid has public edges. For any of the un-migrated grid x, the species will migrate into the grid x with migration probability p_i for the grid x, from the present migrated grid. For fitness k, calculate the number of grids available for the migration, N_k:

$$N_k = \sum_i \sum_j b_{ij}$$

where,

$b_{ij}=1$, if $(v_{ij}=0) \wedge (p_{ij}=p_k) \wedge ((v_{i-1\,j}=1) \vee (v_{i+1\,j}=1) \vee (v_{i\,j-1}=1) \vee (v_{i\,j+1}=1))$ is true;

$b_{ij}=0$, if $(v_{ij}=0) \wedge (p_{ij}=p_k) \wedge ((v_{i-1\,j}=1) \vee (v_{i+1\,j}=1) \vee (v_{i\,j-1}=1) \vee (v_{i\,j+1}=1))$ is false.

And the number of actually migrated grids with fitness k is:

$n_k = \text{INT}\ ((100\ p_k\ N_k + 0.5)/100),\ k=1,2,\ldots,m.$

As a result, n_k actually migrated grids can be a generated among N_k grids. On the other hand, the migrated grid will restore to un-migrated one (species extinction) at the restoration rate r (proportion of grids that can restore from migrated state during unit time):

v_{ij}, if $1-r+R \geq 1$;

$v_{ij}=0$, if $1-r+R<1$.

$i=1,2,\ldots,c;\ j=1,2,\ldots,f.$

where R is a random value, $1 \geq R \geq 0$.

For each unit time, repeat the above iterative process until the scheduled terminal time reaches.

3. Results

3.1. Influence of Restoration on Migration Process

Suppose that there is a region with 15 grids along both the x-axis and y-axis. The terminal time for simulation is 40; there are 4 fitness levels ($m=4$), i.e., $q_1=0$, $q_2=1$, $q_3=2$, $q_4=3$, and the corresponding migration probability is $p_1=0$, $p_2=0.2$, $p_3=0.3$, $p_4=0.5$, respectively. Set the restoration rate $r=0$, 0.05, 0.1, 0.15, 0.2, 0.25, and 0.3, and simulate species migration process using the Java algorithm given in Zhang (2012).

Obviously, if all grids are not restorable, the proportion of migrated grids will increase monotonically with time, and the upper limit of the proportion tends to be a certain value. One reason for the limited upper proportion is that the fitness of some grids is 0 (and migration probability is 0) and the species is not able to migrate into these grids. As the restoration rate increases, the proportion of migrated grids decreases gradually. Once the restoration rate reaches a certain value, the migrated grids will disappear after a period of time. The cellular automata can describe the various types of migration processes, such as the monotonous increase and decline, fluctuation, and periodic oscillation of migration area. For example, the periodic oscillation may occur for $0.10 \geq r \geq 0.05$. For restoration rate r=0.075, the periodic oscillation occurs. As often met in practical situations, such periodic oscillation is not very strict. It has a certain time range, and is not so regular.

The results above indicate that restoration influences, or even determines, the species migration dynamics.

3.2. Influence of Grid Fitness on Migration Process

Fitness corresponds to migration probability. Increased fitness denotes a larger migration probability, and vice versa. In the above example, there are four fitness levels. Set restoration rate r=0, and for the fitness 0, 1, 2, and 3, we fix seven groups of migration probabilities for heterogeneous habitat, (0,0.1,0.2,0.3), (0,0.2,0.3,0.5), (0,0.3,0.4,0.6), (0,0.4,0.5,0.7), (0,0.5,0.6,0.8), and for homogeneous habitat, (0.2,0.2,0.2,0.2) and (0.1,0.1 , 0.1,0.1), and analyze the influence of grid fitness on the migration process.

It is found that migration processes are much different between the heterogeneous and homogeneous habitats. Once a region in the homogeneous habitat has a certain value of fitness, the species would migrate into the entire region in a period of time. If the grid fitness was high, then the upper limit of migration area was high; the species infects more quickly and the time to reach the upper limit was short.

The upper limit of the proportion is about 43.5% to 47.6% in the heterogeneous habitat. In the heterogeneous habitat (or homogeneous habitat) the fitness of grids influence migration processes, but arenot decisive.

3.3. Influence of Border Barriers on Migration Process

Border barriers refer to the continuous grids that the species cannot migrate. The migration probability (and fitness) is 0 for these grids.

We fix four sets of migration probabilities, (0,0.2,0.3,0.5), (0,0.3,0.4,0.6), (0.1,0.2,0.3,0.5), and (0.1, 0.3, 0.4, 0.6), in which the later two sets represent non-border barriers within the region (but it does not mean the habitat is homogeneous because all grids have different s of fitness), and restoration rate r=0, or 0.1. The first two sets have border barriers and the restoration rate r=0.

Simulation results showed that, with or without border barriers, there is a major impact on the migration process. The presence of border barriers can significantly prevent species migration (Zhang and Schoenly, 1999). Border barriers and restoration have decisive impacts on species migration.

4. Discussion

The cellular automata described in the present chapter can be used to analyze the effects of habitat diversity, i.e., mosaic pattern and plant distribution on species migration and population outcome; to determine the influence of species migration and effect on environments, and to optimize the habitat pattern and plant distribution to adjust species migration. The resolution of a grid partition should depend on the understanding of fitness and migration probabilities of grids in the region studied, and the size of the region. A finer partition of the region, that is, more grids, will lead to high simulation accuracy. Methods of generating random numbers are different on different computers, so different computer systems will yield slightly different simulation results. This error can be reduced with the definition of a grid partition.

More variables or parameters are expected in the present cellular automata so that this model can be used in more specific situations.

References

Ballestores F Jr, Qiu ZY. 2012. An integrated parcel-based land use change model using cellular automata and decision tree. *Proceedings of the International Academy of Ecology and Environmental Sciences,* 2(2): 53-69

Batty M, Xie Y, Sun Z. 1999. Modeling urban dynamics through GIS-based cellular automata. *Computers, Environment and Urban Systems,* 23(3): 205-233

Chen Q, Mynett AE. 2003. Effects of cell size and configuration in cellular automata based prey-predator modelling. *Simulation Modelling Practice and Theory*, 11(7-8): 609-625

Cheng J, Masser I. 2004. Understanding spatial and temporal processes of urban growth: Cellular automata modeling. *Environment and Planning B: Planning and Design*, 31(2): 167-194

Clarke KC, Hoppen S, Gaydos L. 1997. A self-modifying cellular automaton model of historical urbanization in the San Francisco bay area. *Environment and Planning B: Planning and Design,* 24(2): 247-261

Griebeler EM. 2011. Are individual based models a suitable approach to estimate population vulnerability? - a case study. *Computational Ecology and Software*, 1(1): 14-24

Jiao J, Boerboom L. 2003. Transition Rule Elicitation Methods for Urban Cellular Automata Models in Innovations. In: *Design & Decision Support Systems in Architecture and Urban Planning* (Van Leeuwen J, Timmermans H, eds). Springer, Netherlands

Li X, Yeh AG. 2002. Neural-network-based cellular automata for simulating multiple land use changes using GIS. *International Journal of Geographical Information Science,* 16(4): 323-343

Li X, Yeh AG. 2004. Data mining of cellular automata's transition rules. *International Journal of Geographical Information Science,* 18(8): 723-744

Liu GH, Zhang WJ. 2011. Computer generation of initial spatial distribution for cell automata. *Computational Ecology and Software*, 1(4): 244-248

Liu XP, Li X, Yeh AG, et al. 2007. Discovery of transition rules for geographical cellular automata by using ant colony optimization. *Science in China, Series D: Earth Sciences,* 50(10): 1578-1588

Ménard A, Marceau DJ. 2005. Exploration of spatial scale sensitivity in geographic cellular automata. *Environment and Planning B: Planning and Design*, 32: 693-714

Moreno N, Wang F, Marceau DJ. 2009. Implementation of a dynamic neighborhood in a land-use vector-based geographic cellular automata model. *Computers, Environment and Urban Systems*, 33: 44-54

Qi YH, Zhang WJ. 2002. A migration model for Species migration in diverse habitat and net work computing software. *Modern Computer*, 133: 16-19

Ward DP, Murray AT, Phinn SR. 2000. A stochastically constrained cellular model of urban growth. *Computers, Environment and Urban Systems*, 24(6): 539-558

White R, Engelen G. 1993. Cellular automata and fractal urban form: A cellular modelling approach to the evolution of urban land-use patterns. *Environment & Planning A*, 25(8): 1175-1199

Wu F. 1996. A linguistic cellular automata simulation approach for sustainable land development in a fast growing region. *Computers, Environment and Urban Systems*, 20(6): 367-387

Zhang WJ, Schoenly KG. 1999. IRRI Biodiversity Software Series. III BOUNDARY: a program for detecting boundaries in ecological landscapes. *IRRI Technical Bulletin No.3.* International Rice Research Institute, Manila, Philippines

Zhang WJ. 2012. *Computational Ecology: Graphs, Networks and Agent-based Modeling.* World Scientific, Singapore

In: Self-organization: Theories and Methods
Editor: WenJun Zhang
ISBN: 978-1-62618-865-5

Chapter 9

ROBUST SELF-ADAPTIVE KALMAN FILTER WITH THE R AND Q ADAPTATIONS AGAINST SENSOR/ACTUATOR FAILURES

***Chingiz Hajiyev**[*1]* ***and Halil Ersin Soken**[2]*
[1]Istanbul Technical University, Aeronautics and Astronautics Faculty,
Istanbul, Turkey
[2]The Graduate University for Advanced Studies (Sokendai),
Department of Space and Astronautical Science,
Sagamihara, Kanagawa, Japan

ABSTRACT

In this chapter a Robust Self-Adaptive Kalman Filter (RSAKF) algorithm with the filter gain correction is developed for the case of sensor/actuator malfunctions. The proposed RSAKF utilizes time variable factors in order to reduce the effect of the faults on the estimation procedure. In this sense, the procedures with single and multiple factors for the adaptation of the filter are presented. In the first case, the filter is adapted by using single adaptive factor as a corrective term on the filter gain while in the second one, an adaptive matrix built of multiple adaptive factors is used to fix the relevant term of the Kalman gain matrix, individually. After chosing the efficient method of adaptation, an overall concept for the RSAKF is proposed. In this concept, the filter detects the type of the fault, either in the sensors or actuators, and after the fault isolation it applies the required adaptation process such that the estimation characteristic is not deteriorated. Effectiveness of the proposed filters is investigated via simulations for the state estimation problem of an UAV. The results of the presented algorithms are compared for different types of sensor/actuator faults and in this context recommendations about their utilization are given.

Keywords: Robust Self-Adaptive Kalman Filter, Unmanned Aerial Vehicle, state estimation, sensor fault, actuator fault

[*] Email: cingiz@itu.edu.tr

1. INTRODUCTION

Nowadays, researchers especially in the military field aim at reducing human factor in flights, so as to keep both human life loss and economic cost as low as possible. As a consequence of this pursuit, there is an overwhelming interest in the Unmanned Aerial Vehicles (UAVs) [1]. Since they reduce the risk of loss of human life, they have been preferred as a part of dangerous, dull and dirty missions and used in many military operations especially beginning in 1990s [2]. However, as an obligation for the successful mission, their control system must be designed perfectly because of the absence of a human pilot who can take initiative in case of an unexpected failure. In this sense, navigation system design of such autonomous vehicles is an important topic that is under discussion [3].

When there are sufficient measurements the necessary states of an autonomous aerial vehicle can be estimated via the Kalman Filter (KF) [4]. Such a procedure is necessary since it is important to know the parameters like velocity, altitude, attitude etc. precisely. In terms of the UAVs, accurate estimation of these parameters is a preliminary obligation for controlling the vehicle successfully. However, both the sensors and actuators should be fault free for such an accurate estimation procedure. Otherwise the filter gives inaccurate results and possibly diverges by time. Therefore, the filter should be built robust in order to achieve fault tolerance in the design of the UAV flight control system.

The KF approach to the state estimation is quite sensitive to any malfunctions. If the real time condition of the system does not correspond to the models used in the synthesis of the filter then these changes resulting from some possible failures at the sensors or actuators significantly decrease the effectiveness of the estimation system. In this case, the Adaptive Kalman Filters (AKFs) can be used for recovering from the possible malfunctions.

The KF can be made adaptive and hence insensitive to the priori sensor or actuator failures by using various different techniques. The basic approaches to the adaptive Kalman filtering problem are the Multiple-Model-based Adaptive Estimation (MMAE) [5-7], Innovation-based Adaptive Estimation (IAE) [6, 8, 9] and Residual-based Adaptive Estimation (RAE) [10, 11].While in the first approach a bank of Kalman filters run in parallel under different models for the filter's statistical information, in the rest the adaptation is done directly to the covariance matrices of the measurement and/or system noises based on the changes in the innovation or residual sequences.

As a drawback, the methods described in [5, 7, 12] assume that the faults are known, and so the designed parallel running KFs are designed for the known sensor/actuator faults. Besides, this approach requires a high computational performance because of several parallel KFs running simultaneously.

For estimating the covariance matrices by the IAE and RAE [6, 8-11] the innovation or residual vectors must be used for *m* epoch. This increases the storage burden and presents the determination of the width of the moving window, *m*, as another problem. Furthermore, the IAE and RAE estimators require that the number, type and distribution of the measurements for all epochs within a window should be consistent. If they do not, the covariance matrices of the measurement noises cannot be estimated based on the innovation or residual vectors.

In [13] an approach to detect and isolate the aircraft sensor/actuator faults affecting the mean of the KF innovation is presented. The effects of the sensor and actuator faults in the innovation of the channels are investigated, and a decision approach to isolate the sensor and actuator faults is proposed. When a KF is used, the decision statistics changes regardless the fault is in the sensors or in the actuators. This paper presents a Robust Kalman Filter (RKF) based on the Doyle-Stein in order to distinguish the sensor and actuator faults. Although it proposes an effective method for the fault isolation, this work examines the sensor and actuator faults affecting only the mean of the innovation, and as another drawback the given measurement noise covariance adaptation procedure is difficult to realize.

There are also documented studies about the adaptation of the KF by using fuzzy logic and neural networks. The AKF presented in [14] has been applied to fuse position signals from the Global Positioning System (GPS) and Inertial Navigation System (INS) for the autonomous mobile vehicles. The Extended Kalman Filter (EKF) and the noise characteristic have been modified using the fuzzy logic adaptive system. In [15], a method of multi-sensor data fusion based on the adaptive fuzzy KF is presented. This method is applied in fusing position and orientation signals from the dead reckoning system and the GPS for landing vehicle navigation. The EKF and characteristics of the measurement noise are modified by using the fuzzy adaptive system, which is based on a covariance matching technique. It has been demonstrated that the fuzzy AKF gives better results (more accurate) than the EKF [14, 15]. In [16], the fuzzy logic-based AKF is used to build adaptive centralized, decentralized, and federated KFs for adaptive multi sensor data fusion. The adaptation is carried out in the sense of adaptively adjusting the measurement noise covariance matrix of each local filter to fit the actual statistics of the noise profiles that is present in the incoming measured data. A fuzzy inference system based on a covariance-matching technique is used as the adaptation mechanism in the paper. The simulation results show that the architectures proposed by the authors are effective in the situations where there are several sensors measuring the same parameters, but each with different measurement dynamic and noise statistics. As an example for the neural network based fuzzy self-adaptive KF, in [17] a combined learning algorithm that uses the fuzzy self-adaptive KF is adopted to train the neural network. Experimental results show that it enables the training process with an overall accuracy and rapid convergence speed. In [18] the radial basis function neural network based on the fuzzy self-adapting KF is applied to control the manipulator motion. The test shows that the automatic control system of neural network is effective.

Although fuzzy logic based AKF algorithms perform well under some specific circumstances, they are knowledge-based systems operating on linguistic variables and these methods, which are based on the human experiences, are not widely applicable to the vital systems such as the aircraft flight control systems.

Another concept is to scale the noise covariance matrix by multiplying it with a time dependent variable. One of the methods for constructing such algorithm is to use a scale factor as a multiplier to the process or measurement noise covariance matrices [19]. When the process noise covariance matrix is scaled (Q adaptation) that means there is an actuator fault to be overcome. Otherwise, when the measurement noise covariance matrix is scaled (R adaptation), a sensor fault is dealt with.

In [20, 21], an adaptation procedure based on single fading factor and in [22] a procedure based on multiple fading factors are presented. The essence of using multiple fading factors is the performance of KF that varies for each variable and actually that is sensible especially for complex multivariable systems where it may be not sufficient to use single fading factor as a multiplier for the covariance matrices.

Nonetheless, as a disadvantage, these papers [20-22] takes only the Q adaptation procedure into consideration and do not examine the R adaptation methods. Per contra, the R adaptation is investigated for possible sensor faults in [23] with a considerably simpler technique, but the Q adaptation is not touched upon. It is shown that if there is a malfunction in the measurement system, the RKF algorithm can be utilized and by the use of a measurement noise scale factor (or adaptive factor) as a multiplier on the measurement noise covariance matrix, insensitiveness of the filter to the current sensor faults can be satisfied.

Moreover there are also published works in the literature which are interested in both the R and Q adaptations [24-27]. Sage-Husa adaptive filter algorithm is available for online simultaneous estimation of statistical characteristics of the system and measurement noises. This filter is based on Maximum a Posterior (MAP) and estimates the measurement noise and system disturbance easily and on-time. But there are many problems, such as it cannot estimate system process and measurement noise covariances at the same time in a way that guarantees the stability and convergence. In these papers an isolation scheme for the fault is absent. These filters can detect whether there is a fault or not in the sensor/actuator, but the type of the fault cannot be determined.

Instead of using a fault isolation scheme a simultaneous adaptation procedure for the R and Q adaptations is performed in these studies. But as indicated in [28], full estimation of the R and Q based on the covariance matching is questionable since these two values are relying on each other in the covariance matching equation. Consequently, these filters cannot estimate the UAV dynamics reliably in the presence of sensor/actuator faults. This problem can be solved through the combined fault isolation and RKF techniques. Besides, the methods proposed in these papers are all based on the single adaptive factor and for complex multivariable systems they may not be sufficiently accurate.

In this study a Robust Self-Adaptive Kalman Filter (RSAKF) algorithm with the multiple adaptive factors is proposed. The RSAKF incorporates the R and Q self-adaptation procedures and adapts itself against the sensor/actuator faults in order to secure the robustness.

The filter first detects the type of the fault (either sensor or actuator) and then isolates it. After that the appropriate adaptation (the R or Q adaptation) is applied such that the estimation characteristic is not deteriorated. Unlike the existing studies, the proposed adaptation methods for both the R and Q adaptations are simplistic and can be applied easily with a small modification to the KF. Besides in this paper it is shown how to incorporate these two different adaptation processes such that the fault is isolated and so the RSAKF algorithm is a complete estimator including the fault detection and isolation processes for the case of sensor/actuator faults. The RSAKF is applied for the state estimation procedure of an UAV and its effectiveness is investigated by the comparisons with the Optimal Kalman Filter (OKF) in case of fault.

The chapter proceeds as follow; in section 2 the flight dynamics model of the UAV is given. In section 3 the OKF is introduced. In section 4, the R adaptation with both single and multiple factors is presented and via simulations, accuracies of these two different adaptation procedures are investigated. In section 5, the Q adaptation method is given. In section 6 an overall scheme for the RSAKF is presented in the light of the results obtained in advance. Section 7 gives a brief summary of the obtained results and the conclusions.

2. Mathematical Model of the UAV Flight Dynamics

The dynamic characteristic of an aircraft must be known in order to build a KF for the state estimation. In general, the equation derivation process for an aircraft may be examined in two steps; the derivation of the rigid body equations of motion and linearization.

In the first step, 6 rigid body equations which consist of 3 force and 3 moment equations, are obtained for the UAV. With the assumptions of;

- the aircraft is a rigid body,
- the mass of the aircraft remains constant for a relatively short duration of time,
- xz plane of the aircraft is the plane of the symmetry,

the following force and moment equations can be written for an aircraft [29]:

$$F_x = m(\dot{u} + qw - rv), \tag{1}$$

$$F_y = m(\dot{v} + ru - pw), \tag{2}$$

$$F_z = m(\dot{w} + pv - qu), \tag{3}$$

$$L = \dot{p}I_{xx} + qr(I_{zz} - I_{yy}) - (\dot{r} + pq)I_{xz}, \tag{4}$$

$$M = \dot{q}I_{yy} - pr(I_{zz} - I_{xx}) + (p^2 - r^2)I_{xz}, \tag{5}$$

$$N = \dot{r}I_{zz} + pq(I_{yy} - I_{xx}) + (qr - \dot{p})I_{xz}. \tag{6}$$

Here, m is the mass of the aircraft; u, v, w are the velocity components and F_x, F_y, F_z are the forces acting on the plane in the body frame along x, y and z directions; p, q, r are the angular rates and L, M, N are the moments about x, y and z axes in the body frame; I_{xx}, I_{yy}, I_{zz} and I_{xz} are the moments of inertia and the product of inertia terms for the related axes.

These nonlinear equations can be linearized by using the small perturbation theory [30]. Hereafter, $\Delta(\cdot)$ term is used for representing the perturbed state.

In order to apply the method, these assumptions are also made:

- The motion of the aircraft consists of small perturbations about the steady flight condition.
- The reference flight condition is symmetric.
- The propulsive forces remain constant.
- x axis is considered as it coincides with the direction of the aircraft velocity vector.
- If $order\ of\,(\Delta q),\ldots,(\Delta v) = \varepsilon$, then neglect the terms with the order of ε^2.

After that, the linearized equations of motion for the aircraft can be derived [29,30]. In general, the equations are examined in two phases; longitudinal and lateral.

Consequently, the linearized longitudinal equations of motion of the UAV in the state space form is,

$$\begin{bmatrix}\Delta\dot{u}\\ \Delta\dot{w}\\ \Delta\dot{q}\\ \Delta\dot{\theta}\\ \Delta\dot{h}\end{bmatrix}=\begin{bmatrix}X_u & X_w & 0 & -g & 0\\ Z_u & Z_w & u_0 & 0 & 0\\ M_u+M_{\dot{w}}Z_u & M_w+M_{\dot{w}}Z_w & M_q+M_{\dot{w}}u_0 & 0 & 0\\ 0 & 0 & 1 & 0 & 0\\ 0 & -1 & 0 & u_0 & 0\end{bmatrix}\begin{bmatrix}\Delta u\\ \Delta w\\ \Delta q\\ \Delta\theta\\ \Delta h\end{bmatrix}$$

$$+\begin{bmatrix}X_{\delta e} & X_{\delta T}\\ Z_{\delta e} & Z_{\delta T}\\ M_{\delta e}+M_{\dot{w}}Z_{\delta e} & M_{\delta T}+M_{\dot{w}}Z_{\delta T}\\ 0 & 0\\ 0 & 0\end{bmatrix}\begin{bmatrix}\Delta\delta_e\\ \Delta\delta_T\end{bmatrix}, \tag{7}$$

and the linearized lateral equations of motion of UAV in the state space form is,

$$\begin{bmatrix}\Delta\dot{\beta}\\ \Delta\dot{p}\\ \Delta\dot{r}\\ \Delta\dot{\phi}\end{bmatrix}=\begin{bmatrix}\frac{Y_\beta}{u_0} & \frac{Y_p}{u_0} & -\frac{u_0-Y_r}{u_0} & \frac{g\cos(\theta_0)}{u_0}\\ L_\beta & L_p & L_r & 0\\ N_\beta & N_p & N_r & 0\\ 0 & 1 & 0 & 0\end{bmatrix}\begin{bmatrix}\Delta\beta\\ \Delta p\\ \Delta r\\ \Delta\phi\end{bmatrix}+\begin{bmatrix}0 & \frac{Y_{\delta r}}{u_0}\\ L_{\delta a} & L_{\delta r}\\ N_{\delta a} & N_{\delta r}\\ 0 & 0\end{bmatrix}\begin{bmatrix}\Delta\delta_a\\ \Delta\delta_r\end{bmatrix}. \tag{8}$$

Here, $\Delta\delta_e$, $\Delta\delta_a$ and $\Delta\delta_r$ are the elevator, aileron and the rudder deflections, $\Delta\delta_T$ is the change in the thrust, $\Delta\theta$ is the pitch angle about y axis, $\Delta\phi$ is the roll angle about x axis, $\Delta\beta$ is the sideslip angle, Δh is the height, θ_0 and u_0 are the values of related terms in the steady state flight, g is the gravity constant and X_u, X_w, $X_{\delta e}, X_{\delta T}, Z_u$, $Z_w, Z_{\delta e}, Z_{\delta T}$, M_u, M_w, M_q, $M_{\dot{w}}$, Y_r, Y_p, Y_β, $Y_{\delta r}$, L_β, L_p, L_r, $L_{\delta a}$, $L_{\delta r}$, $M_{\delta e}$, $M_{\delta T}, N_{\delta a}$, $N_{\delta r}$, N_β, N_p, N_r are the stability derivatives.

3. Optimal Kalman Filter for UAV State Estimation

When the Kalman filter algorithm is built for the UAV state estimation, we worked on combined longitudinal and lateral dynamics of the aircraft. Hence, the state vector to be estimated is formed of $n=9$ states as,

$$x=\begin{bmatrix}\Delta u & \Delta w & \Delta q & \Delta\theta & \Delta h & \Delta\beta & \Delta p & \Delta r & \Delta\phi\end{bmatrix}^T, \tag{9}$$

while the control input vector is,

$$u = \left[\Delta\delta_e \quad \Delta\delta_T \quad \Delta\delta_a \quad \Delta\delta_r\right]^T . \tag{10}$$

Then, let us introduce the UAV process and observation models for the combined dynamics in the state space form as follow:

$$x_k = F_k x_{k-1} + B_k u_{k-1} + G_k w_k \ , \tag{11}$$

$$y_k = H_k x_k + v_k \ , \tag{12}$$

where, F_k is the system dynamics matrix, B_k is the control distribution matrix, y_k is the measurement vector, G_k is the transition matrix of system noises, H_k is the measurement matrix, which is 9×9 identity matrix in case and w_k and v_k are successively, white Gaussian system process and measurement noises;

$$E\left[w_k w_j^T\right] = Q_k \delta_{kj} , \tag{13}$$

$$E\left[v_k v_j^T\right] = R_k \delta_{kj} , \tag{14}$$

$$E\left[w_k v_j^T\right] = 0 . \tag{15}$$

Here, Q_k is the process noise covariance matrix, R_k is the measurement noise covariance matrix and δ_{kj} is the Kronecker delta function as,

$$\delta_{kj} = \begin{cases} 1, & k = j \\ 0, & k \neq j \end{cases} . \tag{16}$$

After that, the OKF for such combined UAV model can be given by these following steps [31]:

State prediction:

$$\tilde{x}_{k/k-1} = F_k \hat{x}_{k-1/k-1} + B_k u_{k-1} . \tag{17}$$

Covariance prediction:

$$P_{k/k-1} = F_k P_{k-1/k-1} F_k^T + G_k Q_k G_k^T . \tag{18}$$

Innovation:

$$\tilde{e}_k = y_k - H_k \tilde{x}_{k/k-1} . \tag{19}$$

Optimal Kalman Gain:

$$K_k = P_{k/k-1}H_k^T\left(H_kP_{k/k-1}H_k^T + R_k\right)^{-1}. \tag{20}$$

State estimation:

$$\hat{x}_{k/k} = \tilde{x}_{k/k-1} + K_k\tilde{e}_k . \tag{21}$$

Covariance estimation:

$$P_{k/k} = \left(I - K_kH_k\right)P_{k/k-1} . \tag{22}$$

Here $\tilde{x}_{k/k-1}$ is the predicted state vector, $P_{k/k-1}$ is the predicted covariance matrix, $\tilde{e}_k$ is the innovation sequence, K_k is the optimal Kalman gain, $\hat{x}_{k/k}$ is the estimated state vector and $P_{k/k}$ is the estimated covariance matrix for discrete step k .

4. R-Adaptation for UAV State Estimation

4.1. RKF with Single Adaptive Factor

In normal operation conditions, where any kind of measurement malfunction is not observed, the OKF (17)-(22) gives sufficiently good estimation results. However, when the measurements are faulty because of malfunctions in the estimation system such as abnormal measurements, sudden shifts or step-like changes in the measurement channel etc. the filter estimation outputs become inaccurate. Therefore, a RKF algorithm, which brings fault tolerance to the filter and secures accurate estimation results in case of faulty measurements without affecting the remaining good estimation characteristic, should be introduced.

The base of the RKF is the comparison of real and theoretical values of the covariance of the innovation sequence [23]. When the operational condition of the measurement system mismatches with the model used in the synthesis of the filter, then the KF gain changes in accordance with the differentiation in the covariance matrix of the innovation sequence. Under these circumstances, covariance matrix of the innovation sequence differs as:

$$P_{e_k} = H_kP_{k/k-1}H_k^T + S_kR_k \tag{23}$$

and so the Kalman gain becomes

$$K_k = P_{k/k-1}H_k^T\left(H_kP_{k/k-1}H_k^T + S_kR_k\right)^{-1} \tag{24}$$

Here S_k is the single measurement noise scale factor or single adaptive factor (SAF).

Due to the approach, the Kalman gain is changed when the predicted observation $H_k\hat{x}_{k/k-1}$ is considerably different from the actual observation y_k because of the significant changes in the operational condition of the measurement system. In other word, if the real value of filtration error exceeds the theoretical error significantly as

$$tr\left\{\tilde{e}_k\tilde{e}_k^T\right\} \geq tr\left\{H_kP_{k/k-1}H_k^T + R_k\right\} \tag{25}$$

then the filter must be run adaptively. Here $tr\{\cdot\}$ denotes the trace of the related matrix.

In order to determine the adaptive factor, S_k, let substitute (23) into the (25) and put in that the adaptation aims at satisfying the equality condition for (25),

$$tr\left\{\tilde{e}_k\tilde{e}_k^T\right\} = tr\left\{H_kP_{k/k-1}H_k^T\right\} + S_k tr\left\{R_k\right\} \tag{26}$$

Then, in the light of $tr\left\{\tilde{e}_k\tilde{e}_k^T\right\} = \tilde{e}_k^T\tilde{e}_k$ equality, S_k can be written as

$$S_k = \frac{\tilde{e}_k^T\tilde{e}_k - tr\left\{H_kP_{k/k-1}H_k^T\right\}}{tr\left\{R_k\right\}} \tag{27}$$

In case of a malfunction in the measurement system, the adaptation of the KF (17)-(22) is performed automatically by correcting the Kalman gain. If the equality condition for (25) is met, then the SAF S_k increases. Higher S_k causes a smaller Kalman gain (24) because of the covariance of the innovation sequence (23) that takes a larger value in the robust case. Consequently, small Kalman gain value reduces the effect of the faulty innovation sequence on the state estimation process (21). In all other cases, where the measurement system operates normally, the SAF takes the value of $S_k = 1$ and so the filter runs optimally.

4.2. RKF with Multiple Adaptive Factors

As it is discussed, it is possible to adapt the filter by using single adaptive factor as a corrective term on the filter gain [23], but that is not a healthy procedure since each term of the innovation sequence has a different effect on the filter performance, especially in a case where one of the sensors is faulty but the others are working properly. [22]. The preferred method is to use an adaptive matrix built of multiple adaptive factors (MAF) to fix the relevant term of the Kalman gain matrix, individually.

In order to determine the scale matrix, again an innovation based process may be followed. It is known that, KF innovation sequence can be determined by (19). Then, as the next step, real and theoretical values of the innovation covariance matrix must be compared. However, this time, different than the procedure for the SAF calculation, we need to perform matrix calculations. Hence, if a scale matrix, S_k, is added into the algorithm as,

$$\frac{1}{\mu}\sum_{j=k-\mu+1}^{k}\tilde{e}_j\tilde{e}_j^T = H_k P_{k/k-1} H_k^T + S_k R_k , \tag{28}$$

then, it can be determined by the formula of,

$$S_k = \left(\frac{1}{\mu}\sum_{j=k-\mu+1}^{k}\tilde{e}_j\tilde{e}_j^T - H_k P_{k/k-1} H_k^T\right) R_k^{-1} . \tag{29}$$

Here, μ is the width of the moving window. In case of normal operation, the scale matrix will be a unit matrix as $S_k = I$. Here I represents the unit matrix. Nonetheless, as μ is a limited number because of the number of the measurements and the computations performed with computer implies errors such as the approximation errors and the round off errors; S_k matrix, found by the use of (29) may not be diagonal and may have diagonal elements which are "negative" or lesser than "one" (actually, that is physically impossible). S_k matrix should be diagonal because only its diagonal terms have significance on the adaptation since each diagonal term corresponds to the noise covariance of each measurement (for the adaptation procedure S_k matrix is multiplied with the diagonal R matrix). Besides the measurement noise covariance matrix must be positive definite (that is why the multiplier S_k matrix cannot have negative terms) and also any term of this matrix cannot decrease in time for this specific problem since there is no possibility for increasing the performance of the onboard sensor (that is why the multiplier S_k matrix cannot have terms less than one). Therefore, in order to avoid such a situation, composing a scale matrix by the following rule is suggested:

$$S^* = diag\left(s_1^*, s_2^*, \ldots, s_n^*\right), \tag{30}$$

where,

$$s_i^* = \max\left\{1, S_{ii}\right\} \; i = 1, n . \tag{31}$$

Here, S_{ii} represents the ith diagonal element of the matrix S . Apart from that point, if the measurements are faulty, S_k^* will change and so affect the Kalman gain matrix;

$$K_k = P_{k/k-1} H_k^T \left(H_k P_{k/k-1} H_k^T + S_k^* R_k\right)^{-1} . \tag{32}$$

Therefore, in case of any kind of malfunctions, related element of the adaptive matrix, which corresponds to the faulty component of the measurement vector increases, and that brings out a smaller Kalman gain, which reduces the effect of the innovation on the state update process. As a result, more accurate estimation results can be obtained.

4.3. Comparison of R-Adaptation Procedures with Single and Multiple Adaptive Factors

In order to compare two different adaptation procedures, with single and multiple adaptive factors, a set of simulations have been realized.

The simulations were run for a period of 100 seconds with 0.1 seconds of sampling time, Δt. As an experimental platform the ZAGI UAV is chosen and the KF applications are performed by using the dynamics and the characteristics of that UAV [32].

The RKF algorithms are tested by implementing a sensor fault to the filter in between 30th and 60th seconds. For this period of 30 seconds a constant term is added to the pitch angle measurement $\Delta\theta$.

In Table 1, the means of the absolute values of error for the estimations in between the 30th and 60th seconds are given for both adaptation methods, with SAF and MAF. As well, same procedure is repeated for the OKF in case of fault and obtained results are included in the table.

As clearly shown in table, the OKF fails in case of measurement fault while the RKF algorithms diminish the effect of the fault and satisfy considerably more accurate estimations. Nonetheless as another fact, the RKF with MAF gives much more accurate estimation results than the RKF with SAF. Hence the theoretical statement given in [22] has been proved via a numerical example. In this sense, since the adaptation with MAF is superior, the overall concept for the RSAKF against sensor/actuator fault is built upon this method.

Table 1.Comparison of the mean absolute values of error for the OKF, RKF with SAF and RKF with MAF in case of sensor fault

Parameter	OKF	RKF (SAF)	RKF (MAF)
$\Delta u(m/s)$	23,0910	1,91025	0,08459
$\Delta w(m/s)$	6,94941	0,33368	0,06365
$\Delta q(rad/s)$	0,78062	0,21531	0,05120
$\Delta\theta(rad)$	6,00792	0,14728	0,02252
$\Delta h(m)$	12,9792	2,66206	0,08202
$\Delta\beta(rad)$	0,02480	0,00828	0,02300
$\Delta p(rad/s)$	0,03142	0,00161	0,02665
$\Delta r(rad/s)$	0,02384	0,00782	0,02194
$\Delta\phi(rad)$	0,04243	0,00898	0,03272

5. Q-Adaptation for UAV State Estimation

When there is an actuator fault in the system that ends up with changes in the control distribution matrix, a procedure for the Q adaptation should be followed.

The essence of the method is again comparing the real and theoretical values of the innovation covariance matrix. When there is an actuator fault, the real error will exceed the theoretical one. Besides, now it is known that using multiple factors for the adaptation gives better estimation accuracy. Hence, if a fading matrix, Λ_k , built of multiple fading factors, is added into the algorithm as,

$$\frac{1}{\mu}\sum_{j=k-\mu+1}^{k}\tilde{e}_k\tilde{e}_k^T = H_k(F_kP_{k-1/k-1}F_k^T + \Lambda_kG_kQ_kG_k^T)H_k^T + R_k \tag{33}$$

then the fading matrix can be determined as,

$$\Lambda_k = H_k^{-1}\left(\frac{1}{\mu}\sum_{j=k-\mu+1}^{k}\tilde{e}_k\tilde{e}_k^T - H_kF_kP_{k-1/k-1}F_k^TH_k^T - R_k\right)\left(G_kQ_kG_k^TH_k^T\right)^{-1}. \tag{34}$$

For a specific case where the all states are measured ($H_k = I$) as in case, (34) reduces to

$$\Lambda_k = \left(\frac{1}{\mu}\sum_{j=k-\mu+1}^{k}\tilde{e}_k\tilde{e}_k^T - F_kP_{k-1/k-1}F_k^T - R_k\right)\left(G_kQ_kG_k^T\right)^{-1}. \tag{35}$$

In a similar manner with the *R*-adaptation, the gained fading matrix should be diagonalized since the Q matrix must be a diagonal, positive definite matrix.

$$\Lambda^* = diag\left(\lambda_1^*, \lambda_2^*, \dots, \lambda_n^*\right), \tag{36}$$

where,

$$\lambda_i^* = \max\{1, \Lambda_{ii}\} \quad i = 1, n. \tag{37}$$

Here, Λ_{ii} represents the ith diagonal element of the matrix Λ. Apart from that point, if there is a fault in the system, Λ_k^* must be put into process as,

$$P_{k/k-1} = F_kP_{k-1/k-1}F_k^T + \Lambda_k^*G_kQ_kG_k^T \tag{38}$$

6. Proposed RSAKF Algorithm for UAV State Estimation

6.1. The Fault Detection and Isolation Process

In this section an integration scheme for the R and Q adaptation methods is proposed such that a RSAKF, which is insensitive against both sensor and actuator faults, is built as a result. Because of the already proved superiority, in both adaptation procedures the multiple factors method is used for the adaptation.

When the adaptation is applied (either R or Q) the covariance of the estimation error of the filter increases compared to the OKF. Therefore, the RSAKF algorithm is operated only when the fault is detected and in all other cases the procedure is run optimally with the regular OKF (17)-(22). The process is controlled by the use of a kind of statistical information. At that point, following two hypotheses may be introduced:

- γ_0; the system is normally operating
- γ_1; there is a malfunction in the estimation system.

To detect the failures a statistical function may be defined as,

$$\beta_k = \tilde{e}_k^T \left[H_k P_{k/k-1} H_k^T + R_k \right]^{-1} \tilde{e}_k \ . \tag{39}$$

This statistical function has χ^2 distribution with s degree of freedom where s is the dimension of the innovation vector. If the level of significance, α , is selected as,

$$P\left\{\chi^2 > \chi^2_{\alpha,s}\right\} = \alpha; ; 0 < \alpha < 1, \tag{40}$$

the threshold value, $\chi^2_{\alpha,s}$ can be found. Hence, when the hypothesis γ_1 is correct, the statistical value of β_k will be greater than the threshold value $\chi^2_{\alpha,s}$, i.e.:

$$\gamma_0 : \beta_k \le \chi^2_{\alpha,s} \ \forall k$$

$$\gamma_1 : \beta_k > \chi^2_{\alpha,s} \ \exists k \, . \tag{41}$$

On the other hand, after detecting the fault in the system the key point is detecting the type of the fault (either a sensor or actuator fault). When a regular KF is used, the decision statistics changes regardless the fault is in the sensors or in the actuators.

If a RKF based on the Doyle-Stein condition is used, it is easy to distinguish the sensor and actuator faults. The KF that satisfies the Doyle-Stein condition is referred in [13,33] as the RKF insensitive to the actuator failures and may be used for fault isolation as well in our study. After that the appropriate adaptation (the R or Q adaptation) may be applied. The Doyle-Stein condition is,

$$K\left(I + H\phi_s K\right)^{-1} = B_s \left(H\phi_s B_s\right)^{-1} . \tag{42}$$

Here K is the filter gain, I is unit matrix, H is the system measurement matrix, B_s is the control distribution matrix in continuous time and $\phi_s = \left(sI - F_s\right)^{-1}$, where F_s is the system matrix in continuous time.

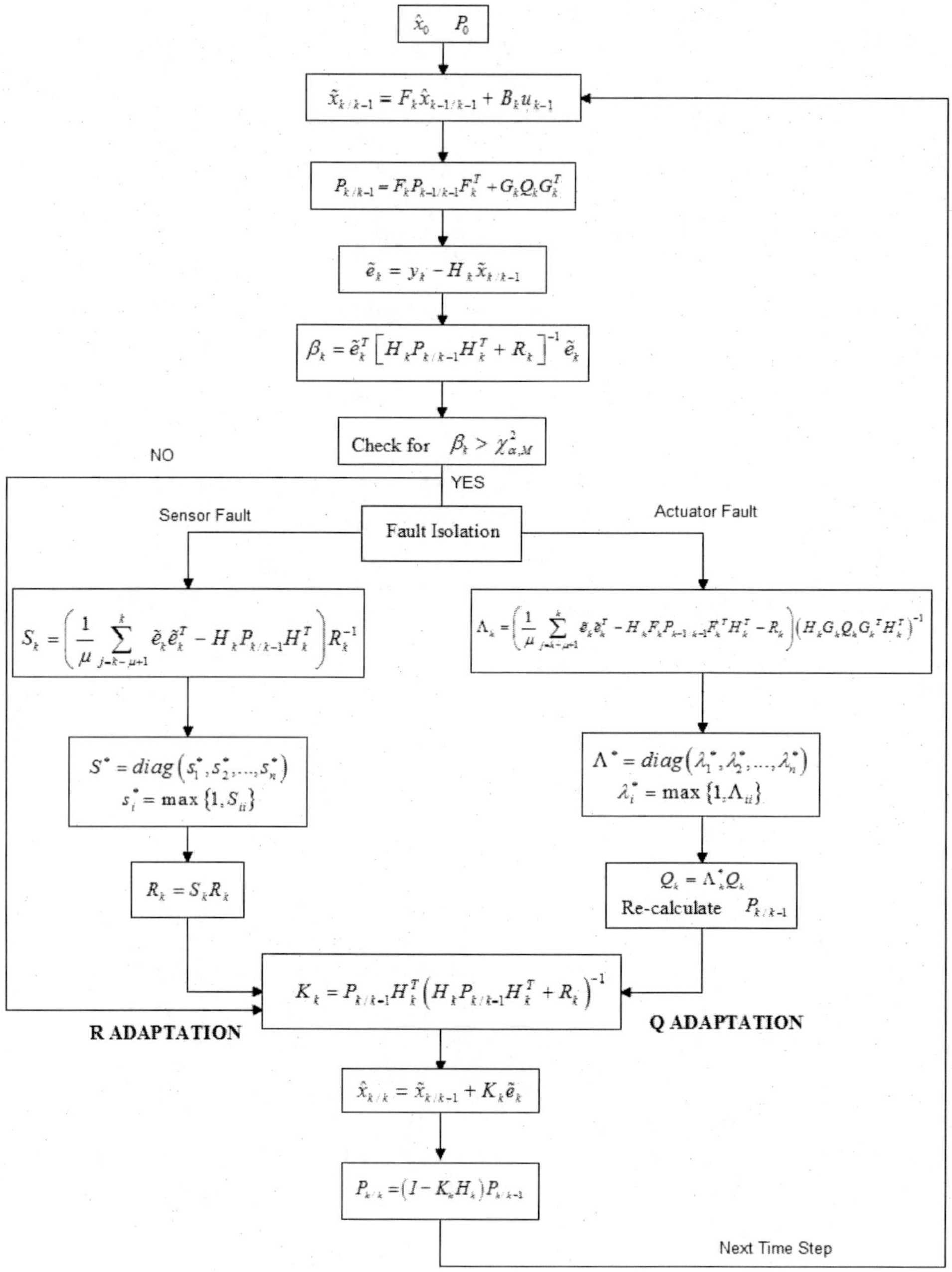

Figure 1. System architecture for the proposed RSAKF algorithm.

The RKF is very useful for isolating the sensor and control surface failures as it is insensitive to the latter failures. If the KF process noise covariance matrix is chosen as,

$$Q(q_r) = Q + q_r^2 B_s V B_s^T \tag{43}$$

then the RKF is obtained. Here Q is the process noise covariance matrix for the nominal plant, q_r is a paremeter that approaches to infinity and V is any positive definite symmetric matrix.

If the sensor fault occurs, the R adaptation is realized; else the procedure is continued with the Q adaptation. The algorithm architecture given in Figure 1 summarizes the integrated R and Q self-adaptation procedures. This new filter, where the robust and self-adaptive KFs are integrated, may be called as the RSAKF.

6.2. Numerical Example

In order to examine the effectiveness of the proposed RSAKF, it has been applied for the state estimation of the ZAGI UAV. Simulations were run for 100 seconds and at some specific period sensor/actuator faults are implemented to the system. Simulations were also repeated with the OKF so as to compare results.

In simulations, $\chi^2_{\alpha,s}$ is taken as 21.7 and this value comes from chi-square distribution when the degree of freedom is 9 and the reliability level is %99. The first top subplot in the figures gives the OKF or RSAKF state estimation results and the actual values in a comparing way. The middle subplot of the figures shows the error of estimation process based on the actual state values of the UAV. The last subplot indicates the variances of the estimations that are obtained as the diagonal elements of the KF estimation error covariance matrix (22).

During the simulations, for testing the fault detection and isolation procedures and also the RSAKF algorithm, two different faults are implemented to the systems. Firstly, the sensor fault is formed by adding a constant term to the measurement of pitch angle (constant term is nearly at same magnitude with the measurement, so the measurement's magnitude is doubled for those instants of time), $\Delta\theta$ in between 30th and 50th seconds. Furthermore, in order to check that fault isolation can be achieved successfully and switching from one adaptation stage to the other one is realized properly, as another fault, actuator fault is simulated by taking the first column elements of longitudinal control distribution matrix nearly zero in between 80th and 90th seconds.

The graph for the statistical values of $\beta(k)$ in case of using the OKF is shown in Figure 2. As seen, except the periods, where the sensor and actuator failures occur, $\beta(k)$ is lower than the threshold.

On contrary, when one of these failures occurs $\beta(k)$ grows rapidly and exceeds the threshold value. Hence γ_1 hypothesis is judged to be true and it is shown that by using such procedure the fault can be detected.

The results for the fault isolation process are given in Figure 3. As mentioned the RKF insensitive to actuator failures is used for this isolation process. The behavior of the statistics $\beta(k)$ in the case of sensor/actuator failures shows that detecting the actuator failure is not possible when the RKF is used since it is insensitive against such faults; on the other hand the sensor failure is detected immediately. Hence the fault isolation can be realized by using such procedure. The RKF, which is insensitive to actuator failures and sensitive to sensor failures, is employed to isolate the sensor and actuator failures in the proposed RSAKF algorithm (See Figure 1).

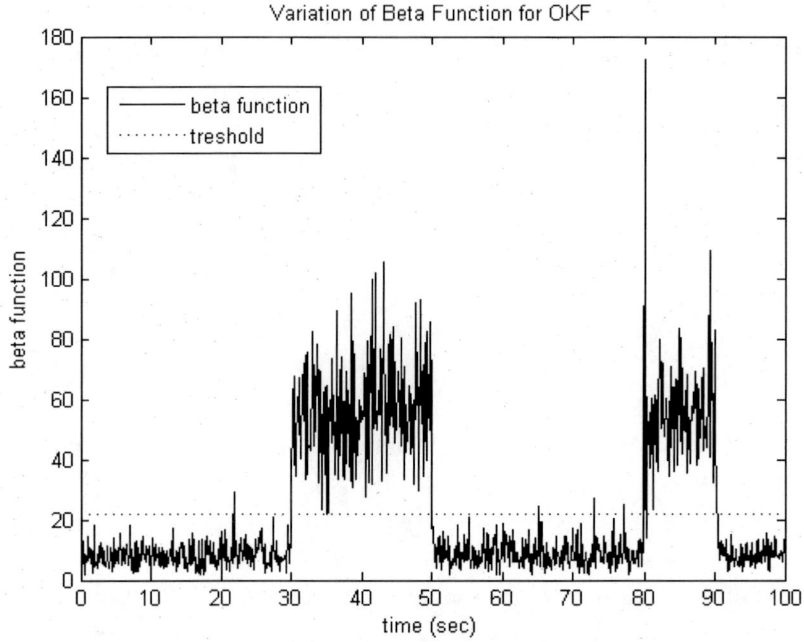

Figure 2. Behavior of the statistics $\beta(k)$ in the case of measurement failure (in between 30th and 50th seconds) and actuator failure (in between 80th and 90th seconds) when the OKF is used.

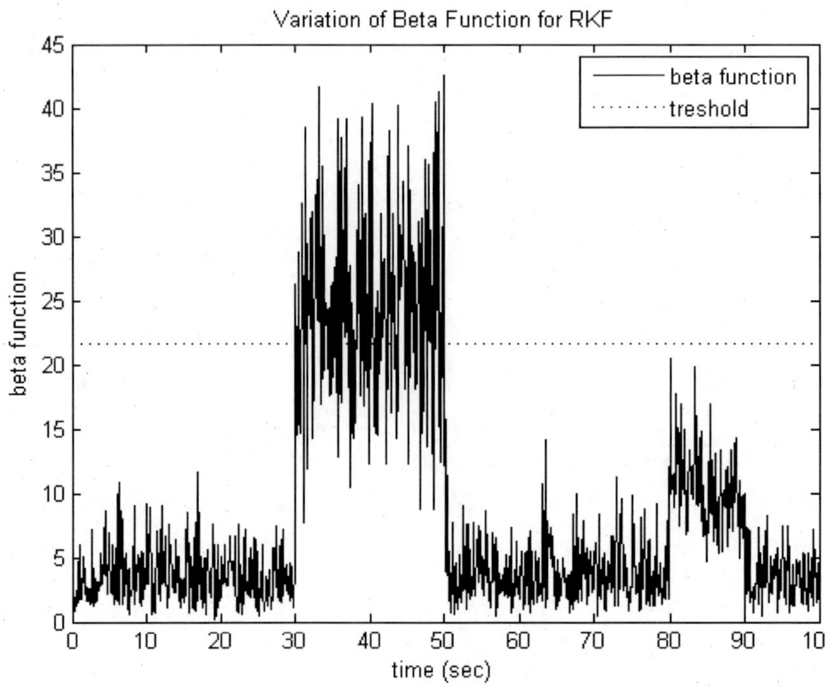

Figure 3. Behavior of the statistics $\beta(k)$ in the case of measurement failure (in between 30th and 50th seconds) and actuator failure (in between 80th and 90th seconds) when the RKF insensitive to the actuator failures is used.

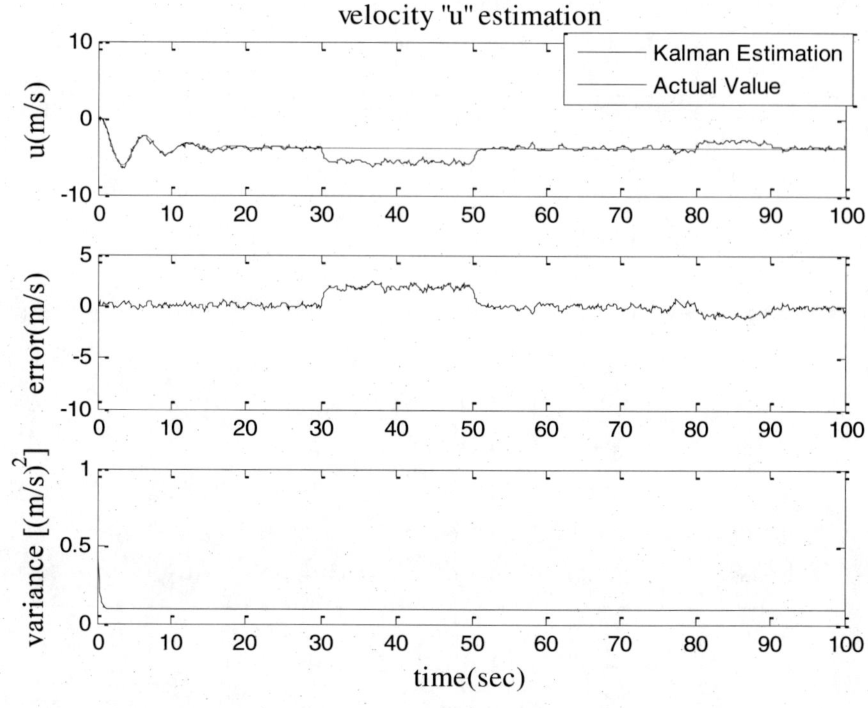

Figure 4. "u" velocity estimation via OKF in case of bias sensor fault (between 30th and 50th s) and actuator fault (between 80th and 90th s).

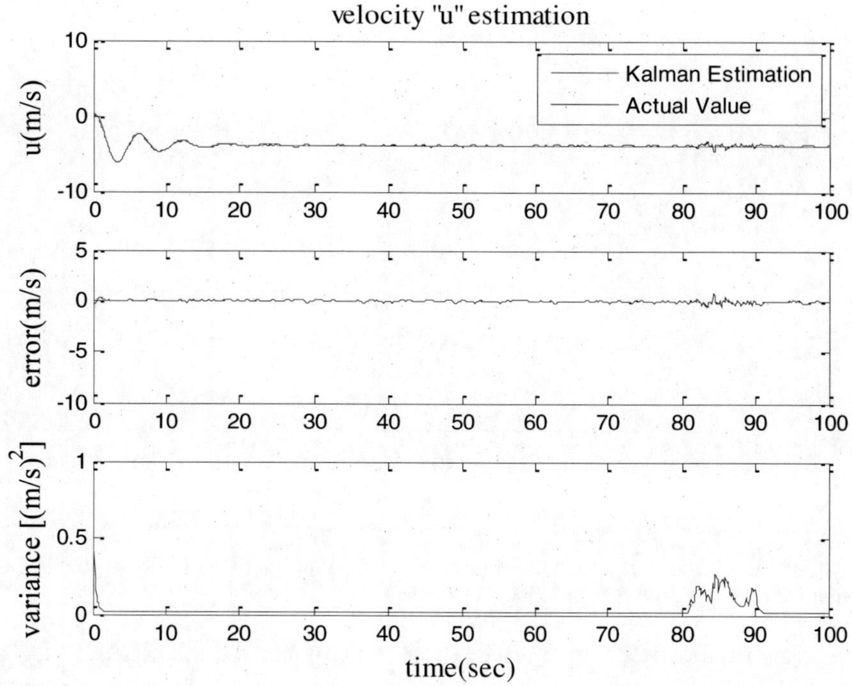

Figure 5. "u" velocity estimation via RSAKF in case of bias sensor fault (between 30th and 50th s) and actuator fault (between 80th and 90th s).

Table 2.Comparison of the mean absolute values of error for the OKF and RSAKF in case of actuator and bias sensor faults

Parameter	Mean Absolute Value of Error for Regular OKF			Mean Absolute Value of Error for RSAKF		
	Sensor Fault	Normally Operating	Actuator Fault	Sensor Fault	Normally Operating	Actuator Fault
$\Delta u(m/s)$	1.8799	0.1891	0.8116	0.0528	0.0578	0.1815
$\Delta w(m/s)$	0.6762	0.0931	1.1939	0.0649	0.0611	0.2678
$\Delta q(rad/s)$	0.0604	0.0621	0.1602	0.0521	0.0542	0.0686
$\Delta\theta(rad)$	0.6497	0.053	0.0796	0.0161	0.0132	0.0325
$\Delta h(m)$	0.5493	0.1059	0.1273	0.0644	0.0636	0.0932
$\Delta\beta(rad)$	0.0319	0.0302	0.0338	0.0123	0.0155	0.0165
$\Delta p(rad/s)$	0.0514	0.0556	0.0538	0.0282	0.0257	0.0304
$\Delta r(rad/s)$	0.0508	0.045	0.0465	0.0115	0.0146	0.0163
$\Delta\phi(rad)$	0.0758	0.0786	0.0805	0.0167	0.0192	0.0349

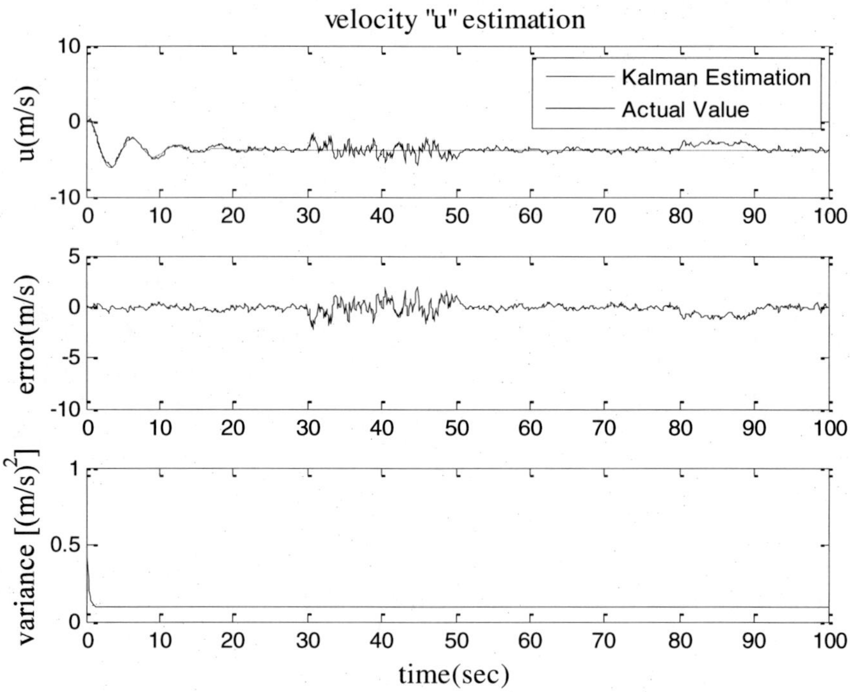

Figure 6. "u" velocity estimation via OKF in case of noise increment sensor fault (between 30th and 50th s) and actuator fault (between 80th and 90th s).

The results of the first fault scenario realized for testing the RSAKF algorithm in the presence of sensor/actuator failures are given in Figures 4 and 5. As it is clear the RSAKF gives superior estimation results than the OKF in both cases.

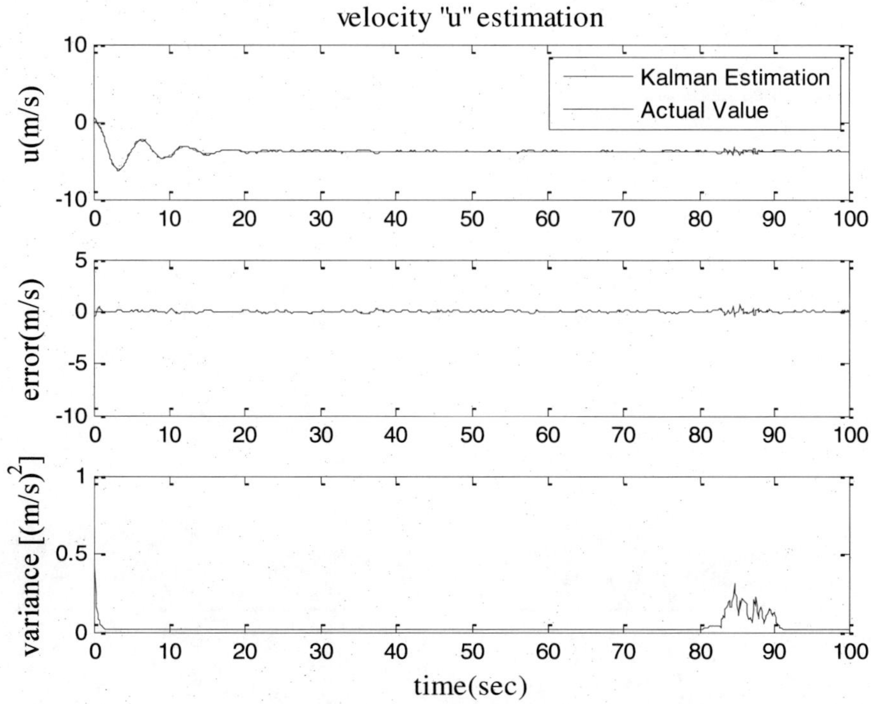

Figure 7. "u" velocity estimation via RSAKF in case of noise increment sensor fault (between 30th and 50th s) and actuator fault (between 80th and 90th s).

The RSAKF precisely detects the type of fault and tunes itself according to the adaptation rule. Hence, it secures accurate estimation results for the whole process while the OKF fails in both sensor and actuator fault conditions. Nonetheless Figure 5 shows that, as it is aforementioned, the covariance of the estimation error of the filter increases compared to the OKF when the filter is adapted. Such increase is more clear in case of the Q adaptation. That signifies the importance of using a fault detection procedure and not using the RSAKF for the whole estimation period. Also, note that similar results are gained for all other estimated parameters. Nevertheless, the performance of both filters; the OKF and RSAKF, in case of faults can be understood more clearly by examining the Table 2. In the table, the mean absolute values of error, which are calculated for 10 seconds, are given for all three conditions; the sensor fault, actuator fault and no fault. It is obvious that, the estimations by the OKF worsen for all of the longitudinal states in case of sensor/actuator fault while the RSAKF keeps accurate estimation behaviour for the whole process. As the second, testing scenario for the RSAKF, this time sensor fault is characterized by multiplying the noise variance of measurement of the velocity component Δu with a constant term in between the 30th and 50th seconds (measurement noise increment) while the actuator fault is not changed. As Figure 6 and Figure 7 show, again the OKF outputs have error while the RSAKF algorithm achieves estimation of the states accurately. Nevertheless, another consequence of the second fault scenario is the good performance of the fault isolation scheme; regardless the type of the sensor fault, the isolation scheme can distinguish the type of the fault effectively and the RSAKF works properly. In order to clarify the results, comparison for the mean absolute values of error is given in Table 3.

Table 3. Comparison of mean absolute values of error for OKF and RSAKF in case of actuator and noise increment sensor faults

Parameter	Mean Absolute Value of Error for Regular OKF			Mean Absolute Value of Error for RSAKF		
	Sensor Fault	Normally Operating	Actuator Fault	Sensor Fault	Normally Operating	Actuator Fault
$\Delta u(m/s)$	0.6235	0.2317	0.8352	0.0819	0.0635	0.1303
$\Delta w(m/s)$	0.2588	0.1482	1.1963	0.0687	0.0626	0.2621
$\Delta q(rad/s)$	0.0611	0.0673	0.1505	0.055	0.0508	0.0711
$\Delta\theta(rad)$	0.0653	0.0621	0.0688	0.0115	0.0155	0.0376
$\Delta h(m)$	0.1101	0.1	0.1323	0.0826	0.0733	0.0785
$\Delta\beta(rad)$	0.0355	0.0392	0.0319	0.0163	0.0149	0.0134
$\Delta p(rad/s)$	0.0497	0.0465	0.0553	0.0308	0.026	0.0265
$\Delta r(rad/s)$	0.0456	0.0443	0.0449	0.0135	0.013	0.0112
$\Delta\phi(rad)$	0.08	0.0694	0.0695	0.0262	0.0196	0.0359

CONCLUSION

In this chapter a RSAKF algorithm is proposed for both cases of faults in system and measurements. The RSAKF incorporates the R and Q self-adaptation procedures and adapts itself against the sensor/actuator faults in order to secure the robustness. The introduced algorithm detects the type of the fault (either sensor or actuator) and isolates it in order to have an accurate estimation result throughout the whole process. After that the appropriate adaptation (the R or Q adaptation) is applied such that the estimation characteristic is not deteriorated. Unlike the existing studies, the proposed adaptation methods for both the R and Q adaptations are simplistic and can be applied easily with a small modification to the KF.

In this chapter it is shown how to incorporate these two different adaptation processes such that the fault is isolated and so the RSAKF algorithm is a complete estimator including the fault detection and isolation processes for the case of sensor/actuator faults. When a regular KF is used, the decision statistics changes regardless the fault is in the sensors or in the actuators, while a RKF based on Doyle-Stein condition is used, it is easy to distinguish the sensor and control surface/actuator faults.

The RSAKF is applied for the state estimation procedure of an UAV and its effectiveness is investigated by the comparisons with the OKF in case of the fault. The proposed RSAKF algorithm may play an important role for the UAV control systems since it gives accurate estimation results despite the sensor or actuator faults. Regarding the harsh environment where the UAVs are generally used, meeting with any kind of fault has a high possibility and so preferring the proposed RSAKF algorithm instead of the regular OKF may bring a significant advantage.

The proposed approach does not require a priori statistical characteristics of the faults and can be used for both linear and nonlinear systems. Furthermore the presented RSAKF algorithm is simple for practical implementations and does not demand a high extra computational performance when compared with the regular OKF.

REFERENCES

[1] Newcome, L.R. (2004). *Unmanned Aviation: A Brief History of Unmanned Aerial Vehicles*, AIAA, Virginia, USA.

[2] Cook, K.L.B. (2007).The silent force multiplier: The history and role of UAVs in warfare, in: Proc. of the 2007 IEEE Aerospace Conference, Montana, USA, pp. 1-7.DOI: 10.1109/AERO.2007.352737.

[3] Parra, S.L., and Angel, J. (2005). Low cost navigation system for UAVs. *Aerospace Science and Technology*, Vol. 9, pp.504-516.DOI: 10.1016/j.ast.2005.06.005.

[4] Wendel, J., Meister, O., Schlaile, C., and Trommer, G. F. (2006). An integrated GPS/MEMS-IMU navigation system for autonomous helicopter. *Aerospace Science and Technology*, Vol. 10, pp.527-533. DOI:10.1016/j.ast.2006.04.002.

[5] Zhang, Y., Li, X.R (1997). Detection and diagnosis of sensor and actuator failures using interacting multiple-model estimator, in: *Proc. of the 36th IEEE Conference on Decision and Control,* Vol. 5, pp. 4475-4480. DOI: 10.1109/CDC.1997.649671.

[6] Maybeck, P.S. (1982). Stochastic Models, Estimation, and Control, Vol.I and II, Academic Press, New York, USA.

[7] White, N.A., Maybeck, P.S., and DcVilbiss, S.L. (1998). MMAE detection of interference/jamming and spoofing in a DGPS-aided inertial system. *IEEE Transactions on Aerospace and Electronic Systems*, Vol. 34, pp. 1208-1217. DOI: 10.1109/7.722708.

[8] Mehra, R.K. (1970). On the identification of variance and adaptive Kalman filtering. *IEEE Transactions on Automatic Control*, Vol. 15, pp.175-184.

[9] Salychev, O.S. (1994). Special studies in dynamic estimation procedures with case studies in inertial surveying, ENGO 699.26 lecture notes, Department of Geomatics Engineering, University of Calgary.

[10] Mohamed, A.H., and Schwarz, K.P. (1999). Adaptive Kalman filter for INS/GPS, *Journal of Geodesy*, Vol. 73, pp. 193-203.

[11] Wang, J., Stewart, M.P., and Tsakiri,M. (1999). Adaptive Kalman filtering for integration of GPS with GLONASS and INS, in: Proc. of the 1999 International Geodesy Symposia (IUGG 99),Birmingham, UK.

[12] Maybeck, P.S. (1999). Multiple model adaptive algorithms for detecting and compensating sensor and actuator/surface failures in aircraft flight control systems. *International Journal of Robust and Nonlinear Control*, Vol. 9, pp. 1051-1070.

[13] Hajiyev, C., and Caliskan, F. (2000). Sensor/actuator fault diagnosis based on statistical analysis of innovation sequence and robust Kalman filtering. *Aerospace Science and Technology*, Vol. 4, pp. 415-422. DOI: 10.1016/S1270-9638(00)00143-7.

[14] Sasiadek, J. Z., and Wang, Q. (1999). Sensor fusion based on fuzzy Kalman filtering for autonomous robot vehicle. *Proc. Of the 1999 IEEE International Conference on Robotics and Automation*, Detroit, Michigan, USA, 4, pp. 2970-2975. DOI: 10.1109/ROBOT.1999.774048.

[15] Zhang, S. T., and Wei, X. Y. (2003). Fuzzy adaptive Kalman filtering for DR/GPS. Proc. of the Second International Conference on Machine Learning and Cybernetics, Xi-an, PRC, 5, pp. 2634-2637.

[16] Escamilla-Ambrosio, P. J., and Mort, N. (2003). Hybrid Kalman filter-fuzzy logic adaptive multisensor data fusion architecture. *Proc. 42nd IEEE Conference on Decision and Control*, Hawaii, USA, 5, pp.5215-5220. DOI: 10.1109/CDC.2003.1272465.

[17] Guo, X.-R., Wang, F.-H., Du, D.-F. and Guo, X.-L.(2009). An Improved Neural Network Based Fuzzy Self-Adaptıve Kalman Fılter And Its Applıcatıon In Cone Pıckıng Robot. *Proc. of the Eighth International Conference on Machine Learning and Cybernetics*, Baoding, 12-15 July 2009, pp.573-577.

[18] Du, D., Yan, M.,Guo, X. and Lu, H. (2009). Application of Fuzzy Self-adapting Kalman Filter in the Control System of an Eradicating Stump Robot. *2009 Second International Symposium on Computational Intelligence and Design*, ISCID 2009, Changsha, Hunan, China, 12-14 December 2009, pp.364-367.DOI 10.1109/ISCID.2009.237.

[19] Hide, C., Moore, T., and Smith, M. (2004). Adaptive Kalman filtering algorithms for integrating GPS and low cost INS, in: *Proc. of the Position Location and Navigation Symposium,* Monterey, USA, pp.227-233.

[20] Hu, C., Chen, W., Chen, Y. and Liu, D. Adaptive Kalman filtering for vehicle navigation. *Journal of Global Positioning Systems*, Vol. 2, pp. 42-47.

[21] Ding, W., Wang, J., and Rizos, C. (2007). Improving adaptive Kalman estimation in GPS/INS integration, *The Journal of Navigation*, Vol. 60, pp. 517–529.DOI: 10.1017/S0373463307004316.

[22] Geng, Y., and Wang, J. (2008). Adaptive estimation of multiple fading factors in Kalman filter for navigation applications. *GPS Solutions*, Vol. 12, pp. 273-279.

[23] Hajiyev, C. (2007). Adaptive Filtration Algorithm with the filter gain correction applied to integrated INS/Radar altimeter. Proc. of the Institution of Mechanical Engineers, Part G: *Journal of Aerospace Engineering,* Vol. 221, pp. 847-885. DOI: 10.1243/09544100JAERO173.

[24] Sage A.P, Husa G.W (1969). Algorithms for sequential adaptive estimation of prior statistics. *IEEE Symp. Adapt. Process*. (8th) Decis. Control, 8, pp. 61-61.

[25] Kim, K.H., Lee, J.G., and Park, C.G. (2006). Adaptive two-stage Kalman filter in the presence of unknown random bias. International *Journal of Adaptive Control and Signal Processing*, Vol. 20, pp. 305-319. DOI: 10.1002/acs.900.

[26] Jwo, D.J., and Weng, T.P. (2008).An adaptive sensor fusion method with applications in integrated navigation. *The Journal of Navigation,* Vol. 61, pp. 705–721.DOI:10.1017/ S0373463308004827.

[27] Jwo, D. J., and Chang, S. C. (2009). Particle swam optimization for GPS navigation Kalman filter. *Aircraft Engineering and Aerospace Technology: An International Journal*, Vol. 81, pp.343-352. DOI: 10.1108/00022660910967336.

[28] Almagbile, A., Wang, J., Ding, W. (2010). Evaluating the performances of adaptive Kalman filter methods in GPS/INS integration, *Journal of Global Positioning Systems* 9, pp.33-40.DOI: 10.5081/jgps.9.1.33.

[29] Yechout, T.R., Morris, S.L., Bossert, D.E., and Hallgren, W.F. (2003). Introduction to Aircraft Flight Mechanics: Performance, Static Stability, Dynamic Stability, and Classical Feedback Control, AIAA Education Series, Virginia, USA.

[30] Nelson, R.C. (1998). Flight Stability and Automatic Control, McGraw-Hill, Massachusetts, USA.

[31] Kalman, R.E. (1960).A new approach to linear filtering and prediction problems. *ASME Journal of Basic Engineering*, pp. 35-45.

[32] Matthews, J.S. (2006). Adaptive Control of Micro Air Vehicles, M.Sc. Thesis, Department of Electrical and Computer Engineering, Brigham Young University, Utah, USA.

[33] Hajiyev Ch and Caliskan F. (2005). Sensor and Control Surface/Actuator Failure Detection and Isolation Applied to F-16 Flight Dynamics. *Aircraft Engineering and Aerospace Technology: An International Journal*, 77(2), pp.152-160.

In: Self-organization: Theories and Methods
Editor: WenJun Zhang

ISBN: 978-1-62618-865-5

Chapter 10

REGULAR APPROXIMATION OF THE STOCHASTIC PUSHDOWN CALCULUS

Marco Carpentieri*
Dipartimento di Matematica ed Informatica, Universitá della Basilicata,
Potenza (Pz), Italy

ABSTRACT

This chapter takes into account the problem of designing regular approximations for the stochastic pushdown computing. I prove that stochastic pushdown automata accepting with finite cut point and for which equivalent non-dirty context-free grammars exist admit arbitrarily accurate regular approximations. I address the problem of designing the uniform representation of the stochastic (context-free/pushdown) computing characterizing it in terms of converging sequences of finite state approximations.

Keywords: Finite state approximation; stochastic pushdown computing

1. INTRODUCTION

Historically, the Stochastic Context Free Grammars (SCFGs) [5, 20, 6, 35] were introduced to facilitate the design and analysis of complex information systems.

Afterwards they have received considerable attention in several research fields such as in *language processing*, *speech or image recognition*, *information retrieval* and *RNA sequencing*.

Fundamental analysis results were introduced by Booth and Thompson [6] (see also [35]) that have considered the problem of characterizing computable probability measures induced over the sentences of the languages associated to weighted grammars [28]. The equivalence between the SCFGs and the Stochastic Pushdown Automata (SPA) was proved by Huang and Fu [20].

* marcar66@virgilio.it.

We introduce a polynomial reduction of the SPA (SCFG) finite cut-point computing to regular (finite state) approximations (SFA) [28].

This allows us to interpret the (efficient) SPA computing [5, 20, 6, 2] by Markovian finite state processes simpler to understand and to analyze [6, 35]. More in detail,

1. we prove that non-dirty SCFGs [6, 35] admit regular representation able to approximate with arbitrary accuracy the associated finite cut point push-down computing;
2. we address the problem of designing uniform descriptions of the (efficient) stochastic context-free or pushdown computing in terms of converging sequences of finite state-approximations; this allows us to get symbolic descriptions of the SCFG calculus and to simplify the statistical analysis of the associated stochastic processes [6, 29, 35].

2. SCFGs

Denote by V^* the Kleene closure of the (finite set) V, namely the set formed by the empty string ε and the words that can be formed by concatenation of elements in V. A Stochastic Context-Free Grammar is a quadruple

$$SCFG = (V_N, V_T, S, P),$$

where

- V_N is a finite set of variables,
- V_T is a finite set of terminal symbols,

$S \in V_N$ is the start symbol,

- P is a finite set of productions, namely stochastic rules

$$\rho : A \rightarrow \pi\ \alpha$$

such that

$$\pi = \pi(\alpha|A)$$

is the probability of replacing the premise $A \in V_N$ with the consequence α ($V = V_N \cap V_T$, $V^+ = V^*\text{-}\{\varepsilon\}$, $\alpha \in V^*$).

We assume that SCFG is normalized, that is, for each production premise $A \in V_N$ it holds that

$$\sum_{\alpha : A \rightarrow \pi\ \alpha \in P} \pi(\alpha|A) = 1.$$

The set

$V = V_N \cap V_T$

is the dictionary of SCFG. A direct derivation

$\beta \Rightarrow \pi(\beta'|\beta)\ \beta'\ (\beta \in V^+,\ \beta' \in V^*)$

states that β' can be obtained from β replacing a premise occurrence (handle) A on β with the consequence α of production ρ with probability

$\pi = \pi(\beta'|\beta).$

An l-length derivation

$$\gamma \Rightarrow^* \pi(\beta_{l+1},\ldots,\beta_2|\beta_1)\ \gamma' \quad (1)$$

($\gamma \in V^+$, $\gamma' \in V^*$) points out that the string γ' can be generated from γ through some sequence of $l \in \mathbf{N}$ direct derivations:

$\beta_1 \Rightarrow \pi(\beta_2|\beta_1)\ \beta_2 \Rightarrow \pi(\beta_3|\beta_2)\ \beta_3 \Rightarrow \ldots. \Rightarrow \pi(\beta_{l+1}|\beta_l)\ \beta_{l+1}$

with $\beta_1 = \gamma$ and $\beta_{l+1} = \gamma'$, where the derivation probability

$\pi(\beta_{l+1},\ldots,\beta_2|\beta_1) = \prod_{r=1,\ldots,l} \pi(\beta_{r+1}|\beta_r)$

is the product of the probabilities of the direct derivation components. Denote by Ω the set consisting of all the derivations. Notation

$\gamma \Rightarrow^* \pi(\gamma'|\gamma)\ \gamma'$

means that the string γ' can be derived from γ with probability $\pi(\gamma'|\gamma)$, where

$$\pi(\gamma'|\gamma) = \sum_{\gamma \Rightarrow^* \pi(\beta_{l+1},\ldots,\beta_2|\beta_1)\ \gamma' \in \Omega} \pi(\beta_{l+1},\ldots,\beta_2|\beta_1)$$

(in other words, $\pi(\gamma'|\gamma)$ is the probability of deriving γ' from γ). The string $\gamma' \in V^*$ is a (sentential) form iff it holds that

$S \Rightarrow^* \pi(\gamma'|S)\ \gamma'$

with probability

$\pi(\gamma'|S) > 0;$

the form is a sentence iff $\gamma' \in V_T^*$. Grammars derivations are usually defined specifying a production application order to the handles of the forms. For example, (1) is derived following a leftmost (rightmost) derivation order iff

$$\beta_i \Rightarrow \pi(\beta_{i+1}|\beta_i)\ \beta_{i+1}$$

is defined by applying a production with premise appearing as leftmost (rightmost) handle on β_i for each i =1,..., l. The stochastic language associated with SCFG is

$$L_{SCFG} = \{w : w \in V_T^*, S \Rightarrow^* \pi(w \mid S)\ w, \pi(w \mid S) > 0\}.$$

Stochastic grammars with the same associated language are said to be equivalent. More specifically, productions with a single variable as premise are context-free linear iff they are of the type

$$A \rightarrow \pi\ w_1 B w_2\ (A, B \in V_N, w_1, w_2 \in V_T^*)$$

and (context-free) non-linear otherwise. Huang and Fu [20] discussed how the productions of any SCFG can be rewritten equivalently in Chomsky normal form:

$$A \rightarrow \pi(BC|A)\ BC \text{ or } A \rightarrow \pi(a|A)\ a\ (A, B, C \in V_N, a \in V_T).$$

3. SCFG Approximation

A fundamental problem consists in asking whether stochastic asymptotically efficient computing [2, 11] admits simplified formal grammar representation. We consider the question in terms of finite state approximations [28, 3, 4, 19] confining our attention to the stochastic pushdown automata (SPA) accepting with finite cut point

$$\lambda \in (0,1).$$

As it is well known, such constrained Turing systems are equivalent to the SCFGs [20]. In this setting, the assumption of acceptance with finite cut point is the basis of the stochastic efficient computing [2] in the sense that, under such assumption, a finite number $O(1/\lambda)$ of i.i.d. repetitions of the computation are required in order to be able to estimate the input acceptance probability with suitable confidence. Characterizing the stochastic context-free calculus in terms of finite state (or regular) approximations allows us to simplify the statistical analysis of the SCFGs; for instance, deciding the consistence of the non-dirty stochastic finite state approximation requires simply to check that the associated Markov chain admits a unique absorbing state [6] (see also [29, 35, 19, 28, 3, 4]).

3.1. Regular Representation

The derivation process of a non-dirty SCFG [35] can be drawn as an oriented derivation process graph. We say that two forms are equivalent if they hold the same number of occurrences of each of the grammar variables. The relation is an equivalence relation and the equivalence classes can be identified by finite patterns

$$A_1^{i(1)} \cdots A_m^{i(m)}$$

of degree

$$dg(A_1^{i(1)} \cdots A_m^{i(m)}) = \sum_{j=1,\ldots,m} i(j)$$

where

$$V_N = \{A_1,\ldots,A_m\}$$

and

$$(i(1),\ldots,i(m)) \in (\mathbf{N} \cap \{0\})^m.$$

The derivation process graph of SCFG has equivalence classes as vertices; an edge

$$A_1^{i(1)} \cdots A_m^{i(m)} \rightarrow \pi\, A_1^{j(1)} \cdots A_m^{j(m)},$$

labeled by the probability π, means that each form in

$$A_1^{i(1)} \cdots A_m^{i(m)}$$

derives directly a form in

$$A_1^{j(1)} \cdots A_m^{j(m)}$$

by the application of a same production

$$A \rightarrow \pi\, \alpha$$

according to some prespecified derivation order ($A_i^0 = \varepsilon \notin V$). The number of vertices of the graph can be finite or infinite.

In particular, the finiteness of the graph size depends on nesting of the non-linear productions that are the only ones able to produce derivations of forms belonging to patterns with strictly increasing degrees.

We say that a finite state (or regular) (δ,λ)-approximation of SCFG ($\delta \in [0, 1)$, $\lambda \in (0,1)$) is a stochastic grammar SG whose derivation process graph has a finite number of vertices (states) and such that any sentence $w \in L_{SCFG}$, derived with probability

$\pi(w) > \lambda$,

is derived by SG with (relative) probability error less than or equal to δ. The following theorem states that stochastic pushdown automata accepting with (finite) cut point $\lambda \in (0,1)$ and for which equivalent non-dirty SCFGs exist admit arbitrarily accurate finite state (δ,λ)-approximations.

To fix the ideas we shall assume that the SCFGs are expressed in Chomsky normal form [20, 19], where at least one linear production

$A \to\pi\ \alpha\ (\pi \neq 0)$

exists for each other non-linear one

$A \to\pi'\ BC$

with the same premise (A). The result can be generalized since each non-dirty grammar can be redefined equivalently in such a way that its production set contains at least one linear production with associated non-null probability (and consequence in V_T^{*}) for each premise A $\in V_N$.

Theorem 1

Non-dirty SCFGs equivalent to SPA computing with finite cut point have regular (δ,λ)-approximations for every $\delta \geq 0$ and $\lambda > 0$.

Proof

Let

$SCFG = (V_N, V_T, S, P)$

whose productions are in Chomsky normal form. Consider the derivation process graph restricted to the forms generated applying non-linear productions at most a finite number of times (following a leftmost derivation order). Denote the graph state-patterns

$A_1^{i(1)} \cdots A_m^{i(m)}\ (A_1 = S)$

by distinct auxiliary variables X_i and assume that their degree is less than or equal to d. Construct a new set P' of productions with variables X_i and terminal symbols in V_T consisting of the edges

$X_i \to\pi(\alpha|A)\ X_j$

iff it holds that

$\alpha \in V_N^{*}$

and

$X_i \to \pi(a|A)\ aX_j\ (a \in V_T)$,

iff

$X_i \to \pi(a|A)\ X_j$

is an edge of (the restriction of) the derivation process graph. Auxiliary variables X'_i (distinct from the X_is) are used to map any transition entering a state of degree greater than d to some other state of degree less than d + 1. Since the number of distinct patterns consisting of variables and having degree not greater than d is finite, the number of distinct auxiliary variables X_i, X'_i is finite and our construction

$SG = (\{X_i, X'_i\}, V_T, S, P')$

defines a stochastic regular (δ,λ)-approximation of SCFG.

Example 1

Consider the stochastic grammar

SG = SCFG

with productions

$S \to \pi(AB|S)\ AB$, $S \to \pi(h|S)\ h$, $A \to \pi(BC|A)\ BC$, $A \to \pi(a|A)\ a$,

$B \to \pi(b|B)\ b$, $B \to \pi(bB|B)\ bB$, $C \to \pi(c|C)\ c$, $C \to \pi(cS|C)\ cS$,

where the productions with the premises B and C may be rewritten as required by using auxiliary variables D and E, namely we can replace them with

$B \to \pi(b|B)\ b$, $B \to \pi(DB|B)\ DB$, $D \to b$,

$C \to \pi(c|C)\ c$, $C \to \pi(ES|C)\ ES$, $E \to c$.

The derivation process graph of the regular $(0,\lambda)$-approximation restricted to the states specified by the patterns of degree less than four has vertices labeled by

S, B, C, $AB = X_1$, $BC = X_2$, $B^2C = X_3$,

where the transitions are defined by

$S \to \pi(AB|S)\ X_1$, $S \to \pi(h|S)\ h$ ($S \to \pi(h|S)\ hL_{SG}$), $B \to \pi(b|B)\ b$

(B $\rightarrow \pi(b|B)$ bL_{SG}), B $\rightarrow \pi(bB|B)$ bB, C $\rightarrow \pi(c|C)$ c (C $\rightarrow \pi(c|C)$ cL_{SG}),

C $\rightarrow \pi(cS|C)$ cS, $X_1 \rightarrow \pi(BC|A)$ X_3, $X_1 \rightarrow \pi(a|A)$ aB, $X_2 \rightarrow \pi(b|B)$ bC,

$X_2 \rightarrow \pi(bB|B)$ bX_2, $X_3 \rightarrow \pi(b|B)$ bX_2, $X_3 \rightarrow \pi(bB|B)$ bX_3.

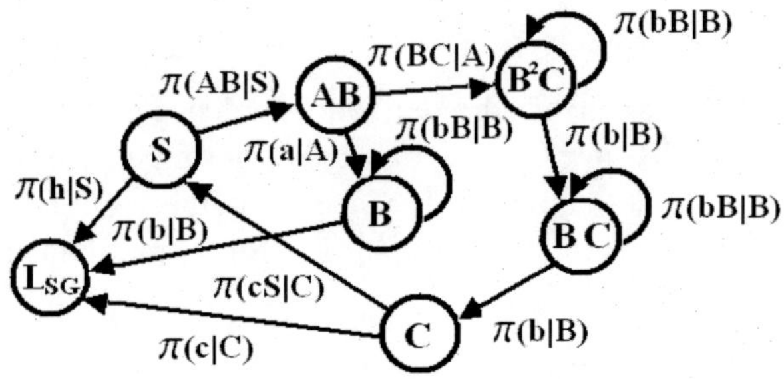

Figure 1. Finite state (0,λ)-approximation diagram for SCFG following (leftmost) lexicographical derivation order.

The (0,λ)-approximation diagram for the SCFG is reported in Figure 1. Note SCFG has non-linear productions forming finite length derivations; consequently it is equivalent to a stochastic finite state grammar whose productions link finite degree states.

The error bound δ of the finite state approximation (specified by cutting derivations formed by nesting finitely non-linear productions) depends on the derivation order used to generate sentences. For instance, assume that the states of the approximation are identified by sequences of grammar variable occurrences taken in the order they appear on the derived forms.

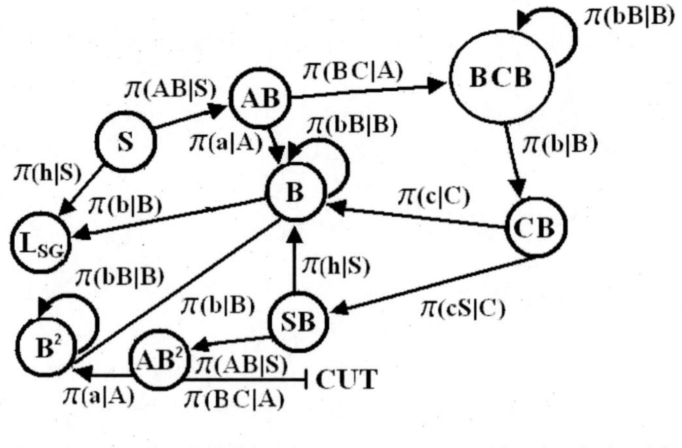

Figure 2. Approximation diagram for SCFG with states represented by the derived forms (the productions are applied following a leftmost derivation order).

Figure 2 displays the (approximated) SCFG derivation diagram drawn applying productions to such new kind of patterns (having degree less than four). The diagrams reported in figures 1 and 2 evidence that the choice of the state-patterns and of the derivation order are crucial to get minimal approximation representation (see [22] about the minimization of stochastic automata, where non-expanding productions with a same premise should be applied first).

Example 2

The SCFG' with productions

$S \rightarrow \pi(aSb|S)\ aSb,\ S \rightarrow \pi(ab|S)\ ab$

generates the stochastic non-regular language

$L_{SCFG'} = \{a^i b^i : S \Rightarrow^* \pi^{i-1}(aSb|S)\pi(ab|S)\ a^i b^i,\ i \geq 1\}.$

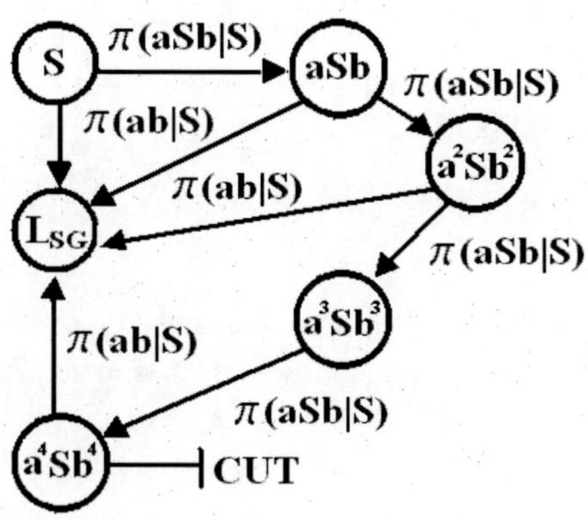

Figure 3. Derivation expansions of length less than five for SCFG'.

A finite state approximation diagram can be drawn directly by expanding derivatons of finite length like those depicted in Figure 3.

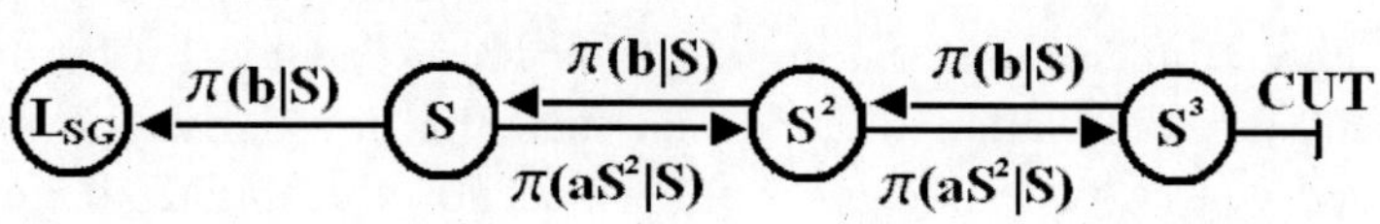

Figure 4. Approximated derivation diagram for SCFG''.

Example 3

A finite state approximation diagram for the SCFG'' with productions

$S \rightarrow \pi(aS^2 \mid S)\ aS^2,\ S \rightarrow \pi(b \mid S)\ b$

and associated language

$L_{SCFG''} = \{wb : S \Rightarrow^* \pi^i(aS^2|S)\pi^{i+1}(b|S)\ wb,\ \#_a(w) = \#_b(w) \geq 0\},$
where $\#_x(w)$ is the number of occurrences of x in w, can be traced from graphs like that reported in Figure 4.

3.2. Uniform Representation

Theoretically every stochastic context-free language recognized by a SPA computing with finite cut point can be accepted with error ≈ 0 by a (stochastic) finite state automaton. Moreover, the equivalence between SCFGs and SPA [20] implies that (finite) time complexity of any SPA, initialized with probability distribution

$$< \pi_0, \Sigma^* >$$

and accepting some language L_{SPA}, is the (worst-case) architecture complexity (number of control states) of the finite state (δ,λ)-approximation accepting L_{SPA} with bounded error.

Finite state approximations may be used to characterize the uniform representation of languages L_{SCFG} in terms of converging sequences as the length l of the derivations formed by nesting non-linear productions tends to ∞. Let

$$\{FSA_n\}_{n\in \mathbf{N}} \tag{2}$$

be the sequence of finite state (δ_l,λ)-approximations for some SCFG drawn by ensuring that derivations are obtained by applying at most l stochastic non-linear productions. If a finite l_0 exists such that

$$\delta_l = 0$$

for each

$$l \geq l_0,$$

then the sequence admits a uniform regular limit-representation for the language L_{SCFG}; this happens when some derivation order exists such that the derivation process diagram consists of bounded degree pattern-states from which no other higher degree state can be reached. An example of this situation is displayed in Figure 1 of Subsection 3.1.

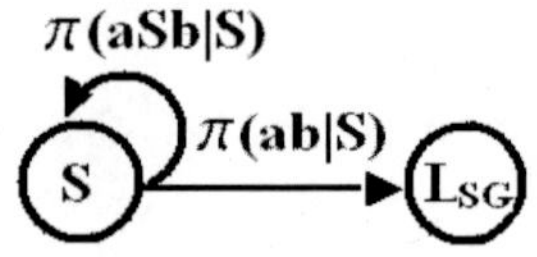

Figure 5. Uniform finite state diagram for SCFG'.

Example 2 in the same subsection (3.1) introduces a non-regular SCFG' for which we are not able to devise finite state automata approximating $L_{SCFG'}$ with error $\delta = 0$. Nevertheless, as $l \to \infty$, the forms of the finite state approximations may be clustered into a finite number of (equivalence) classes mapping the derivation process to the finite state automaton reported in Figure 5. According to such view, the limit of the sequence (2) denotes a uniform representation of the (stochastic) context-free calculus meaning that SCFG' can be approximated linearly with error ≈ 0 for each sufficiently large l (note that non-dirtiness ensures that deterministic non-linear productions may produce only finite length derivations).

The uniform representation, defined in terms of converging sequences of Markovian SFSG approximations, is helpful to describe compactly derivation processes; it allows us to simplify qualitative analysis of the SCFGs making conceivable symbolic higher-level calculus.

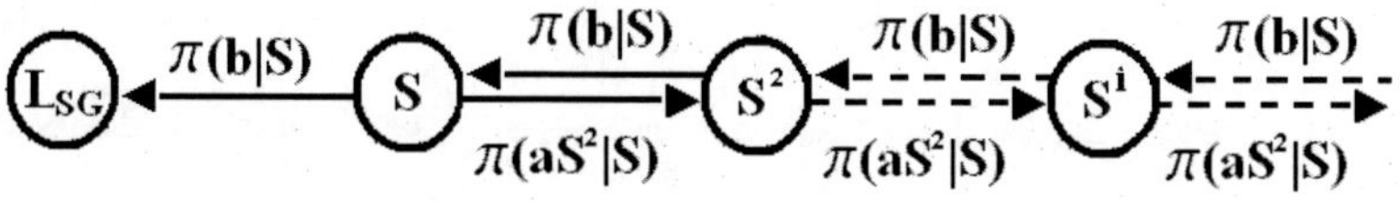

Figure 6. Uniform finite state diagram for SCFG''.

Unfortunately, the uniform regular representation can be traced readily only when the pattern-states have finitely bounded degree. The SCFG'' in Example 3 (Subsection 3.1) is an instance that seems to require a more powerful symbolic derivation notation in order to be able to draw an associated uniform finite state representation (see Figure 6).

CONCLUSION

We have proved that the (efficient) stochastic pushdown automata computing (with finite cut point acceptance) can be approximated with arbitrary accuracy in terms of stochastic finite state automata. This allows us to simplify the semantic and the statistical analysis [6,29,35] of the stochastic (efficient) pushdown automata computing.

We have interpreted and characterized the uniform representation of the SCFG calculus as the limit of converging sequences of finite state approximations. Such view of (SCFG or SPA) computing allows us to explain as symbolic descriptions arise by simple approximation sequences, making possible high level formalisms. The well known formal language and computational complexity analysis results [19, 6, 1, 12] can be applied so making possible to perform the symbolic (pushdown) calculus and to simplify the statistical analysis.

In particular, finite state approximations suggest that the SPA calculus could be performed by a system formed by at least two, highly synchronized, levels (see Figure 7): the higher order level is required to extract main qualitative properties of the represented phenomena and the lower level serves to learn fundamental characteristics by examples. The top level is highly associative and suitable to get conceptual descriptions, while the lower level is required to check by examples adequacy of the conceptual hypotheses. In this setting non-linear uniform descriptions (such as that reported in Figure 6) deserve further research work and analysis to be better understood. The reader interested in alternative representation techniques and main open questions regarding the stochastic calculus is referred to the enclosed bibliography (in particular to [17, 25, 18, 3, 4, 7, 34, 8–11]), where the approximation relationship existing between the physically realizable and the non-deterministic (efficient) computing was stated by Goemans and Williamson in [17] (see also [18]).

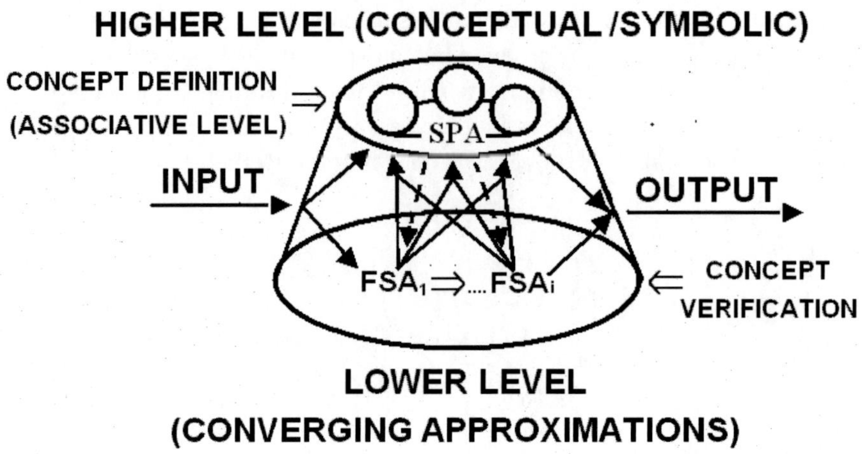

Figure 7. Representation of the pushdown automata calculus.

References

A. V. Aho, R. Sethi, J. D. Ullman, Compilers: Principles, Techniques and Tools, *Addison-Wesley,* 1988.

J. L. Balkazar, J. Diaz, J.Gabarro, Structural Complexity, *Springer-Verlag,* 1988.

A. Bertoni, M. Carpentieri, Analogies and Differences between Quantum and Stochastic Automata, Theoretical Computer Science, Vol. 262 (1-2), pp. 69-81, *Elsevier Science* B. V, 2001.

M. Carpentieri, A. Bertoni, Regular Languages Accepted by Quantum Automata, Information and Computation, Vol. 165, pp. 174-182, *Elsevier Science* B. V., Academic Press, 2001.

T. L. Booth, Probability Representation of Formal Languages, *IEEE Tenth Annual Symposium on Switching and Automata Theory,* 1969.

T. L. Booth, R. A. Thompson, Applying Probabilistic Measures to Abstract Languages, *IEEE Transactions on Computers*, Vol. C-22, n. 5, 1973.

M. Carpentieri, On the Simulation of Quantum Turing Machines, Theoretical Computer Science, Vol. 304, pp. 103-128, *Elsevier Science* B. V., Amsterdam, The Netherlands, 2003.

M. Carpentieri, A Genetic System Based on Simulated Crossover of Sequences of Two-Bit Genes, Theoretical Computer Science, Vol. 367, n. 3, pp. 324-335, *Elsevier Science,* 2006.

M. Carpentieri, A Genetic System Based on Simulated Crossover: Stability Analysis and Relationships with the Neural Nets, Evolutionary Computation, Vol. 17, n. 2, *Mit Press,* 2009.

M. Carpentieri, On Fundamental Genetic Techniques to Solve Randomized Hard Problems, submitted to Evolutionary Computation, 2012.

M. Carpentieri, Syntactic Models of Stochastic Computing, Internal Report, Universitá della Basilicata, Italy, 2012.

F. Charniak, Statistical Language Learning, *MIT Press,* Cambridge-Mass., 1993.

F. Charniak, Statistical Parsing with a Context-Free Grammar Word Statistics, *Proceedings of the 14th National Conference on Artificial Intelligence* (AAAI 9 7), 1997.

N. Chomsky, Context-free Grammars and Pushdown Storage, *Quarterly Prog. Rept.*, n. 65, *Research Laboratory of Electronics*, Massachusetts Institute of Technology, pp. 187-194, 1962.

J. Evey, The Theory and Application of Pushdown Machines, *Doctoral Thesis,* Harward University, 1963.

S. German, M. Johnson, Probabilistic Grammars and their Applications, International Encyclopedia of the Social and Behavioral Sciences, *Pergamon,* Oxford, pp. 12075-12082, 2002.

M. X. Goemans, D. P. Williamson, Improved Approximation Algorithms for Maximum Cut and Satisfiability Problems Using Semidefinite Programming, *Journal of the Association for Computing Machinery,* Vol. 42, n. 6, 1995.

J. Hastad, Some Optimal Inapproximability Results, Proceedings of the Twenty-Ninth ACM Symposium on the Theory of Computation, pp. 1-10, *ACM Press*, New York, NY, 1997.

J. E. Hopcroft, J. D. Ullman, Introduction to Automata Theory, Languages and Computation, *Addison-Wesley*, 1979.

T. Huang, K. S. Fu, On Stochastic Context-Free Languages, Information Sciences, Vol. 3, pp. 201-224, 1971.

C. D. Manning, H. Shutze, Foundations of Statistical Natural Language Processing, *MIT Press.*

P. Mateus, D. Qiu, L. Li, On the Complexity of Minimizing Stochastic and Quantum Automata, Information and Computation, Vol. 218, pp. 36-52, 2012.

A. M. Mood, F. A. Graybill, D. C. Boes, Introduction to the Theory of Statistics, *McGraw-Hill Companies*, 1974.

M. Mohri, M. J. Nederhof, Regular Approximation of CFG through Transformation, Robustness in Language and Speech Technology, *Kluwer-Academic,* pp. 251-261, 2000.

M. Mosbah, Probabilistic Graph Grammars, Fundamentae Informatica, Vol. 26, pp. 341-360, 1996.

M. J. Nederhof, Practical Experiments with Regular Approximation of Context-Free Languages, CL, Vol. 26 (1), pp. 17-44, 2000.

H. Ney, Stochastic Grammars and Pattern Recognition, Speech Recognition and Understanding, P. Laface and R. De Mori eds., pp. 319-344, Springer, 1992.

A. Paz, Introduction to Probabilistic Automata, New York: Academic, 1971.

D. V. Pynadath, M. P. Wellman, Generalized Queries on Probabilistic Context-Free Grammars, *IEEE Transactions on Pattern Analysis and Machine Intelligence,* Vol. 20, n. 1, pp. 65-77, 1998.

M. O. Rabin, Probabilistic Automata, Sequential Machines, *Addison-Wesley,* MA, 1964.

Y. Sakakibara, M. Brown, R. C. Underwood, I. S. Mian, D. Haussler, Stochastic Context-Free Grammars for Modeling RNA, *Proc. 27th Hawai Int'l Conf. System Sciences,* pp. 284-293, 1995.

A. Salomaa, Probabilistic and Weighted Grammars, Information and Control, Vol. 15, pp. 529-544, 1969.

G. W. Stewart, Introduction to Matrix Computations, Computer Science and Applied Mathematics Textbook, Werner Rheinboldt Editor, *Academic Press,* New York, 1972.

D. Suciu, N. Dalvi, Foundation of Probabilistic Answers to Queries, Tutorial, http://www.cs.washington.edu/ homes/suciu, 2005.

C. S. Wetherel, Probabilistic Languages: A Review and Some Open Questions, Computing Surveys, Vol. 12, n. 4, 1980.

INDEX

A

B

C

D

E

F

G

H

I

J

K

L

M

N

O

P

R

S

T

U

V

W

Y